Results and Problems in Cell Differentiation

45

Dietmar Richter
Center for Molecular Neurobiology
University Medical Center Hamburg-Eppendorf (UKE)
University of Hamburg
Martinistrasse 52
20246 Hamburg
Germany
richter@uke.uni-hamburg.de

Henri Tiedge
The Robert F. Furchgott Center for Neural and Behavioral Science
Department of Physiology and Pharmacology
Department of Neurology
SUNY Health Science Center at Brooklyn
Brooklyn, New York 11203
USA
htiedge@downstate.edu

Series Editors
D. Richter, H. Tiedge

Günter Schäfer, Harvey S. Penefsky (Eds.)

Bioenergetics
Energy Conservation and Conversion

 Springer

GÜNTER SCHÄFER
Institute of Biochemistry
University of Lübeck
Ratzeburger Allee 160
23538 Lübeck
Germany
ggw.schaefer@web.de

HARVEY S. PENEFSKY
International Center for Public Health
Public Health Research Institute
225 Warren Street
Newark, NJ 07103
USA
penefshs@umdnj.edu

ISSN 0080-1844
ISBN-13 978-3-540-78621-4 Springer Berlin Heidelberg New York
DOI 10.1007/978-3-540-78622-1

Library of Congress Control Number: 2008922553

This work is subject to copyright. All rights are reserved, whether the whole or part of the material is concerned, specifically the rights of translation, reprinting, reuse of illustrations, recitation, broadcasting, reproduction on microfilm or in any other way, and storage in data banks. Duplication of this publication or parts thereof is permitted only under the provisions of the German Copyright Law of September 9, 1965, in its current version, and permission for use must always be obtained from Springer. Violations are liable for prosecution under the German Copyright Law.

Springer is a part of Springer Science+Business Media

springer.com

© Springer-Verlag Berlin Heidelberg 2008
Printed in Germany

The use of registered names, trademarks, etc. in this publication does not imply, even in the absence of a specific statement, that such names are exempt from the relevant protective laws and regulations and therefore free for general use.

Cover design: WMXDesign GmbH, Heidelberg
Typesetting and Production: LE-TEX Jelonek, Schmidt & Vöckler GbR, Leipzig

Printed on acid-free paper   39/3180/YL – 5 4 3 2 1 0

Introduction

The fermentation of sugar by cell-free yeast extracts was demonstrated more than a century ago by E. Buchner (Nobel Prize 1907). Buchner's observations put an end to previous animistic theories regarding cellular life. It became clear that metabolism and all cellular functions should be accessible to explication in chemical terms. Equally important for an understanding of living systems was the concept, explained in physical terms, that all living things could be considered as energy converters [E. Schrödinger (Nobel Prize 1933)] which generate complexity at the expense of an increase in entropy in their environment.

Bioenergetics was established as an essential branch of the biochemical sciences by the investigations into the chemistry of photosynthesis in isolated plant organelles [O. Warburg (Nobel Prize 1931)] and by the discovery that mitochondria were the morphological equivalent that catalyzed cellular respiration. The field of bioenergetics also encompasses a large variety of additional processes such as the molecular mechanisms of muscle contraction, the structure and driving mechanisms of microbial flagellar motors, the energetics of solute transport, the extrusion of macromolecules across membranes, the transformation of quanta of light into visual information and the maintenance of complex synaptic communications. There are many other examples which, in most cases, may perform secondary energy transformations, utilizing energy stored either in the cellular ATP pool or in electrochemical membrane potentials.

The recognition that primary energy conservation can indeed occur via formation of electrochemical potential gradients formed by small ions, the chemiosmotic mechanism, has fundamentally revolutionized the understanding of these processes [P. Mitchell (Nobel Prize 1978)]. Oxygenic photosynthesis in chloroplasts from green plants and algae on the one hand, and cellular respiration in mitochondria or aerobic bacteria on the other, are the best-known processes of that type. The general importance of electrochemical potential gradients well deserves the dedication of a special volume within this series of Results and Problems of Cell Differentiation. The further importance of ion gradients is illustrated by the fact that they may be subject to defects and diseases and may also be useful targets for herbicides and drugs. The basic mechanistic principles of these molecular energy converters may provide models for the design of bionic devices of importance to the development of novel energy sources utilizing solar light. The most essential bioenergetic process on earth—the oxidation of water—is catalyzed by a 4-metal-cluster that might be synthetically reconstructed to achieve a non-biological splitting of water. Together with fuel cells such devices might in the future help to solve mankind's major energy problems in an ecologically acceptable manner.

Nature has had billions of years to develop these efficient devices. Present knowledge, based upon enormous progress in the understanding of molecular and structural biology during the past half-century, has made it evident that the principle of primary energy conservation as well as that of the utilizing secondary machinery, the ATP-synthase, was established very early in cellular evolution. This point is highlighted in the first chapter of this volume which provides insight into the phylogenetic relations of the vast number of organisms exhibiting the genetic instructions for heme–copper oxidases, the key elements in cellular oxygen respiration. The question may even be raised as to which came first, photosynthesis or respiration. In addition, the cyclic tetrapyroles, the metal hosts which are the catalytically active key structures, must be at least of similar evolutionary age. Interestingly, prophyrin-like metal-chelators are not restricted to heme–Cu enzymes but are also found in similar, catalytically highly active metal complex structures such as chlorophylls and cobalamines. In addition to oxygen respiration, other forms of respiratory chains are widely distributed in facultative or strictly anaerobic microbes. Sulfur respiration is an example as well as branched bacterial respiratory chains with a variety of possible electron acceptors.

Chapters Two and Three discuss light-driven energy converters. These systems, comprising both anoxygenic and oxygenic photosynthesis, generate an ion-motive force across organelle or cellular membranes. The membrane-resident anoxygenic photosystem-I of bacteria was the first to be structurally elucidated on an atomic scale [H. Michel, J. Deisenhofer, R. Huber (Nobel Prize 1988)]. It should be considered a milestone in modern structural biology, opening up not just access to an understanding of the molecular mechanisms of biological energy transduction but also to novel methods for the crystallization of membrane proteins. Subsequent substantial progress also enabled the crystallization of photosystem-II, including the water-splitting device as reviewed in this volume. In addition to such chlorophyll-based mechanisms, completely different light-harvesting systems evolved which were first detected in extremophilic archaea, the retinal-dependent ion pumps. The present review makes clear how minor modifications in rhodopsins, utilizing the same retinal cofactor, can permit these molecules to serve not only as ion pumps but also as sensory devices. Considering that rhodopsins are also found in visual sensory receptors of higher animals one may ask, from the point of view of developmental biology, whether this device independently evolved twice, or was adopted via endosymbiotic events or early lateral gene transfer. The answer presently remains enigmatic.

Two prominent examples of membrane-linked energy conservation during anaerobic fermentation are presented in subsequent chapters and illustrate the ability of various methanogenic archaea to derive energy from fermentative reactions characterized by a rather low free energy change. In addition, the peculiarities of archaeal A_OA_1-type ATP-synthases are reviewed. The latter exhibit significant similarities to vacuolar V_OV_1-ATPases and are likely to be evolutionarily related to them. The membrane-resident steps of these metabolic pathways are coupled to primary transformation of chemical potential into chemiosmotic potentials (using either protons or sodium ions). This point also applies to another unique example presented in the accompanying chapter which indicates that fermentation in conjunction with a decarboxylation reaction can serve as an energy source for the formation of an ion gradient across microbial cell membranes.

The following chapters deal with two membrane-bound protein complexes which share strong similarities in some of their subunits while assuming entirely different physiological functions: complex-I of respiratory chains and hydrogenases. The proton translocating NADH dehydrogenases of the respiratory chain (complex-I) may comprise between 14 (bacterial) and 42 (eucaryotic) different protein subunits. It is the only major complex of the respiratory chain that has not as yet been crystallized. Thus, high-resolution structural data are not available, a problem shared with the equally complex hydrogenases. Nevertheless, the apparent evolutionary relationship of both membrane-residing protein assemblies is obvious and illustrates the evolutionary reuse of functional modalities once developed. It also becomes obvious that the generation of ion conduction at the expense of redox energy is not necessarily conserved in all types of these enzymes.

In contrast, crystal structures are available for terminal oxidases of bacteria and mammalia as well as for complex III of aerobic respiratory chains, for example, the quinol-cytochrome-c-reductases. This module of energy transduction is of special interest because it occurs in a very similar structure and with identical function as a member of the photosynthetic electron transport chains. The complex not only catalyzes a rather interesting proton translocation, including disproportionation of charge which allows the transition from a 2-electron to a 1-electron-conducting redox pathway, but also demonstrates that inherent large-scale movements of protein domains assume a key function directly mediating the distribution of electrons. Accordingly, a specific domain of the iron–sulfur protein in complex-III exerts a definite pivoting movement in order to pick up electrons at a donor site and deliver them to a distant acceptor site in its molecular environment.

The final chapter concentrates on the universal secondary energy transducer, the F_OF_1-type ATP-synthase. As to its basic composition and function it may serve as a model also for the above-mentioned A_OA_1-type ATP-synthases from archaea and even for the vacuolar ATPases which, however, can work only as ion pumps but not in the reverse direction as ATP-synthesizing devices. The essential parts of its three-dimensional atomic structure have been established [J. Walker, P. Boyer (Nobel Prize 1998)]. They confirmed the previous proposal of a rotary molecular engine which utilizes the flux of ions across a membrane as a driving force for the cyclic structural rearrangements that permit the catalysis of ATP formation from ADP and Pi. The intramolecular stepwise rotation of the central subunit within this macromolecular protein assembly was originally visualized by the author of the present contribution. The ATP synthase is the master example of secondary energy transducers in cellular biology. It is the best-investigated member of a family of reversibly functioning, membrane-residing ATPases spread throughout all evolutionary kingdoms. The central aspect discussed in this volume is the problem of regulation of such devices in response to the energetic demands within living cells.

In summary, the articles presented in this issue review present knowledge of selected examples of biological energy conservation machinery. The machinery is clearly highly variable, particularly that which is distributed throughout microorganisms, but all of these devices rely universally on one unique underlying physico-chemical principle.

January 2008

Günter Schäfer
Harvey S. Penefsky

Contents

**Diversity of the Heme–Copper Superfamily in Archaea:
Insights from Genomics and Structural Modeling**
James Hemp, Robert B. Gennis 1
1 Introduction 1
2 Introduction to the Heme–Copper Superfamily 2
3 Classification of the Superfamily 3
4 Heme-Copper Family Properties 5
4.1 Oxygen Reductases 5
4.2 Nitric Oxide Reductases 6
5 Heme–Copper Oxygen Reductases in Archaea 8
5.1 A-Family 8
5.2 B-Family 19
5.3 C-Family 21
5.4 Families D Through H 21
6 Distribution and Evolution
 of the Heme–Copper Superfamily 26
References 28

Structure of Photosystems I and II
Petra Fromme, Ingo Grotjohann 33
1 Overview on Oxygenic Photosynthesis 33
2 Photosystem I 35
2.1 General Structure of PS I 35
2.2 Protein Subunits of PS I 36
2.3 Electron Transfer Chain in PS I 45
2.4 The Antenna System of PS I 48
3 Photosystem II 50
3.1 General Structure of PS II 50
3.2 Arrangement of Protein Subunits in PS II 52
3.3 Electron Transport Chain of PS II 59
3.4 Antenna System of PS II 62
References 65

Microbial Rhodopsins:
Scaffolds for Ion Pumps, Channels, and Sensors
JOHANN P. KLARE, IGOR CHIZHOV, MARTIN ENGELHARD 73
1 Introduction 73
2 Microbial (Type 1) Rhodopsins 75
2.1 Archaeal Rhodopsins 75
2.2 Eubacterial Rhodopsins 79
2.3 Eukaryotic Type 1 Rhodopsins 82
2.4 Structure of the Rhodopsins 85
2.5 Evolution of Type I Rhodopsins 89
3 Ion Transfer and Signal Transfer Mechanisms 91
3.1 Photocycle 91
3.2 Mechanism of Signal and Ion Transfer 95
4 Phototaxis 101
4.1 The Transducer Proteins 102
4.2 Two-Component Systems 104
4.3 Mechanisms of Signal Transduction 106
4.4 Comparison Between Phototaxis and Chemotaxis 110
5 Outlook 110
References 111

Life Close to the Thermodynamic Limit:
How Methanogenic Archaea Conserve Energy
UWE DEPPENMEIER, VOLKER MÜLLER 123
1 Introduction 123
2 The Process of Methanogenesis 125
3 Reactions and Compounds
 of the Methanogenic Electron Transport Chains 128
4 Structure and Function of Ion-Translocating Enzymes 133
4.1 $F_{420}H_2$ Dehydrogenase 134
4.2 F_{420}-nonreducing Hydrogenase 135
4.3 Ech Hydrogenase 136
4.4 Heterodisulfide Reductase 136
4.5 Structure and Function
 of the Methyltetrahydromethanopterin:
 Coenzyme M Methyltransferase,
 a Primary Sodium Ion Pump 137
5 ATP Synthesis in Methanogens 139
5.1 Structure and Function
 of the A_1A_O ATP Synthase from Methanogens 140
5.2 The Unique Membrane-Embedded Rotor
 of A_1A_O ATP Synthases 142
6 Concluding Remarks 144
References 145

ATP Synthesis by Decarboxylation Phosphorylation
Peter Dimroth, Christoph von Ballmoos 153
1 Introduction 153
2 Fermentation Pathways
 with Na^+-Transport Decarboxylases (NaT-DC) 154
2.1 Fermentation of Citrate 154
2.2 Fermentation of Succinate or Lactate 156
2.3 Fermentation of Malonate 157
2.4 Fermentation of Glutarate 158
3 Structure and Mechanism of the NaT-DC Enzymes 159
3.1 Oxaloacetate Decarboxylase Na^+ Pump 159
4 ATP Synthesis Energized
 by an Electrochemical Na^+ Ion Gradient 162
4.1 H^+- and Na^+-Translocating ATP Synthases 162
4.2 The F_1 Motor 165
4.3 The F_0 Motor 166
4.4 Subunit C Structures 167
4.5 Structure of the C Ring From *I. tartaricus* 168
5 Mechanism of the F_0 Motor 171
5.1 The Proton Motor 171
5.2 The Sodium Motor 173
References 177

The Three Families of Respiratory NADH Dehydrogenases
Stefan Kerscher, Stefan Dröse, Volker Zickermann,
Ulrich Brandt 185
1 Introduction 185
2 "Alternative" or NDH-2-Type NADH Dehydrogenases 187
2.1 Characteristics 187
2.2 Physiological Roles 188
2.3 Other Enzymes
 Related to Alternative NADH Dehydrogenases 191
2.4 Structural Model and Mechanistic Implications 194
3 Sodium-Pumping NADH Dehydrogenases (Nqr) 196
3.1 Occurrence, Cellular Functions
 and Physiological Significance in Bacteria 196
3.2 Subunit Composition and Cofactors 198
3.3 Functional Properties and Catalytic Mechanism 201
4 Proton-Pumping NADH:Ubiquinone Oxidoreductase
 (Complex I) 205
4.1 Occurrence and General Features 205
4.2 Can Bacterial Complex I Act as a Sodium Pump? 206
4.3 The Functional Modules of Complex I 209

| 4.4 | Mechanism of Proton Translocation | 213 |

References 213

Hydrogenases and H^+-Reduction in Primary Energy Conservation
PAULETTE M. VIGNAIS 223

1	Introduction	223
2	Diversity and Classification of Hydrogenases	224
2.1	The [NiFe]hydrogenases	225
2.2	The [FeFe]hydrogenases	233
2.3	Hydrogenases and Complex I	236
3	Modes of Energy Conservation by Hydrogenases	238
3.1	Energy Conservation via Energy-Transducing Electron Transport Chains by Respiratory [NiFe]hydrogenases (Group 1)	238
3.2	Energy-Conservation by Proton/Na^+ Translocation by Membrane-Bound, H_2 Evolving [NiFe]hydrogenases (Group 4)	240
3.3	Disposal of Excess Reducing Equivalents	241
4	Conclusions and Perspectives	243

References 244

A Structural Perspective on Mechanism and Function of the Cytochrome bc_1 Complex
CAROLA HUNTE, SOZANNE SOLMAZ, HILDUR PALSDÓTTIR, TINA WENZ 253

1	Introduction	253
2	Structural Characterization of cyt bc_1	255
2.1	Redox-Active Subunits	258
2.2	Tight Binding of Phospholipids to the cyt bc_1	260
3	Mechanistic Considerations	261
3.1	Ubihydroquinone Oxidation at Center P	262
4	Interaction of cyt c with cyt bc_1	268
5	Respiratory Supercomplexes	269

References 270

Regulatory Mechanisms of Proton-Translocating F_OF_1-ATP Synthase
BORIS A. FENIOUK, MASASUKE YOSHIDA 279

1	Introduction	279
2	Structure and Rotary Catalysis: a Brief Summary	280
2.1	Structure	280
2.2	Catalytic Mechanism	281
3	ADP-Inhibition: a Common Regulatory Mechanism	284
3.1	Mechanism of ADP-Inhibition	285
3.2	Factors Affecting ADP-Inhibition	286
4	Subunit ϵ in Bacterial and Chloroplast Enzyme	291

4.1	Structure of Subunit ϵ	291
4.2	Inhibition of ATP Hydrolysis by Subunit ϵ	292
4.3	Conformational Transitions of Subunit ϵ C-Terminal Domain	293
4.4	The Role of βDESLEED Region in Inhibition Mediated by Subunit ϵ	294
5	Thiol Regulation in Chloroplast Enzyme	295
6	Mitochondrial Inhibitor Protein IF_1	297
7	Conclusions	298
References		299

Subject Index 309

Contributors

Christoph von Ballmoos
Institute of Microbiology, ETH Zürich, Wolfgang-Pauli-Strasse 10, 8093 Zürich, Switzerland

Ulrich Brandt
Molecular Bioenergetics Group, Centre of Excellence "Macromolecular Complexes", Johann Wolfgang Goethe-Universität, 60590 Frankfurt am Main, Germany

Igor Chizhov
Medizinische Hochschule Hanover, Carl-Neuberg-Straße 1, 30623 Hannover, Germany

Uwe Deppenmeier
Department of Biological Sciences, University of Wisconsin-Milwaukee, 3209 N. Maryland Ave, Milwaukee, WI 53211, USA

Peter Dimroth
Institute of Microbiology, ETH Zürich, Wolfgang-Pauli-Strasse 10, 8093 Zürich, Switzerland

Stefan Dröse
Molecular Bioenergetics Group, Centre of Excellence "Macromolecular Complexes", Johann Wolfgang Goethe-Universität, 60590 Frankfurt am Main, Germany

Martin Engelhard
Max-Planck-Institute for Molecular Physiology, Otto Hahn Str. 11, 44227 Dortmund, Germany

Boris A. Feniouk
ATP System Project, Exploratory Research for Advanced Technology, Japan Science and Technology Corporation (JST), 5800-3 Nagatsuta, Midori-ku, 226-0026 Yokohama, Japan
Chemical Resources Laboratory, Tokyo Institute of Technology, 4259 Nagatsuta, Midori-ku, 226-8503 Yokohama, Japan

Petra Fromme
Department of Chemistry and Biochemistry, Arizona State University, Box 871604, Tempe, AZ 85287-1604, USA

Robert B. Gennis
Department of Chemistry, University of Illinois, Urbana, IL 61801, USA
Department of Biochemistry, University of Illinois, 600 S. Mathews Avenue, Urbana, IL 61801, USA

Ingo Grotjohann
Department of Chemistry and Biochemistry, Arizona State University, Box 871604, Tempe, AZ 85287-1604, USA

James Hemp
Center for Biophysics and Computational Biology, University of Illinois, Urbana, IL 61801, USA
Department of Chemistry, University of Illinois, Urbana, IL 61801, USA

Carola Hunte
Dept. Molecular Membrane Biology, Max Planck Institute of Biophysics, 60438 Frankfurt, Germany

Stefan Kerscher
Molecular Bioenergetics Group, Centre of Excellence "Macromolecular Complexes", Johann Wolfgang Goethe-Universität, 60590 Frankfurt am Main, Germany

Johann P. Klare
Fachbereich Physik, University Osnabrück, Barbarastraße 7, 49069 Osnabrück, Germany

Volker Müller
Molecular Microbiology & Bioenergetics, Institute of Molecular Biosciences, Goethe University Frankfurt am Main, 60438 Frankfurt/Main, Germany

Hildur Palsdóttir
Dept. Molecular Membrane Biology, Max Planck Institute of Biophysics, 60438 Frankfurt, Germany

Sozanne Solmaz
Dept. Molecular Membrane Biology, Max Planck Institute of Biophysics, 60438 Frankfurt, Germany

Paulette M.Vignais
CEA Grenoble, Laboratoire de Biochimie et Biophysique des Systèmes Intégrés, UMR CEA/CNRS/UJF no. 5092, Institut de Recherches en Technologies et Sciences pour le Vivant (iRTSV), 17 rue des Martyrs, 38054 Grenoble cedex 9, France

Tina Wenz
Dept. Molecular Membrane Biology, Max Planck Institute of Biophysics, 60438 Frankfurt, Germany

Masasuke Yoshida
Chemical Resources Laboratory, Tokyo Institute of Technology, 4259 Nagatsuta, Midori-ku, 226-8503 Yokohama, Japan
ICORP ATP-Synthesis Regulation Project (Japanese Science and Technology Agency), National Museum of Emerging Science and Innovation, 2-41 Aomi, Koto-ku, 135-0064 Tokyo, Japan

Volker Zickermann
Molecular Bioenergetics Group, Centre of Excellence "Macromolecular Complexes", Johann Wolfgang Goethe-Universität, 60590 Frankfurt am Main, Germany

Diversity of the Heme–Copper Superfamily in Archaea: Insights from Genomics and Structural Modeling

James Hemp[1,2] · Robert B. Gennis[1,2,3] (✉)

[1]Center for Biophysics and Computational Biology, University of Illinois, Urbana, IL 61801, USA

[2]Department of Chemistry, University of Illinois, Urbana, IL 61801, USA

[3]Department of Biochemistry, University of Illinois, 600 S. Mathews Avenue, Urbana, IL 61801, USA
r-gennis@uiuc.edu

Abstract Recent advances in DNA sequencing technologies have provided unprecedented access into the diversity of the microbial world. Herein we use the comparative genomic analysis of microbial genomes and environmental metagenomes coupled with structural modelling to explore the diversity of aerobic respiration in Archaea. We focus on the heme–copper oxidoreductase superfamily which is responsible for catalyzing the terminal reaction in aerobic respiration—the reduction of molecular oxygen to water. Sequence analyses demonstrate that there are at least eight heme–copper oxygen reductase families: A-, B-, C-, D-, E-, F-, G-, and H-families. Interestingly, five of these oxygen reductase families (D-, E-, F-, G-, and H-families) are currently found exclusively in Archaea. We review the structural properties of all eight families focusing on the members found within Archaea. Structural modelling coupled with sequence analysis suggests that many of the oxygen reductases identified from thermophilic Archaea have modified proton channel properties compared to the currently studied mesophilic bacterial oxygen reductases. These structural differences may be due to adaptation to the specific environments in which these enzymes function. We conclude with a brief analysis of the phylogenetic distribution and evolution of Archaeal heme–copper oxygen reductases.

1
Introduction

All life on Earth is currently divided into three Domains based on phylogenetic and genomic analyses: Bacteria, Archaea, and Eukaryota (Woese and Fox 1977). Early work on Archaea characterized them as predominantly inhabiting extreme environments (hyperthermophiles, halophiles, acidophiles) or as performing novel metabolisms (methanogens). Environmental sampling utilizing rDNA as a phylogenetic marker has revealed an enormous diversity of microbial life (DeLong and Pace 2001; DeSantis et al. 2006), with the vast majority belonging to uncultured species (Rappe and Giovannoni 2003). Recently, it has become apparent that Archaea are not limited to extreme environments, but are common in almost every environment studied (Robertson et al. 2005). In fact the diversity of Archaea is greatest in cold environments, where they form a significant fraction of the biosphere (Cavicchioli 2006). However, very little is known about the physiology and ecology of these new Archaeal groups, creating a fascinating opportunity for the characterization of the bioenergetic properties of these organisms.

Recent advancements in DNA sequencing and analysis applied to microbial genomes and communities (Riesenfeld et al. 2004) have vastly improved our understanding of microbial diversity and physiology. Herein, we review the comparative genomics and structural diversity of the heme–copper superfamily in Archaea. Members of the superfamily play key roles in aerobic and anaerobic respiration. Analysis of the DNA sequences from more than 950 genomes from Archaea and Bacteria along with 15 metagenomic projects has identified thousands of new members of the heme–copper superfamily. The superfamily can currently be classified into eight oxygen reductase and five nitric oxide reductase families. It is very interesting that five of the oxygen reductase families are currently only found in Archaea. This leads to questions concerning the evolution and ecology of Archaeal members of the superfamily.

2
Introduction to the Heme–Copper Superfamily

The majority of Eukaryota are aerobic heterotrophs, oxidizing organic compounds to CO_2 while reducing O_2 to water via aerobic respiration within mitochondria. Archaea and Bacteria have much greater physiological diversity, extracting energy from their environments utilizing a wide range of metabolisms including; aerobic and anaerobic respirations, fermentation, and photosynthesis (Madigan and Martinko 2006). In aerobic eukaryotes, respiration is carried out within the inner mitochondrial membrane by a linear electron transfer chain with cytochrome c oxidase as the terminal electron acceptor. In Archaea and Bacteria, respiration is performed by a series of soluble and integral membrane protein complexes found within cytoplasmic or thylakoid membranes (Madigan and Martinko 2006). Archaea and Bacteria usually have complex branched respiratory chains with alternative terminal electron acceptors (Poole and Cook 2000) (aerobic and anaerobic), allowing them to adjust their metabolisms to fit the availability of substrates in their environments.

The heme–copper superfamily is a large and extremely diverse superfamily of integral membrane proteins that contains members which play crucial roles in both aerobic and anaerobic respiration (Garcia-Horsman et al. 1994; Pereira et al. 2001). Members of the superfamily are oxidoreductases, which currently are divided into two classes based on the chemical reaction that they perform: the oxygen reductases and the nitric oxide reductases.

Oxygen reductases $\quad O_2 + 4H_{chem}^+ + 4e^- + 4H_{in}^+ \rightarrow 2H_2O + 4H_{out}^+$,
Nitric oxide reductases $\quad 2NO + 2H^+ + 2e^- \rightarrow N_2O + H_2O$.

The oxygen reductase members of the superfamily are terminal oxidases in the aerobic respiratory chains of mitochondria and many Archaea and Bacteria. These enzymes catalyze the reduction of molecular oxygen to water with the concomitant transfer of protons across the membrane, contributing to the generation of the electrochemical gradient used for membrane transport and ATP synthesis (Michel et al. 1998; Hosler et al. 2006). The oxygen reductases have a broad phylogenetic distribution and are found in all three domains of life: Archaea, Bacteria, and Eukaryotes. The nitric oxide reductase (NOR) members of the superfamily catalyze the reduction of nitric oxide to nitrous oxide in Archaea and Bacteria capable of anaerobic denitrification starting with nitrate or nitrite (Zumft 2005). Nitric oxide reductase can also play a nitric oxide detoxifying role in some pathogenic bacteria (Philippot 2005). Unlike the oxygen reductases, the currently identified nitric oxide reductases are not electrogenic and are unable to pump protons (Reimann et al. 2007). Nitric oxide reductases have only been found in Archaea and Bacteria, however recently the unicellular eukaryotic foraminifer *Globobulimina pseudospinescens* has been shown to perform denitrification (Risgaard-Petersen et al. 2006). The genes responsible for denitrification in this organism have not been characterized, so it is not clear whether a heme–copper nitric oxide reductase or an alternative nitric oxide reducing enzyme performs the nitric oxide reduction step.

Structurally, both the oxygen and nitric oxide reductases are multi-subunit, integral membrane protein complexes. The complexes are composed of a main subunit, which is the functional core of the enzyme complex, along with secondary subunits that serve a variety of functions, such as interacting with the mobile electron donors (e.g., cytochrome c or quinol) or regulatory functions. The main subunit (subunit I) contains all of the amino acids and cofactors necessary for both catalysis and proton translocation and it is this core subunit that defines the superfamily (Garcia-Horsman et al. 1994). Only this subunit is homologous between the oxygen and nitric oxide reductases, with the secondary subunits having independent evolutionary histories.

3
Classification of the Superfamily

We have analyzed DNA sequences from microbial genomes and metagenomic projects to identify thousands of new members of the heme–copper superfamily (Hemp 2007). To characterize this diversity we proposed an updated classification system in which the superfamily is first divided into classes, and then further into families, and subfamilies (see Pereira et al. 2001 for the original classification scheme). Classes are defined by the reaction that the enzymes catalyze. The superfamily is currently divided into two classes: the oxygen reductases and the nitric oxide reductases. It is possible that some enzymes of one class may also have low rates of reaction for the other class

(Giuffre et al. 1999). In these cases enzymes were placed into classes based on the primary reaction that they catalyze. It should be noted that the class is not a phylogenetic property in that the reaction types could have evolved independently multiple times within the superfamily.

Families were delineated based on phylogenetic, genomic, sequence, and structural information. Members of a given family were defined to perform the same chemical reaction, share common protein complex subunit architecture, have similar proton channel structures, and form separate sequence clusters. Different families clearly have different physiological functions, are paralogous to each other, and can be objectively defined by sequence analyses. Phylogenetic analysis of the whole superfamily using Bayesian (Ronquist and Huelsenbeck 2003), Neighbor-Joining, and Maximum Parsimony methods all produced 13 phylogenetic clusters irrespective of the method used. The same 13 families were identified using all-vs-all BLAST sequence clustering with a 35% sequence identity cutoff. The intra-family sequence identity was at least 35%, whereas inter-family sequence identities were usually less than 20%. This shows that the differences between families is quite large. Structural analysis (active-site structure and channel properties) and subunit architecture further support the delineation of the currently identified superfamily members into 13 families.

Table 1 Family properties of heme–copper superfamily members

Family	Class	Active-site residue	Cross-linked cofactor	Proton channels[a]	Proton pumping	Subunit II fold[b]
A	Oxygen reductase	Y	+	2[c]	+	Cu_A
B	Oxygen reductase	Y	+	1	+	Cu_A
C	Oxygen reductase	Y	+	1	+	Heme c
D	Oxygen reductase	Y	+*	1	+	Other
E	Oxygen reductase	Y	+*	1	+	Cu_A
F	Oxygen reductase*	Y	+*	0	−*	Cu_A
G	Oxygen reductase*	Y	+*	1	+*	Cu_A
H	Oxygen reductase*	Y	+*	1	+*	Cu_A
cNOR	Nitric oxide reductase	E	−	0	−	Heme c
qNOR	Nitric oxide reductase	E	−	0	−	−
sNOR	Nitric oxide reductase*	N	−	0	−*	Cu_A
eNOR	Nitric oxide reductase*	Q	−	1	+/−	Cu_A
gNOR	Nitric oxide reductase*	D	−	0	−*	Cu_A

[a] Number of conserved proton channels determined by sequence analysis and structural modelling
[b] Cu_A-Subunit II has cupredoxin fold, homologous to subunit II of mitochondrial cytochrome c oxidases. Heme c-Subunit II has cytochrome c fold
[c] Some members of the A-family have modified D- and/or K-channels
* Predicted

Eight of the families identified belong to the oxygen reductase class: the A-, B-, C-, D-, E-, F-, G-, and H-type oxygen reductases. Two of the families belong to the nitric oxide reductase class: the cNOR and qNOR nitric oxide reductases. The other three families have not been previously described. On the basis of expression data (Cho et al. 2006) and structural analysis we predict that all three families belong to the nitric oxide reductase class: putatively the sNOR, eNOR, and gNOR nitric oxide reductases. Table 1 lists the heme–copper families along with their associated properties.

4
Heme-Copper Family Properties

Many of the recently identified oxygen reductase families are currently only found in Archaea: the D-, E-, F-, G-, and H-families. These new families have unique structural features in comparison to the previously studied A-, B-, and C-families. It is also common for Archaea to contain members of the A- and B-families that have modified structural features. The reason that Archaea have modified heme–copper superfamily members is not yet clear. It may be related to the structural properties of their membranes, or to the environments in which they are found. To identify the structural features that are modified in Archaea we first review the structural properties of oxygen reductases already elucidated from the study of bacterial enzymes.

4.1
Oxygen Reductases

The oxygen reductases are structurally diverse. They vary in the type of electron donor (cytochrome c, quinol, or high potential iron-sulfur protein), the types of heme present [hemes B, O, O_{P1}, O_{P2}, A, or A_S (Lübben and Morand 1994)], and the number of subunits present in the enzymatic complex (ranging from 2, in some bacteria, to as many as 13 in mammalian mitochondria). X-ray crystal structures have been reported for members of both the A-type (Ostermeier et al. 1997; Abramson et al. 2000; Yoshikawa et al. 2000; Svensson-Ek et al. 2002) and B-type (Soulimane et al. 2000) oxygen reductase families. These structures show that the protein component of the main subunit, subunit I, has 12 transmembrane helices arranged in a pseudo three-fold rotational symmetry with the symmetry axis perpendicular to the membrane. Subunit I also contains three redox-active metals; a low-spin heme and a binuclear heme–copper catalytic site. The six-coordinate low-spin heme accepts electrons from the electron donor specific for the particular complex and transfers the electrons to the catalytic site. The heme–copper catalytic site is composed of a five coordinate high-spin heme and a copper ion ligated to three conserved histidines. In three of the oxygen reductase families (A-, B-, and C-families) a novel His-Tyr cross-linked cofactor has been shown by X-ray crystallography (A- and B-families) or mass spectrometry (A-, B-, and C-families) (Hemp et al. 2006; Rauhamaki et al. 2006) to be present in the active site. This cross-linked pair of residues forms a cofactor which plays a critical role in oxygen reduction. Since the other five families (D-, E-, F-, G-, and H-families) conserve the residues which form the cross-link it is likely that they also contain this cofactor.

Mutagenesis studies performed on bacterial oxygen reductase members of the superfamily have identified proton input channels necessary for proton pumping and the delivery of protons from the cytoplasm to the active site. In the A-type oxygen reductases, two proton input channels have been identified (Hosler et al. 1993). One channel, the K-channel, leads from the interface of subunits I and II on the cytoplasmic side of the membrane (equivalent to the matrix side of the mitochondrial enzymes) to the cross-linked tyrosine at the active-site. The K-channel has been shown to be important in the A-type oxidases for the delivery of chemical protons to the catalytic active site (Konstantinov et al. 1997). The second channel, the D-channel, has been implicated in the transfer of both chemical protons to the active site and protons which are pumped across the membrane. The D-channel leads from an aspartate on the cytoplasmic surface of subunit I to a gating residue near the active site which is a branch point from which protons are directed either to the active site, where they are consumed to generate water, or are pumped across the membrane and released on the opposite side as pumped protons, contributing to the electrochemical gradient. Both the K- and D-channels contain highly conserved hydrophilic residues which help to stabilize and orient the water in conformations facilitating proton transfer (Wraight 2006).

In the B-type and C-type oxygen reductases, a channel analogous to the K-channel has been identified through modelling studies (B- and C-families) (Sharma et al. 2006; Hemp et al. 2007), X-ray crystallography (B-family) (Soulimane et al. 2000) or site-directed mutagenesis (C-family) (Hemp et al. 2007). The X-ray structure of the ba_3-type oxidase (B-family) from *Thermus thermophilus* has been used to postulate two additional proton input channels, one analogous to the D-channel, and another called the Q-channel (Soulimane et al. 2000). There are no experimental studies to support a functional role of these putative channels, and sequence analysis indicates that the D- and Q-channels are not conserved in the B-family of oxygen reductases. Similar analyses indicates that there is only one conserved channel in the C-family (Hemp et al. 2007). Hence, it is highly likely that there is only one proton input channel in the B- and C-families of oxidases.

4.2
Nitric Oxide Reductases

The nitric oxide reductases are also structurally diverse and vary in the number of subunits present and the type of electron donor. There are currently two established families (Hendriks et al. 2000; Zumft 2005) cNOR and qNOR, and our recent work indicates three additional putative families of nitric oxide reductases: sNOR, eNOR, and gNOR families (Stein et al. 2007; Sievert et al. 2007). All five nitric oxide reductase families have a core subunit that is homologous to the main subunit in the oxygen reductases. The cNOR family forms a two-subunit complex. The main subunit (NorB) is homologous to the core subunit in the oxygen reductases and contains the catalytic site, whereas the second subunit (NorC) contains heme c and acts as an electron shuttle between a mobile cytochrome c and the core subunit. The enzymes in the qNOR family contain a single protein subunit (NorZ), which appears to be a fusion of the core subunit and a second subunit (Cramm et al. 1997). The qNOR's do not contain cytochrome c nor use cytochrome c as an electron donor. Instead, the qNOR enzymes receive electrons from membrane-bound quinols. There have been reports of a hybrid type of nitric oxide reductase which can receive electrons from either quinol or cytochrome c (Suharti et al. 2001). However, no sequence data is available to allow this enzyme to be classified.

There are no X-ray structures of a nitric oxide reductase. However, the nitric oxide reductases are homologous to the oxygen reductases and homology modelling coupled with spectroscopic data and metal analysis has allowed structural models to be inferred (Zumft 2005; Reimann et al. 2007). In the cNOR and qNOR families the main subunit (NorB or NorZ, respectively) contains a six-coordinate low spin heme, which accepts electrons from the electron donor (cytochrome c or quinol, respectively), and a binuclear active site, where catalysis occurs. The binuclear center in the nitric oxide reductases does not contain copper, as in the oxygen reductases, but instead contains an iron ion (Girsch and de Vries 1997; Hendriks et al. 1998). The three conserved histidine residues which ligate to Cu_B in the active site of the oxygen reductases are also present in the NO reductases. However, the ligation state of the iron is not clear, electronic structure calculations (Blomberg et al. 2006) and small molecule active-site models (Collman et al. 2006) suggest that a conserved glutamate acts as a fourth ligand to the iron, whereas in whole protein models it appears only the three histidine ligands are present (Zumft 2005; Reimann et al. 2007). The NO reductases do not have the His-Tyr cross-linked cofactor in the active site and do not perform oxygen reduction at an appreciable rate. It appears that all the members of the heme–copper superfamily that contain the His-Tyr cofactor are oxygen reductases whereas those without it are nitric oxide reductases.

The cNOR and qNOR families of nitric oxide reductases are not electrogenic, i.e., charges do not cross the membrane bilayer concomitant with enzyme catalysis (Reimann et al. 2007). The protons required for chemistry are delivered from the periplasmic side of the membrane and not, as in the case of the oxygen reductases, from the cytoplasmic side. Hence, there is no need for proton input channels from the cytoplasmic side of the enzymes. Indeed, there is no pattern of conservation of residues that define any proton channel from the cytoplasmic surface of the NO reductases (Reimann et al. 2007). Instead a proton input channel that leads to the active site, and which contains a pair of conserved glutamate residues near the periplasmic surface (Reimann et al. 2007; Thorndycroft et al. 2007), has been postulated.

5
Heme–Copper Oxygen Reductases in Archaea

There are many excellent reviews covering the biochemical properties of heme–copper oxygen reductases in Archaea (Schäfer et al. 1999; Schafer 2004), so we will focus on the structural properties determined by sequence analysis coupled with structural modelling. These analyses demonstrate that the Archaea harbor seven different families of heme–copper oxygen reductases, many of which have only one conserved proton channel (Tables 2 and 3). Comparative studies of these oxidases will be crucial to determine the mechanism of proton pumping.

5.1
A-Family

The A-family is the largest and best studied heme–copper family. The enzyme complex typically contains three main subunits, subunits I, II, and III, which are homologues of the subunits of the prototypical A-family oxidase, the mitochondrial cytochrome c oxidase. Many of the A-family complexes contain additional subunits (Fig. 1A). Homologues of subunit III are only found in the A-family oxygen reductases. In Archaea it is common for subunit III to be fused to the C-terminus of subunit I. On the basis of sequence divergence, approximately 20 subfamilies have been defined within the A-family of oxygen reductases. Of these 20 subfamilies, seven are exclusively found in Archaea. There are also a number of unclassified A-family oxidases identified within Archaea. These are identified in Table 3.

The most-studied A-family member from Archaea is the SoxM supercomplex from *Sulfolobus acidocaldarius* (Schafer 2004), listed as Subfamily XIV in Table 2. The SoxM supercomplex from *Sulfolobus acidocaldarius* is a fusion of an Archaeal bc_1 homolog (SoxFG) and an A-type oxygen reductase (SoxM) (Komorowski et al. 2002). The SoxFG bc_1 homolog oxidizes caldariella quinol and transfers the electrons via sulfocyanin (SoxE) to Cu_A, which is located in subunit II of the oxygen reductase. The core subunit of SoxM is a fusion of subunits I+III, and is unique in that the active site contains heme b. This is the only known example of a heme b incorporated into an A-type oxygen reductase active-site, forming a heme b_3-Cu_B binuclear center. A-family oxygen reductases have also been isolated in *Metallosphaera sedula* (Kappler et al. 2005), *Aeropyrum pernix* (Ishikawa et al. 2002), *Pyrobaculum oguniense*

Table 2 Phylogenetic distribution of Archaeal heme–copper and bd oxidase superfamily members. Organisms are classified by domain, phylum, class, and order. Organisms in **bold** are from completed genomes, those in plain font are from incomplete genomes, while those in *italics* are from individual submissions to GenBank. The number of heme–copper and bd oxidase members from each family that an organism has is given in the respective column. bd oxidases underlined have two subunit I genes encoded instead of subunits I and II

	bd	A	B	C	D	E	F	G	H	qNOR	eNOR
Crenarchaeota											
Thermoprotei											
Caldisphaerales											
Cenarchaeales											
Cenarchaeum symbiosum		1									
Desulfurococcales											
Aeropyrum pernix (K1)		1	1								
Hyperthermus butylicus (DSM 5456)		<u>1</u>									
Ignicoccus (Kin4-I)											
Staphylothermus marinus (F1)											
Marine archaeal group 1											
Nitrosopumilus maritimus (SCM1)		1									
Sulfolobales											
Acidianus ambivalens					1						
Metallosphaera sedula (DSM 5348)		1			2	1	1				
Sulfolobus acidocaldarius (DSM 639)		1			1	1					
Sulfolobus metallicus							1				
Sulfolobus solfataricus (P2/ATCC 35092/DSM 1617)		1			1	1	1				
Sulfolobus tokodaii (7/JCM 10545)		1			1	1	1			1	
Thermoproteales											
Caldivirga maquillingensis (IC-167)	<u>1</u>							1		1	
Pyrobaculum aerophilum (IM2/ATCC 51768/DSM 7523)	1		1							1	

Table 2 (continued)

	bd	A	B	C	D	E	F	G	H	qNOR	eNOR
Pyrobaculum arsenaticum (9DSM 13514)	1									1	
Pyrobaculum calidifontis (JCM 11548)	1	1								1	
Pyrobaculum islandicum (DSM 4184)	1	1									
Pyrobaculum oguniense		1									
Thermofilum pendens (Hrk 5)	1										
Euryarchaeota											
Archaeoglobi											
Archaeoglobales											
Archaeoglobus fulgidus (DSM 4304/ATCC 49558)	1										
Halobacteria											
Halobacteriales											
Haloarcula marismortui (ATCC 43049)		2	1							1	
Halobacterium salinarum (NRC-1/ATCC 700922)		4	1								
Halobacterium salinarum (R1/ATCC 19700)		2	1								
Haloferax volcanii (DS2/ATCC 29605)		1	1							1	
Halorubrum lacusprofundi (ATCC 49239)			1							1	
Haloquadratum walsbyi (DSM 16790)			1								
Natronomonas pharaonis (DSM 2160)			1								1
Methanobacteria											
Methanobacteriales											
Methanosphaera stadtmanae (DSM 3091)											
Methanothermobacter thermautotrophicum (Delta H)											
Methanococci											
Methanococcales											
Methanocaldococcus jannaschii (DSM 2661/ATCC 43067)											
Methanococcus aeolicus (Nankai-3)											

Table 2 (continued)

	bd	A	B	C	D	E	F	G	H	qNOR	eNOR
Methanococcus maripaludis (C5)											
Methanococcus maripaludis (C7)											
Methanococcus maripaludis (S2/LL)											
Methanococcus vannielii (SB)											
Methanomicrobia											
Methanomicrobiales											
Methanobrevibacter smithii (ATCC 35061)											
Methanocorpusculum labreanum (Z)											
Methanoculleus marisnigri (JR1)											
Methanoregula boonei (6A8)											
Methanospirillum hungateii (JF-1)											
Methanosarcinales											
Methanococcoides burtonii (DSM 6242)											
Methanosaeta thermophila (PT/DSM 6194)											
Methanosarcina acetivorans (C2A/ATCC 35395/DSM 2834)		1									
Methanosarcina barkeri (fusaro)		1									
Methanosarcina mazei (Go1/ATCC BAA-199/DSM 3647)											
Methanopyri											
Methanopyrales											
Methanopyrus kandleri (AV19/DSM 6324/JCM 9639)											
Thermococci											
Thermococcales											
Pyrococcus abyssi (GE5/Orsay)											
Pyrococcus furiosus (DSM 3638/ATCC 43587)											
Pyrococcus horikoshii (OT3)											
Thermococcus kodakaraensis (KOD1)											

Table 2 (continued)

	bd	A	B	C	D	E	F	G	H	qNOR	eNOR
Thermoplasmata											
Thermoplasmatales											
Ferroplasma (type II)		1									
Ferroplasma acidarmanus (fer1)	2	1									
Picrophilus torridus (DSM 9790/ATCC 70027)	2	1									
Thermoplasma acidophilum (DSM 1728)	2										
Thermoplasma volcanium (GSS1/DSM 4299/JCM 9571)	2										
Thermoplasmatales archaeon (Gp1)	2										
Unclassified											
Methanogenic archaea (Rice Cluster I)									1		
Korarchaeota											
Nanoarchaeota											
Nanoarchaeum											
Nanoarchaeum equitans (Kin4-M)											

Table 3 Archaeal members of the heme-copper superfamily. This table lists all of the currently known members of the heme-copper superfamily found within Archaea, including those from environmental samples. The organism, identification number, operon structure, and type of electron acceptor (subunit II) are given for each member. The NCBI identification number is given for environmental samples

A-family oxygen reductases			
Subfamily XII			
Aeropyrum pernix (K1)	5010020	II I+III Sco1	Cu$_A$
Subfamily XIII			
Haloarcula marismortui (ATCC 43049) [1]	5020319a	III [I] II	Cu$_A$
Haloarcula marismortui (ATCC 43049) [3]	5020319c	[II] [I] [III]	Cu$_A$
Halobacterium salinarum (NRC-1/ATCC 700922)	5020320	II [III] I	Cu$_A$
Halobacterium salinarum (R1/ATCC 19700) [1]	5020321	II I III	Cu$_A$
Haloferax volcanii (DS2/ATCC 29605) [2]	5020322b	[II CtaB] [I] [III]	Cu$_A$
Haloquadratum walsbyi (DSM 16790)	5020323	[II CtaB] [I] [III]	Cu$_A$
Natronomonas pharaonis (DSM 2160)	5020447	CtaB II III I	Cu$_A$
Subfamily XIV			
Metallosphaera sedula (DSM 5348)	5010392	[SoxF SoxG II SoxI] SoxE I+III	Cu$_A$
Sulfolobus acidocaldarius (DSM 639)	5010718	[SoxF SoxG II SoxI] SoxE I+III	Cu$_A$
Sulfolobus solfataricus (P2/ATCC 35092/DSM 1617)	5010719	[SoxF SoxG II SoxI] SoxE I+III	Cu$_A$
Sulfolobus tokodaii (7/JCM 10545)	5010720	[SoxF SoxG II SoxI] SoxE I+III	Cu$_A$
Subfamily XV			
Pyrobaculum aerophilum (IM2/ATCC 51768)	5010543	II I+III X CtaB	Cu$_A$
Pyrobaculum calidifontis (JCM 11548)	5010893	II I+III X CtaB	Cu$_A$
Pyrobaculum oguniense	501X004	II I+III	Cu$_A$
Subfamily XVI			
Ferroplasma (type II)	5020274	I+III Sulfocyanin II CtaB	Cu$_A$
Ferroplasma acidarmanus (fer1)	5020275	I+III Sulfocyanin II CtaB	Cu$_A$
Picrophilus torridus (DSM 9790/ATCC 70027)	5020496	I+III Sulfocyanin II CtaB	Cu$_A$

Table 3 (continued)

Subfamily XVII			
Cenarchaeum symbiosum	5010160	X II I Cyanin CtaA	Cu$_A$
Nitrosopumilus maritimus (SCM1)	5011037	X II I Cyanin CtaA	Cu$_A$
	AACY020042774		
	AACY020165265		
	AACY020271526		
	AACY020383948		
	AACY020408154		
	AACY021781753		
	AACY022062082		
	AACY022206108		
	AACY022690326		
	AACY023085447		
	AACY023776262		
	AACY024074640		
Subfamily XVIII			
Haloarcula marismortui (ATCC 43049) [2]	5020319b	II I+III IV	Cu$_A$
Haloferax volcanii (DS2/ATCC 29605) [1]	5020322a	II I+III IV	Cu$_A$
Halorubrum lacusprofundi (ATCC 49239)	5020995	II I+III IV	Cu$_A$
Unclassified subfamily			
Euryarchaeota (Predicted from genomic context)	AACY01004662	II I III	Cu$_A$
	AACY020067637		
	AACY020071257		
	AACY020098741		
	AACY020109206		
	AACY020112219		
	AACY020116897		
	AACY020207114		

Table 3 (continued)

AACY020242961
AACY020255478
AACY020321485
AACY020327424
AACY020330951
AACY020335240
AACY020338338
AACY020402825
AACY020499439
AACY020546099
AACY020561279
AACY021070130
AACY022154585
AACY022225003
AACY022384987
AACY022690422
AACY022694601
AACY022699076
AACY022700678
AACY023388866
AACY023397990
AACY023399007
AACY023410494
AACY023454892
AACY023474211
AACY023519752
AACY023526653
AACY024102295

Table 3 (continued)

	DU733775		
B-family oxygen reductases			
Subfamily IV			
Haloarcula marismortui (ATCC 43049)	5020319	CbaE Halocyanin/IIa II I	Cu_A
Halobacterium salinarum (NRC-1/ATCC 700922)	5020320	CbaE Halocyanin/IIa II I	Cu_A
Halobacterium salinarum (R1/ATCC 19700)	5020321	CbaE Halocyanin/IIa II I	Cu_A
Haloferax volcanii (DS2/ATCC 29605)	5020322	CbaE IIa II I	Cu_A
Halorubrum lacusprofundi (ATCC 49239)	5020995	CbaE IIa II I	Cu_A
Natronomonas pharaonis (DSM 2160)	5020447	CbaE IIa II I	Cu_A
Subfamily V			
Aeropyrum pernix (K1)	5010020	[IIa II scoI] [I]	Cu_A
Pyrobaculum aerophilum (IM2/ATCC 51768)	5010543	IIa II I scoI	Cu_A
Pyrobaculum calidifontis (JCM 11548)	5010893	IIa II I scoI	Cu_A
Pyrobaculum oguniense	501X004	IIa II I scoI	Cu_A
D-family oxygen reductases			
Acidianus ambivalens	501X001	doxB doxC doxE	quinol
Metallosphaera sedula (DSM 5348) [1]	5010392	doxB doxC doxE	quinol
Sulfolobus acidocaldarius (DSM 639)	5010718	doxB doxC doxE	quinol
Sulfolobus solfataricus (P2/ATCC 35092/DSM 1617)	5010719	doxB doxC doxE	quinol
Sulfolobus tokodaii (7/JCM 10545)	5010720	doxB doxC doxE	quinol
Operon code:			
doxB – subunit I homolog			
doxC – subunit II homolog (very distant)			
E-family oxygen reductases			
Subfamily I			
Metallosphaera sedula (DSM 5348)	5010392	soxA soxB soxC soxD	No Cu_A
Sulfolobus acidocaldarius (DSM 639)	5010718	soxA soxB soxC soxD	No Cu_A

Table 3 (continued)

Sulfolobus solfataricus (P2/ATCC 35092/DSM 1617)	5010719	soxA soxB soxC soxD	No Cu$_A$
Sulfolobus tokodaii (7/JCM 10545) [1]	5010720a	soxA soxB soxC soxD	No Cu$_A$
Subfamily II			
Sulfolobus tokodaii (7/JCM 10545) [2]	5010720b	soxA soxB	Cu$_A$
Operon code:			
soxB – subunit I homolog			
soxA – subunit II homolog			
F-family oxygen reductases			
Metallosphaera sedula (DSM 5348) [1]	5010392a	foxB foxC foxD foxE foxF foxJ foxG foxH [foxI foxA foxA]	Cu$_A$
Metallosphaera sedula (DSM 5348) [2]	5010392b	foxG foxH foxI foxJ foxE foxF foxD]	Cu$_A$
Sulfolobus metallicus	501X002	[foxB foxC foxD]	
Sulfolobus tokodaii (7/JCM 10545)	5010720	foxA foxB foxC foxD foxE foxF [foxG foxH foxI foxJ]	Cu$_A$
G-family oxygen reductases			
Caldivirga maquilingensis (IC-167)	5010896	[II soxE] I bc1	Cu$_A$
H-family oxygen reductases			
Thermoplasmatales archaeon (Gpl)	5020905	II I CtaB	Cu$_A$
qNOR-family nitric oxide reductases			
Caldivirga maquilingensis (IC-167)	5010896		
Haloarcula marismortui (ATCC 43049)	5020319		
Haloferax volcanii (DS2/ATCC 29605)	5020322		
Halorubrum lacusprofundi (ATCC 49239)	5020995		
Pyrobaculum aerophilum (IM2/ATCC 51768/DSM 7523)	5010543		

Table 3 (continued)

Pyrobaculum arsenaticum (9DSM 13514)	5010940	
Pyrobaculum calidifontis (JCM 11548)	5010893	
Sulfolobus solfataricus (P2/ATCC 35092/DSM 1617)	5010719	
eNOR-family nitric oxide reductases		
Natronomonas pharaonis (DSM 2160)	5020447	IIa II I Cu_A

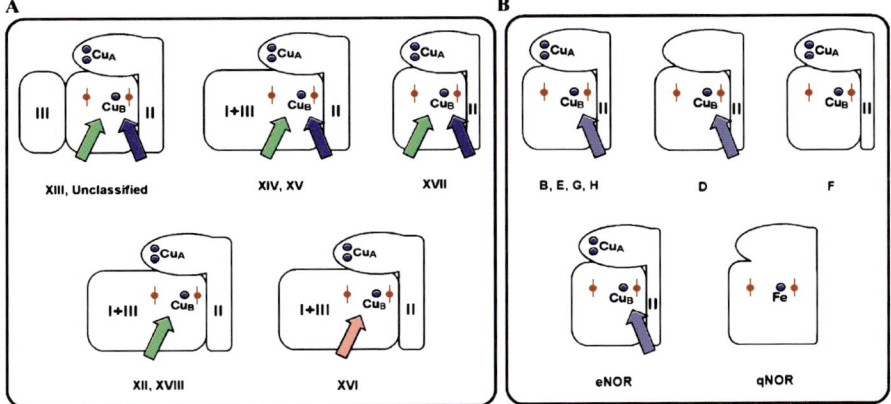

Fig. 1 Core subunit architecture of the heme–copper superfamily. **A** The structural diversity within the A-family in Archaea. **B** The other oxygen reductase and nitric oxide reductase families found in Archaea. Conserved channels are indicated by *arrows*. The K-channel is *blue*. The D-channel is *green*, while the modified D-channels are shown in *red*. The family and subfamily identifications are given in Table 3

(Nunoura et al. 2005), *Halobacterium halobium* (Fujiwara et al. 1989), and *Haloferax volcanii* (Tanaka et al. 2002).

The prototypical A-type oxygen reductases have two conserved proton channels, the D- and K-channels, for the transport of chemical and pumped protons. However, a few subfamilies that are unique to Archaea have modified D- or K-channels. Table 4 shows the key conserved residues that define these two channels. In three subfamilies (XII, XVI, and XVIII) most of the residues defining the K-channel are not conserved (Fig. 2), and in Subfamily XVII two of the residues are replaced by residues that could conceivably also function to assist proton transfer (e.g., *R. sphaeroides* K362 is replaced by a histidine). Subfamily XVI, in addition to missing conserved residues in the K-channel, has a modified D-channel in which one of the conserved asparagines (N207) is replaced by an aspartate (Table 4). The N207D mutant of the *R. sphaeroides* oxidase has been shown to be decoupled from the proton pump (Han et al. 2006). Therefore, the properties of this oxidase are clearly of interest. Figure 1A summarizes the features of the channels in the Archaeal A-family oxygen reductases.

5.2
B-Family

The B-type oxygen reductase complex consists of three subunits (Subunits I, II, and IIa). Subunits I and II are homologous to subunits I and II of the A-type family. The third subunit (IIa) is a single transmembrane helix which is structurally analogous to the second transmembrane helix in subunit II

Table 4 Channel properties of Archaeal heme-copper superfamily members. Conserved channel residues from known crystal structures (*Rhodobacter sphaeroides* [1M56] and *Thermus thermophilus* [1XME]) are given. Non-standard channel residues are indicated with an *asterisk*. Many Archaeal A-family oxidases are missing important K-channel residues

Family	A	A	A	A	A	A	A	A	A	B	B	B	D	E	F	G	H	sNOR	eNOR
Subfamily	IV	XII	XIII	XIV	XV	XVI	XVII	XVIII	UNC	III	IV	V							
PDB	1M56									1XME									
K-channel	Y288	Y	Y	Y	Y	Y	Y	Y	Y	Y237	Y	Y	Y	Y	Y	Y	Y	N*	Q*
	T359	–	S	S	S	S	T	S	–	T	S309	S	S	S/T	S	–	S	S	–	S
	K362	–	K	K	K	K	–	H*	–	K	T312	T	T	T	T	–	T	T	–	H*
	S299	–	S	–	–	S	–	Y*	–	S	Y248	Y	Y	Y	Y	–	Y	Y	Y	Y
	E101	–	E	E	E	E	–	E	–	E	E15	E	E	E	E	–	E	E	–	E
D-channel	E286	E	E	E	E	E	E	E	E										
	Y33	Y	Y	Y	Y	Y	Y	Y	Y										
	N139	N	N	N	N	N	N	N	N										
	N121	N	N	N	–	N	N	N	N										
	N207	N	N	N	E*	N	N	N	N										
	D132	D	D	D	D	D	D	D	D										

UNC Unclassified subfamily
* Non-standard amino acid substitutions

Fig. 2 Sequence alignment of representatives from each Archaeal heme–copper family. ▶ Residues important for channel formation are indicated below the alignment. Sequence identifiers are given in Table 3. Sequences for known crystal structures are also included for reference. (*Rhodobacter sphaeroides* [A1200559a] is an A-family oxidase, while *Thermus thermophilus* [B1110763] is a B-family oxidase. Both species are members of the Bacteria domain.) For space considerations non-homologous regions have been removed from the alignment

from the A-type oxygen reductases (Soulimane et al. 2000). However, subunit IIa is not homologous to the second helix. The B-family of oxygen reductases can be subdivided into five subfamilies (I, II, III, IV, and V) based on groupings with greater than 50% sequence identity in the core subunit. Subfamilies IV and V are found in Archaea exclusively (Table 4). Archaeal members of the B-family have been studied in *Aeropyrum pernix* (Ishikawa et al. 2002), *Pyrobaculum oguniense* (Nunoura et al. 2003), and *Natronobacterium pharaonis* (Mattar and Engelhard 1997).

Mapping of conserved residues (Fig. 2) onto the crystal structure for the *Thermus thermophilus* ba_3 oxidase shows that there is only one conserved proton input channel, a modified K-channel analog (Fig. 1B). The residues which were previously identified to form an analogous D-channel and an alternative proton channel (Q-channel) (Soulimane et al. 2000) are not conserved across the family. The pattern of conserved residues, therefore, suggests that the B-type oxygen reductases have only one proton channel, analogous to the K-channel in the A-family oxygen reductases. This is being tested experimentally.

5.3
C-Family

The C-family of oxygen reductases, also known as the cbb_3 oxidases, is the second largest oxygen reductase family (24%) after the A-family (71%) found in genomic and community DNA sequences. There are no C-family oxygen reductases found in the Archaea in the sequences collected to date. It is noted that two subunits in the C-family oxygen reductases contain heme c. The ability to synthesize c-type cytochromes is unusual and is sporadically distributed through the Archaea (Bertini et al. 2007). Hence, most Archaea do not have the machinery needed to correctly assemble the proteins in the C-family of oxygen reductases.

5.4
Families D Through H

The classification scheme introduced by Pereira et al. (2001) and colleagues would combine the enzymes represented in the D-, E-, F-, G-, and H-families

```
              ------Helix XIII------       ------             ------Helix IX------                                          ------Helix X------                                           ------Helix XI------
A1200559a     VIAVPT--GIKIFSWIATM---TPMLWALGF-LFLFTV-GGVTGIV-LSQASVDRYHDTYVVAHFHYVMSLGAVFGIFAGIYFWIGKM---WAGKLHFWMMFVGANLTFFPQHFLGRQ-GMPRR---TW
A5010020      LISIPF--EMAVMSFIFTL---VPMLFAVGA-LLNFII-GGSTGVV-LGSIAIDRGFR-TYWVVAHFHYILVGTVTLGLIAGLYYWWPKI---RLGKIHFALAMLGVALTLPQFALMDM--PRR---PL
A5010160      AAVPAS--AMHVFNFVATM---SPMMWSIGG-IALFFS-AGAGGVL-NAAMPLDFTTHDTYWVGHFHLFVMGTIAFGSIGFFYIMFPFV---KLGKIHFVSFVGAVLLFFTQHILGIY-GMPRR---AM
A5010718      AIAIPS--AMHVFNWVAIM---TPTILLISF-IVMFLL-GGIFGVF-FPLVPIDYALNGTYFVVGHFHMYA-ILYALLGALFYYFPFW---DLGKTGAILLVAGTFLTATGMSIAGIL-GMPRR---PL
A501X004      AVAIPS--GVKVFNWVATL---APMLFMLTF-LLFFIV-GGVTGVF-FPVVFLDLHLQDTYFVLGHFHYIV-NAIAFGALGALLYFPHL---RLAVLSWGLIITVGGFMAVTLMSAAGIL-GSPRR---PY
A5020319b     GISLFF--DLMVFSLIYTM---TPFLFSLGA-LILFII-GGITGVF-LGAIVLDIQFRGTYWVVAHFHYVMVAGAT-ALFGGLYYWYFKI---TLGKIQFGVYIFLGFNLLYFPMFIAWET--PRR---IW
A5020320      AISIPS--AVKVFNWITTM---APMLFCIGF-VQNFII-GGVTGVF-LAVIPIDLILHDTYVVAHFHFTVYGAIGFALFAASYYWFPMV---RLAHAHFWTALVGSNATFLAMLWLGYG-GMPRR---TA
A5020496      TISIPT--SMLIVGLIATM---TPILFALAA-STSFMI-GGVSGVT-QSSLAIDTIVHGDYYVVAHFHYMVGAVMGLMGNLLYYYFFKF---KLGRIHWAFAYPGVNILYFPMEFLYDL--PRR---IF
AAACY01004662 LVAVPT--GIKIFNWLMTM---THTLWALGF-LVTFTL-GGISGMF-FPSIAMDTHLHESYFVVAHFHVLVGGTVFGFFAGIYYWWFKA---KLGVLHFLTAFVSYNALFWFPMERLGVW-GMTRR---GW
B1110763      FVAVPS--LMTAFTVAASL---NPAFVAPVLGLLGFIP-GGAGGIV-NASFTLDYVHNTAWPGHFHLQVASLVTLTAMGSLYWLLPNL---RGLAVVWLWFLGMMIMAVGLHWAGLL-NVPRR---VP
B501X004      LVAVPS--FITAPNVIATL---DPVFTGTFAIILFGI-AGFSGVV-NASFNVNYNVHNTAWIVGHFHLTVGGGVTLTFIATSFLLVPLL---KFLIAVPTLWFVGQLIFGISYHIVGLL-HAPRR---PW
B5020320      MLLLPS--LLTAFTVASL---KPSFSAMALAGLVFAA-GGFSGMI-NAGMNINYLIHNTINVPGHFHLTVSGGVTLTFIATSFLLVPLL---TLAAIQPFWVLGMTLMSNAMHRGGLA-GLPRR---EM
D501X001      AVVVPS--MMTFLNLWATA---NVITAFVATSFAGAIA-AGVTGIA-NATIAFDSIVHNSMWVGHFHAMILLSIVPAAMAVLYFMIPMM---KMAWIHYIGYTIGASILIIGFEMIGFY-GVVRR---FA
E5010718      ILATGS--GLTVLNLGLTI---DPVGMGALISLIGFIL-AGAQALV-LPENSINPLFHNSYYVVGHFHLMIWTLIIMGTTVFLDMLRTS---KWMRLGMIWWTAPFMGVGYAMSVGYL-GFLRR---PY
F5010720      ALIAIP--FAAFWLLFFTI---DTGLAPFLYAAAIWNI I-GGLAPLV-AGAQALV-DPYMESLIA-MFANLE-AGYSGVA-QATVAWNLFHMTLWVAHFHTMALLNIAMVAIAILYYIVFKA---KLSWLHFWCWQAGMLFVTASAIGFY-GLIRR---PY
G5010896      GISIVS--AVSVFALLAMM---DPYMESLIA-MFAMLE-AGYSGVA-QATVAWNLILHNTLWVAHFHTMALLNIAMVAIALYYIVEKA---KWSKVHFWMTLVGGFGVNTWFAEGVL-GVPRR---NL
H5020905      LVVIPS--AITIFNYIASL---TPMLFVNG-IFDFII-GGITGVM-QSMLVVNIQVHGTYWVT GHFHFFMGITTGILFASVYMLWPTI---RLATWHMALTAIGSFIMSMGWTIGGFL-GMPRY---VY
sNOR1070753   AVGVPAFFVTIVGSLGLL---APVLFILAG-LWGWTV-QGMAAVI-DCTIAINNVMHNTLWVPAHFHSYYLFGAAGYTWAYLFHRLSGV--ALARRLAWLYAIGAGFVLAFFLSGLQ-SVPRR---GL
eNOR5020447   GAAFAS--MIHAFAIPAGL---NPVFSATIFSIILFGFMGGITGVL-MGQLQLNMTWHNTFATVGHFHTVALGTLAFMGLVFFVIRTM---KLASLVPFYYAGAMGIAVLMMMYVGLLYVPRR---PL
K-channel                                *
D-channel                                                                                                    *  *
Heme ligands             *  *

              ------Helix XII------
A1200559a     NFVSSLGAFLSFASFLFFLGVIFY
A5010020      NQLSTLGAFIFGGSMAIGLVNFLY
A5010160      NQIATVGAMLIGVGMAIFLGNMIH
A5010718      QFMASVGAVLTGIGLFILAGVLVH
A501X004      HVGMIFTTIILAAGLVIYFANLLY
A5020319b     HSMATVGGFILGASFLLMFYNLFV
A5020320      HRLATVGAFLIGVSTLIWLFNMAT
A5020496      NEIATAGAFIFGLSFLIMFANFAW
AAACY01004662 NMFITVSGFIFFSNFIMVFNMIV
B1110763      MVFNVLAGIVLLVALLLFIYGLFS
B501X004      LQLGAIGGVIFALGGALFLLLTFA
B5020320      QLQIAIGGTLLFVSLALFLAVMVG
D501X001      ENLATAGALIIAEIATLVWFVNIVA
E5010718      NLVESPLAEIGIPGLLLTLFVGMF
F5010720      YQLLMIGGWVAGFSLLVFTFNLIL
G5010896      DLIGVIFALILGIAQGIFAVKLFN
H5020905      QDIAIAGGVIIGIGQLFFIANLIK
sNOR1070753   AQAAVPFVLLLVIAFGGLLLMALR
eNOR5020447   FLILGPAAVLAIIAGALFVVVAVL
K-channel
D-channel
Heme ligands
```

with those in the B-family. We have separated these into distinct families based on the extreme divergence that is evidenced by their sequences. These five families are unique to Archaea, and some are represented by only a single sequence entry. As additional sequences become available, it is likely that the number of highly divergent oxygen reductases will increase substantially.

D-Family

The family of oxygen reductases related to *DoxB* from *Acidianus ambivalens* (Purschke et al. 1997) has been reclassified as the D-family. The enzyme was originally isolated from *Acidianus ambivalens* as a complex of five proteins encoded by two operons, DoxBCE and DoxDA. Recently, it has been shown that these two operons in fact encode two separate complexes; a thiosulfate:caldariella quinone oxidoreductase (DoxDA) and a caldariella quinol:oxygen reductase (DoxBCE) (Muller et al. 2004). The complexes together couple the oxidation of thiosufate to the reduction of O_2.

Subunit II of the oxygen reductase complex is very divergent and does not contain residues required for the formation of the Cu_A site. This is not surprising since the complex oxidizes caldariella quinol and other oxygen reductases which oxidize quinols are, as a rule, also missing the residues which bind Cu_A. It has been suggested that a caldariella quinone may be permanently bound to subunit I as a redox active cofactor, functionally taking the place of the Cu_A site. This cofactor would be capable of sequentially transferring two electrons to the low spin heme a (Schafer 2004).

Structural modelling and sequence analysis of subunit I have identified the presence of a modified K-channel, identical to the one found in the B-family (Tables 1 and 4 and Figs. 1B and 2). No other conserved channels can be defined based on the pattern of residue conservation. It was previously suggested that the D-family has a D-channel analog, with E80 functionally replacing the glutamate (E286 in Rhodobacter) at the top of the D-channel in the A-family (Gomes et al. 2001). However, this glutamate (E80) is not conserved in any of the other D-family members. This observation, combined with the fact that most of the other residues implicated in the formation of D- and Q-channel analogs (Victor et al. 2004) are not conserved, strongly suggests that the D-family only has one functional proton input channel.

E-Family

The E-family members are related to the *SoxB* oxygen reductase from *Sulfolobus acidocaldarius* (Lübben et al. 1992). The proteins encoded by the SoxABCD operon form a supercomplex with SoxLN which is an Archaeal equivalent of a bc_1 complex. It appears that caldariella quinone is the ultimate electron donor, with the electrons being passed from the quinone to the SoxLN complex, which transfers them directly to the SoxABCD complex (Schafer 2004). This is similar to the situation with the SoxM supercomplex (also found in *Sulfolobus acidocaldarius*), except in this case the oxygen reductase component (*SoxB*) is an E-family oxygen reductase. Structural analysis shows that the E-family contains a conserved K-channel similar to the one found in the B-family (Tables 1 and 4 and Figs. 1B and 2).

F-Family
The oxygen reductase genes related to *FoxA* from *Sulfolobus metallicus* (Bathe and Norris 2007) have been classified as the F-family. These genes form part of a gene cluster that is highly expressed when the cells are grown on ferrous iron or pyrite but not sulfur. This implies that this oxygen reductase family is part of a respiratory chain that specifically couples the oxidation of ferrous iron [Fe(II) oxidized to Fe(III)] to the reduction of O_2.

Structurally, the F-family is a two subunit complex with subunit II containing a Cu_A binding domain. Analysis of subunit I shows that it does not have any conserved proton input channels (Tables 1 and 4 and Figs. 1B and 2). Unless non-conserved residues comprise the proton delivery channel, this suggests that the F-family of oxygen reductases cannot be electrogenic since both the electrons and protons would have to be derived from the same side of the membrane, and no proton channels are available for proton pumping. This situation is similar to that of the NO reductases. The putative lack of proton pumping may not be surprising, considering the driving force available from Fe(II) oxidation and the acidic environment in which these organisms grow. However, one might expect at the least that protons would be utilized from the cytoplasm to maintain a neutral internal pH. This is another example in which biochemical studies are needed to test the validity of predictions inferred by sequence analysis.

Interestingly, the acidophilic bacteria *Thiobacillus ferrooxidans* contains a modified A-type oxygen reductase that is expressed when grown on ferrous iron (Yarzabal et al. 2004). This modified A-type oxygen reductase is also missing the residues which form the K-channel, though the D-channel appears to be present (Ingledew 2004). Possibly, this bacterial A-type oxidase also lacks the ability to pump protons but, in this case, protons are utilized from the cytoplasm to form water at the enzyme active site. This would be an example of the same evolutionary solution (lack of proton channels) to an environmental problem (low pH) being implemented in two different families.

G- and H-Families
Members of the G- and H-families are recent discoveries and, so far, represented by only one sequence entry for each. Neither family has been biochemically or genetically characterized. The G-family sequence was identified in the Crenarchaeota *Caldivirga maquilingensis (IC-167)* in an operon encoding homologs to subunits I and II along with genes for sulfocyanin, polyferredoxin, and *SoxL* and *SoxC* homologs. These proteins could form a supercomplex similar to the one found in the E-family. The H-family was identified in *Thermoplasmatales archaeon (Gpl)* assembled from acid mine drainage

(Tyson et al. 2004). It is located within an operon encoding subunit I and II homologs along with a CtaB homolog. *Caldivirga maquilingensis (IC-167)* and *Thermoplasmatales archaeon (Gpl)* do not appear to encode any other heme–copper oxygen reductases. This implies that both of the enzymes representing these families are functional in their respective organisms for oxygen respiration.

In these two enzymes, subunit II has the residues required for the formation of a Cu_A binding site and subunit I has the residues needed for the formation of a modified K-channel, analogous to the one found in the B-family of oxygen reductases (Tables 1 and 4 and Figs. 1B and 2). Genomic and community sequencing projects in progress along with biochemical analyses will be required to further elucidate the properties of these new families.

6
Distribution and Evolution of the Heme–Copper Superfamily

The phylogenetic and family distribution of heme–copper superfamily members found within sequenced Archaeal genomes is given in Tables 1 and 2. The Archaea encode members from seven of the eight currently identified heme–copper oxygen reductase families, whereas the Bacteria only have members from three; the A-, B-, and C-families. The D-, E-, and F-families have so far only been found in the Sulfolobales class of the Crenarchaeota. Figure 3 shows the phylogenetic positions of the newly identified heme–copper families. Phylogenetic analyses strongly suggest that the heme–copper oxygen reductases originated within the Bacteria and members were later transferred to the Archaea. This leads to the question: Why is there so much sequence divergence within Archaeal members of the superfamily after they have acquired them from Bacteria? One answer may be that Archaeal membranes select for different sequence characteristics than do Bacterial membranes, leading to rapid divergence. Another might be that since most of the oxygen reductases in Archaea have only one channel there is less selection pressure on most of the protein, and hence a higher rate of evolution.

The nitric oxide reductase members of the superfamily which are found in Archaea are predominantly from the qNOR family. Very few Archaea utilize cytochrome c as an electron carrier, which would exclude the use of the cytochrome c-requiring cNOR family. The qNOR family has a very sporadic and phylogenetically discordant distribution with members found in both the Crenarchaeota and Euryarchaeota. Phylogenetic analysis of the qNOR family

Fig. 3 Phylogenetic analysis of new Archaeal heme–copper families. Sequence identifiers ▶ are given in Table 3. Identifiers starting with a 1 represent Bacteria sequences

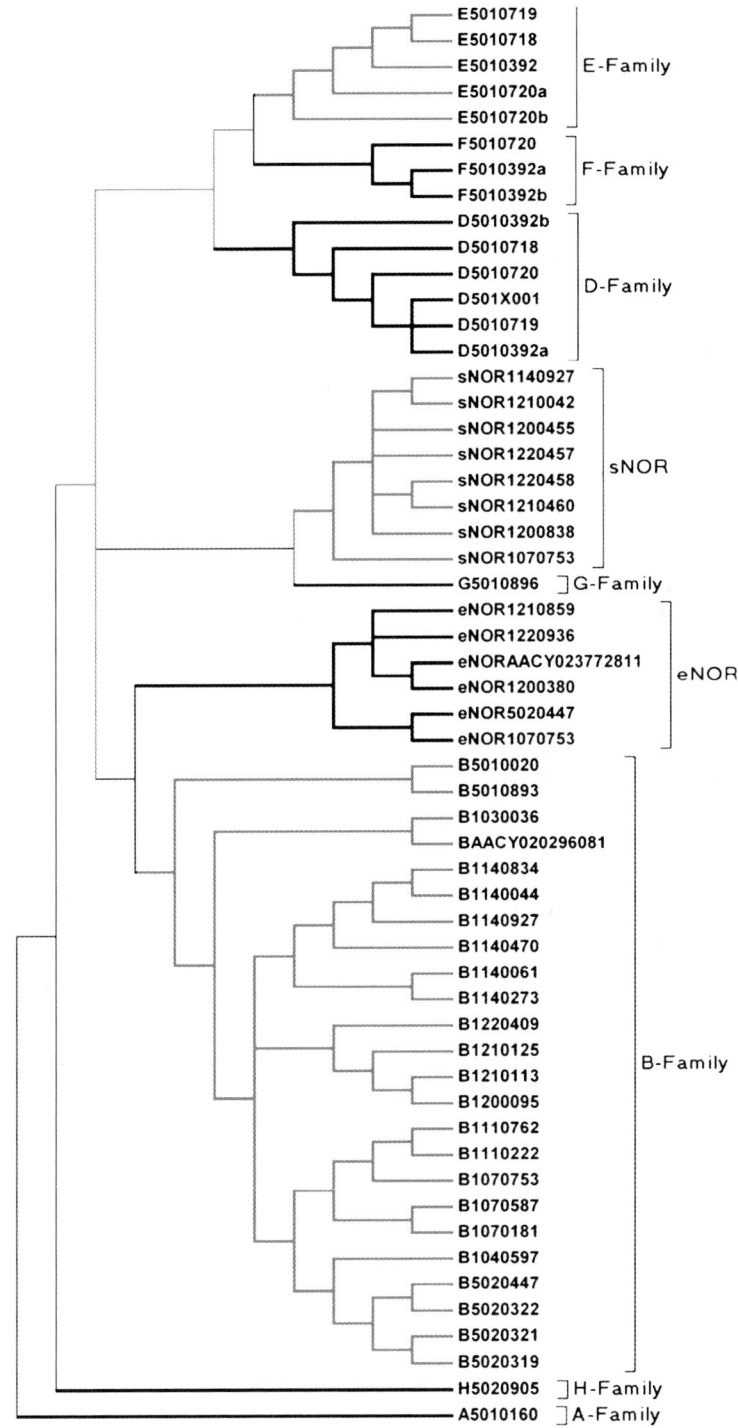

also shows that it is very likely that the Archaea acquired them via lateral gene transfer from Bacteria.[1]

References

Abramson J, Riistama S, Larsson G, Jasaitis A, Svensson-Ek M, Laakkonen L, Puustinen A, Iwata S, Wikström M (2000) The Structure of the Heme–Copper Oxidase from *Escherichia coli* and its Binding Site for Ubiquinone. Nat Struct Biol 7(10):910–917

Bathe S, Norris PR (2007) Ferrous iron- and sulfur-induced genes in Sulfolobus metallicus. Appl Environ Microbiol 73(8):2491–2497

Bertini I, Cavallaro G, Rosato A (2007) Evolution of mitochondrial-type cytochrome c domains and of the protein machinery for their assembly. J Inorg Biochem 101(11–12):1798–1811

Blomberg LM, Blomberg MR, Siegbahn PE (2006) Reduction of nitric oxide in bacterial nitric oxide reductase-a theoretical model study. Biochim Biophys Acta 1757(4):240–252

Cavicchioli R (2006) Cold-adapted archaea. Nat Rev Microbiol 4(5):331–343

Cho CM, Yan T, Liu X, Wu L, Zhou J, Stein LY (2006) Transcriptome of a Nitrosomonas europaea mutant with a disrupted nitrite reductase gene (nirK). Appl Environ Microbiol 72(6):4450–4454

Collman JP, Yan YL, Lei J, Dinolfo PH (2006) Active-site models of bacterial nitric oxide reductase featuring tris-histidyl and glutamic acid mimics: influence of a carboxylate ligand on Fe(B) binding and the heme Fe/Fe(B) redox potential. Inorg Chem 45(19):7581–7583

Cramm R, Siddiqui RA, Friedrich B (1997) Two isofunctional nitric oxide reductases in Alcaligenes eutrophus H16. J Bacteriol 179(21):6769–6777

DeLong EF, Pace NR (2001) Environmental diversity of bacteria and archaea. Syst Biol 50(4):470–478

DeSantis TZ, Hugenholtz P, Larsen N, Rojas M, Brodie EL, Keller K, Huber T, Dalevi D, Hu P, Andersen GL (2006) Greengenes, a chimera-checked 16S rRNA gene database and workbench compatible with ARB. Appl Environ Microbiol 72(7):5069–5072

Fujiwara T, Fukumori Y, Yamanaka T (1989) Purification and properties of *Halobacterium halobium* cytochrome aa3 which lacks CuA and CuB. J Biochem (Tokyo) 105(2):287–922

Garcia-Horsman JA, Barquera B, Rumbley J, Ma J, Gennis RB (1994) The Superfamily of Heme–Copper Respiratory Oxidases. J Bacteriol 176(18):5587–5600

Girsch P, de Vries S (1997) Purification and initial kinetic and spectroscopic characterization of NO reductase from *Paracoccus denitrificans*. Biochim Biophys Acta 1318(1–2):202–216

Giuffre A, Stubaueer G, Sarti P, Brunori M, Zumft WG, Buse G, Soulimane T (1999) The Heme–copper Oxidases of *Thermus thermophilus* Catalyze the Reduction of Nitric Oxide: Evolutionary Implications. PNAS 96:14718–14723

Gomes CM, Backgren C, Teixeira M, Puustinen A, Verkhovskaya ML, Wikstrom M, Verkhovsky MI (2001) Heme–copper oxidases with modified D- and K-pathways are yet efficient proton pumps. FEBS Lett 497(2–3):159–164

---

[1] Sequences used in this study were retrieved from the NCBI, JGI, JVCI, and TIGR databases. Since many genomes were not annotated an inhouse numbering system was used to provide a unique identifier for each sequence. Sequences and alignments are available from the authors at request.

Han D, Namslauer A, Pawate A, Morgan JE, Nagy S, Vakkasoglu AS, Brzezinski P, Gennis RB (2006) Replacing Asn207 by aspartate at the neck of the D channel in the aa3-type cytochrome c oxidase from Rhodobacter sphaeroides results in decoupling the proton pump. Biochemistry 45(47):14064–14074

Hemp J (2007) The Heme–Copper Oxidoreductase Superfamily: Genomics and Structural Analyses. PhD Thesis, University of Illinois at Urbana-Champaign

Hemp J, Han H, Roh JH, Kaplan S, Martinez TJ, Gennis RB (2007) Comparative genomics and site-directed mutagenesis support the existence of only one input channel for protons in the C-family (cbb3 oxidases) heme–copper oxygen reductases. Biochemistry 46(35):9963–9972

Hemp J, Robinson DE, Ganesan KB, Martinez TJ, Kelleher NL, Gennis RB (2006) Evolutionary Migration of a Post-Translationally Modified Active-Site Residue in the Proton-Pumping Heme–Copper Oxygen Reductases. Biochemistry 45(51):15405–15410

Hendriks J, Oubrie A, Castresana J, Urbani A, Gemeinhardt S, Saraste M (2000) Nitric oxide reductases in bacteria. Biochim Biophys Acta 1459(2–3):266–273

Hendriks J, Warne A, Gohlke U, Haltia T, Ludovici C, Lubben M, Saraste M (1998) The active site of the bacterial nitric oxide reductase is a dinuclear iron center. Biochemistry 37(38):13102–13109

Hosler JP, Ferguson-Miller S, Calhoun MW, Thomas JW, Hill J, Lemieux L, Ma J, Georgiou C, Fetter J, Shapleigh J, Tecklenburg MMJ, Babcock GT, Gennis RB (1993) Insight into the Active-Site Structure and Function of Cytochrome Oxidase by Analysis of Site-Directed Mutants of Bacterial Cytochrome aa_3 and Cytochrome bo. J Bioenerg Biomembr 25(2):121–136

Hosler JP, Ferguson-Miller S, Mills DA (2006) Energy transduction: proton transfer through the respiratory complexes. Annu Rev Biochem 75:165–187

Ingledew WJ (2004) Fe(II) Oxidation by *Thiobacillus ferrooxidans*: The Role of the Cytochrome c Oxidase in Energy Coupling. Springer, Berlin Heidelberg New York

Ishikawa R, Ishido Y, Tachikawa A, Kawasaki H, Matsuzawa H, Wakagi T (2002) *Aeropyrum pernix* K1, a strictly aerobic and hyperthermophilic archaeon, has two terminal oxidases, cytochrome ba3 and cytochrome aa3. Arch Microbiol 179(1):42–49

Kappler U, Sly LI, McEwan AG (2005) Respiratory gene clusters of *Metallosphaera sedula* – differential expression and transcriptional organization. Microbiology 151(Pt 1):35–43

Komorowski L, Verheyen W, Schafer G (2002) The archaeal respiratory supercomplex SoxM from *S. acidocaldarius* combines features of quinole and cytochrome c oxidases. Biol Chem 383(11):1791–1799

Konstantinov AA, Siletsky S, Mitchell D, Kaulen A, Gennis RB (1997) The Roles of the Two Proton Input Channels in Cytochrome *c* Oxidase from *Rhodobacter sphaeroides* Probed by the Effects of Site-Directed Mutations on Time-Resolved Electrogenic Intraprotein Proton Transfer. Proc Natl Acad Sci USA 94:9085–9090

Lübben M, Kolmerer B, Saraste M (1992) An Archaebacterial Terminal Oxidase Combines Core Structures of Two Mitochondrial Respiratory Complexes. EMBO J 11(3):805–812

Lübben M, Morand K (1994) Novel Prenylated Hemes as Cofactors of Cyhtochrome Oxidases. J Biol Chem 269(34):21473–21479

Madigan MT, Martinko JM (2006) Brock Biology of Microorganisms. Pearson/Prentice Hall, Upper Saddle River, NJ

Mattar S, Engelhard M (1997) Cytochrome ba3 from *Natronobacterium pharaonis*—an archaeal four-subunit cytochrome-c-type oxidase. Eur J Biochem 250(2):332–341

Michel H, Behr J, Harrenga A, Kannt A (1998) Cytochrome c oxidase: structure and spectroscopy. Annu Rev Biophys Biomol Struct 27:329–356

Muller FH, Bandeiras TM, Urich T, Teixeira M, Gomes CM, Kletzin A (2004) Coupling of the pathway of sulphur oxidation to dioxygen reduction: characterization of a novel membrane-bound thiosulphate:quinone oxidoreductase. Mol Microbiol 53(4):1147–1160

Nunoura T, Sako Y, Wakagi T, Uchida A (2003) Regulation of the aerobic respiratory chain in the facultatively aerobic and hyperthermophilic archaeon *Pyrobaculum oguniense*. Microbiology 149(Pt 3):673–688

Nunoura T, Sako Y, Wakagi T, Uchida A (2005) Cytochrome aa3 in facultatively aerobic and hyperthermophilic archaeon *Pyrobaculum oguniense*. Can J Microbiol 51(8):621–627

Ostermeier C, Harrenga A, Ermler U, Michel H (1997) Structure at 2.7 Å Resolution of the *Paracoccus denitrificans* Two-Subunit Cytochrome c Oxidase Complexed with an Antibody F_V Fragment. Proc Natl Acad Sci USA 94:10547–10553

Pereira MM, Santana M, Teixeira M (2001) A Novel Scenario for the Evolution of Haem-copper Oxygen Reductases. Biochim Biophys Acta 1505:185–208

Philippot L (2005) Denitrification in pathogenic bacteria: for better or worst? Trends Microbiol 13(5):191–192

Poole RK, Cook GM (2000) Redundancy of Aerobic Respiratory Chains in Bacteria? Routes, Reasons and Regulation. Adv Microb Physiol 43:165–224

Purschke WG, Schmidt CL, Petersen A, Schäfer G (1997) The Terminal Quinol Oxidase of the Hyperthermophilic Archaeon *Acidianus ambivalens* Exhibits a Novel Subunit Structure and Gene Organization. J Bacteriol 179(4):1344–1353

Rappe MS, Giovannoni SJ (2003) The uncultured microbial majority. Annu Rev Microbiol 57:369–394

Rauhamaki V, Baumann M, Soliymani R, Puustinen A, Wikstrom M (2006) Identification of a histidine-tyrosine cross-link in the active site of the cbb3-type cytochrome c oxidase from Rhodobacter sphaeroides. Proc Natl Acad Sci USA 103(44):16135–16140

Reimann J, Flock U, Lepp H, Honigmann A, Adelroth P (2007) A pathway for protons in nitric oxide reductase from *Paracoccus denitrificans*. Biochim Biophys Acta 1767(5):362–373

Riesenfeld CS, Schloss PD, Handelsman J (2004) Metagenomics: genomic analysis of microbial communities. Annu Rev Genet 38:525–552

Risgaard-Petersen N, Langezaal AM, Ingvardsen S, Schmid MC, Jetten MS, Op den Camp HJ, Derksen JW, Pina-Ochoa E, Eriksson SP, Nielsen LP, Revsbech NP, Cedhagen T, van der Zwaan GJ (2006) Evidence for complete denitrification in a benthic foraminifer. Nature 443(7107):93–96

Robertson CE, Harris JK, Spear JR, Pace NR (2005) Phylogenetic diversity and ecology of environmental Archaea. Curr Opin Microbiol 8(6):638–642

Ronquist F, Huelsenbeck JP (2003) MrBayes 3: Bayesian phylogenetic inference under mixed models. Bioinformatics 19(12):1572–1574

Schäfer G (2004) Respiratory Chains in Archaea: From Minimal Systems to Supercomplexes. In: Zannoni D (ed) Advances in Photosynthesis and Respiration, vol 16. Respiration in Archaea and Bacteria: Diversity of Prokaryotic Respiratory Systems. Springer, Berlin Heidelberg New York, pp 1–33

Schäfer G, Engelhard M, Müller V (1999) Bioenergetics of the Archaea. Microbiol Mol Biol Rev 63(3):570–620

Sharma V, Puustinen A, Wikstrom M, Laakkonen L (2006) Sequence analysis of the cbb3 oxidases and an atomic model for the Rhodobacter sphaeroides enzyme. Biochemistry 45(18):5754–5765

Sievert SM, Scott KM, Klotz MG, Chain PSG, Hauser LJ, Hemp J, Hugler M, Land M, Lapidus A, Larimer FW, Lucas S, Malfatti SA, Meyer F, Paulsen IT, Ren Q, Simon J

(2007) The genome of epsilon-proteobacterial chemolithoautotroph *Sulfurimonas denitrificans*. Appl Environ Microbiol (in press)

Soulimane T, Buse G, Bourenkov GP, Bartunik HD, Huber R, Than ME (2000) Structure and Mechanism of the Aberrant ba_3-cytochrome c Oxidase from *Thermus thermophilus*. EMBO J 19(8):1766–1776

Soulimane T, Than ME, Dewor M, Huber R, Buse G (2000) Primary structure of a novel subunit in ba3-cytochrome oxidase from *Thermus thermophilus*. Protein Sci 9(11):2068–2073

Stein LY, Arp DJ, Berube PM, Chain PSG, Hauser L, Jetten MSM, Klotz MG, Larimer FW, Norton JM, Op den Camp HJM, Shin M, Wei X (2007) Whole-genome analysis of the ammonia-oxidizing bacterium, *Nitrosomonas eutropha* C91: implications for niche adaptation. Environ Microbiol 9(12):2993–3007

Suharti, Strampraad MJ, Schroder I, de Vries S (2001) A novel copper A containing menaquinol NO reductase from *Bacillus azotoformans*. Biochemistry 40(8):2632–2639

Svensson-Ek M, Abramson J, Larsson G, Tornroth S, Brzezinski P, Iwata S (2002) The X-ray Crystal Structures of Wild-type and EQ (I-286) Mutant Cytochrome c Oxidases from *Rhodobacter sphaeroides*. J Mol Biol 321:329–339

Tanaka M, Ogawa N, Ihara K, Sugiyama Y, Mukohata Y (2002) Cytochrome aa(3) in *Haloferax volcanii*. J Bacteriol 184(3):840–845

Thorndycroft FH, Butland G, Richardson DJ, Watmough NJ (2007) A new assay for nitric oxide reductase reveals two conserved glutamate residues form the entrance to a proton-conducting channel in the bacterial enzyme. Biochem J 401(1):111–119

Tyson GW, Chapman J, Hugenholtz P, Allen EE, Ram RJ, Richardson PM, Solovyev VV, Rubin EM, Rokhsar DS, Banfield JF (2004) Community structure and metabolism through reconstruction of microbial genomes from the environment. Nature 428(6978):37–43

Victor BL, Baptista AM, Soares CM (2004) Theoretical identification of proton channels in the quinol oxidase aa3 from *Acidianus ambivalens*. Biophys J 87(6):4316–4325

Woese CR, Fox GE (1977) Phylogenetic structure of the prokaryotic domain: the primary kingdoms. Proc Natl Acad Sci USA 74(11):5088–5090

Wraight CA (2006) Chance and design-proton transfer in water, channels and bioenergetic proteins. Biochim Biophys Acta 1757(8):886–912

Yarzabal A, Appia-Ayme C, Ratouchniak J, Bonnefoy V (2004) Regulation of the expression of the *Acidithiobacillus ferrooxidans* rus operon encoding two cytochromes c, a cytochrome oxidase and rusticyanin. Microbiology 150(Pt 7):2113–2123

Yoshikawa S, Shinzawa-Itoh K, Tsukihara T (2000) X-ray Structure and the Reaction Mechanism of Bovine Heart Cytochrome c Oxidase. J Inorg Biochem 82:1–7

Zumft WG (2005) Nitric oxide reductases of prokaryotes with emphasis on the respiratory, heme–copper oxidase type. J Inorg Biochem 99(1):194–215

Structure of Photosystems I and II

Petra Fromme (✉) · Ingo Grotjohann

Department of Chemistry and Biochemistry, Arizona State University, Box 871604, Tempe, AZ 85287-1604, USA
pfromme@asu.edu

Abstract Photosynthesis is the major process that converts solar energy into chemical energy on Earth. Two and a half billion years ago, the ancestors of cyanobacteria were able to use water as electron source for the photosynthetic process, thereby evolving oxygen and changing the atmosphere of our planet Earth. Two large membrane protein complexes, Photosystems I and II, catalyze the primary step in this energy conversion, the light-induced charge separation across the photosynthetic membrane. This chapter describes and compares the structure of two Photosystems and discusses their function in respect to the mechanism of light harvesting, electron transfer and water splitting.

Keywords Light harvesting · Membrane proteins · Photosynthesis · Photosystem I · Photosystem II · Water oxidation · X-ray structure analysis

1
Overview on Oxygenic Photosynthesis

All higher life on Earth depends on the process of oxygenic photosynthesis, which is catalyzed by cyanobacteria, algae and plants. In the process of photosynthesis, the light energy from the sun is converted into chemical energy. Fossil records show that the first primitive forms of photosynthetic algae may have evolved on Earth as early as 3.8 billion years ago. 2.5 billion years ago ancestors of cyanobacteria, which already contained two distinct photosystems, were able to use water as source for the electron transport chain, thereby "inventing" oxygenic photosynthesis.

1.5 billion years ago, the first eukaryotic photosynthetic cells developed by the process of endosymbiosis, but the primary steps of photosynthesis are still very similar and have been conserved over the 2.5 billion years of evolution.

The energy conversions catalyzed by photosynthetic organisms are functionally divided into the light reactions, which convert the light energy into an electrochemical gradient that is used to synthesize the high-energy substrates ATP and NADPH, and the dark reactions that consume ATP and NADPH for the production of carbohydrates by CO_2 fixation. The light reactions take place in the photosynthetic membrane, and are catalyzed by four large membrane protein complexes: Photosystems I and II, the cytochrome b_6f complex, and ATP synthase.

The content of this chapter focuses on the structure and function of Photosystems I and II (abbreviated as PS I and PS II respectively), which catalyze the central step of the energy conversion process, the transmembrane charge separation. PS I and PS II contain internal antenna systems consisting of chlorophylls and carotenoids, which capture the light energy of the sun and use this energy to perform a charge separation across the thylakoid membrane. In addition, both photosystems are functionally connected to peripheral antenna complexes, thereby increasing the cross section of light capturing. The peripheral antenna proteins have evolved after the endosymbiotic event, so they differ significantly between different photosynthetic species. In plants and green algae, membrane integral antenna protein complexes that contain chlorophyll a and b, are associated with PS I and II. Large membrane associated phycobillisomes serve as peripheral antenna complexes in cyanobacteria. Some marine cyanobacteria contain membrane intrinsic antenna complexes that belong to the Pscb protein family. Under iron deficiency, an 18-fold ring of IsiA surrounds PS I and serves as a peripheral antenna.

Photosystem II provides the electrons for the whole photosynthetic electron transport chain. It captures the light with an internal antenna system consisting of 35–36 chlorophylls and 10–12 carotenoids. The excitation energy is transferred to the center of the complex where charge separation takes place. Four electrons are extracted in four subsequent charge separation events from 2 water molecules bound to the oxygen evolving complex. Thereby 4 protons are released into the thylakoid lumen and oxygen is evolved as a by-product of photosynthesis. The electrons are transferred to a mobile plastoquinone (PQ), which, after double reduction, binds two protons and leaves the binding site as PQH_2. The PQH_2 serves as mobile electron and proton carrier and is in constant exchange with the PQ pool in the membrane. It tansfers two electrons and two protons to the cytochrome b_6f complex, which releases 2 protons to the inside of the thylakoids (lumen), and subsequently reduces 2 molecules of plastocyanin (Bialek-Bylka et al.) or cytochrome c_6 (cyt c_6). In addition, the b_6f complex pumps an additional proton across the membrane in a process known as the Q-cycle, thereby further contributing to the establishment of a proton gradient across the thylakoid membrane. The structure of the b_6f complex has recently been determined both from green algae (Stroebel et al. 2003) and cyanobacteria (Kurisu et al. 2003; Yamashita et al. 2007).

PC and cytochrome c_6 serve as mobile electron carriers between the cytochrome b_6f complex and PS I. Photosystem I catalyzes the light driven electron transfer from PC or Cyt c_6 at the lumenal site to the soluble electron carrier ferredoxin or favodoxin at the stromal site of the membrane. The Electron transport chain consists of 6 chlorophylls, 2 phylloquinones, and 3 4Fe4S clusters. In addition to the cofactors of the ETC, PS I contains a large core antenna system consisting of 90 chlorophylls and 22 carotenoids. The light-driven electron transfer in PS I provides the electrons for the final reduction of protons to hydrogen, which is stored in form of NADPH. Ferredoxin docks to PS I at the stromal site and brings the electrons over to FNR (ferredoxin:NADP+-reductase). Under iron deficiency, flavodoxin replaces ferredoxin as mobile electron carrier. The electron transport chain is completed by the reduction of $NADP^+$ to NADPH by FNR. Two electrons are required for the reduction of $NADP^+$, thereby two subsequent electron transfer events take place before NADPH can be formed. The formation of NADPH further reduced the proton concentration in the stroma. The photosynthetic electron transport chain establishes a proton motive force ($\Delta pH + \Delta \psi$) in the form of an electrochemical potential across the membrane, which drives synthesis of ATP by the ATP-Synthase.

In this chapter, we will focus on the structures of PS I and II.

2
Photosystem I

2.1
General Structure of PS I

Photosystem I exists in different oligomeric forms: in plants it is a monomer surrounded by LHC I, whereas it is a trimer in cyanobacteria. The trimeric PS I from *Thermosynechococcus elongatus* is shown in Fig. 2 and has a molecular weight of 1 080 000 Da. Photosystem I represents the largest and most complex membrane protein that has been crystallized (Fromme 1998) and for which a high resolution structure has been determined (Jordan et al. 2001). The outer shape of the trimeric complex of PS I can be described as a large disc with a diameter of 220 Å (see Figs. 1 and 2B). Most of the protein is located in the membrane, but each monomer contains a large membrane extrinsic stromal hump, which provides the docking site for ferredoxin. The hump extends the membrane about 40 Å at the stromal site. The lumenal site of Photosystem I extends the membrane only about 15 Å. Here, the docking site for PC/cyt c6 is located. One monomeric unit of cyanobacterial PS I consists of 12 different proteins to which 127 cofactors are non-covalently bound. The large proteins PsaA and PsaB form a joint reaction center and core antenna complex and are located in the center of PS I. The PsaA/B core is surrounded by seven small membrane intrinsic proteins (PsaF, PsaI, PsaJ, PsaK, PsaL, PsaM and PsaX).

Three stromal subunits (PsaC, PsaD and PsaE) form the stromal hump of PS I.

Plant PS I lacks PsaM and PsaX but contains at least four additional subunits PsaG, PsaH, PsaO and PsaN. The structure of the supercomplex of plant PSI with the LHC I peripheral antenna has been determined at 4.4 Å resolution (Ben-Shem et al. 2003) and was recently improved to 3.4 Å (Amunts et al. 2007).

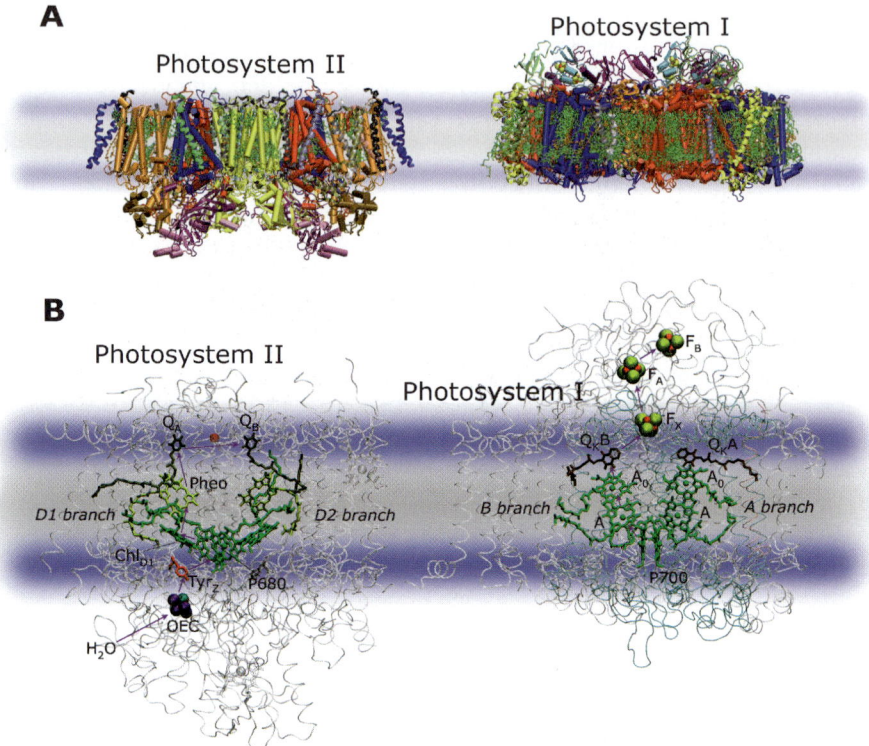

Fig. 1 Structural overview of the structures of PS I and II **A** Structural model of the dimeric PS I and trimeric PS I and their proposed location in the membrane. The model is based on the pdb files 1JB0 (Jordan et al. 2001) and 2 AXT (Loll et al. 2005). Note that PS I contains three extrinsic subunits at the stromal side, which are responsible for the docking of ferredoxin, while the extrinsic subunits of PS II are located at the lumenal side and stabilize the oxygen evolving complex. **B** The comparison of the electron transport chain in Photosystem II and I. The chlorophylls are depicted in *green*, Pheophytin in *yellow*, quinones in *orange*. Fe atoms are shown as red spheres, sulfur atoms as yellow spheres, Mn atoms as purple spheres and the Ca atom as a cyan sphere. The redox active tyrosine Tyr$_Z$ is highlighted in red. The backbone of the protein is shown as grey hollow ribbons. All model images were done using the graphic program VMD (Humphrey et al. 1996)

2.2
Protein Subunits of PS I

2.2.1
Core of PS I: Large Subunits PsaA and PsaB

The core of PS I is formed by a heterodimer of the two large subunits, PsaA and PsaB. The heterodimeric nature of PS I may be a relatively recent

Structure of Photosystems I and II

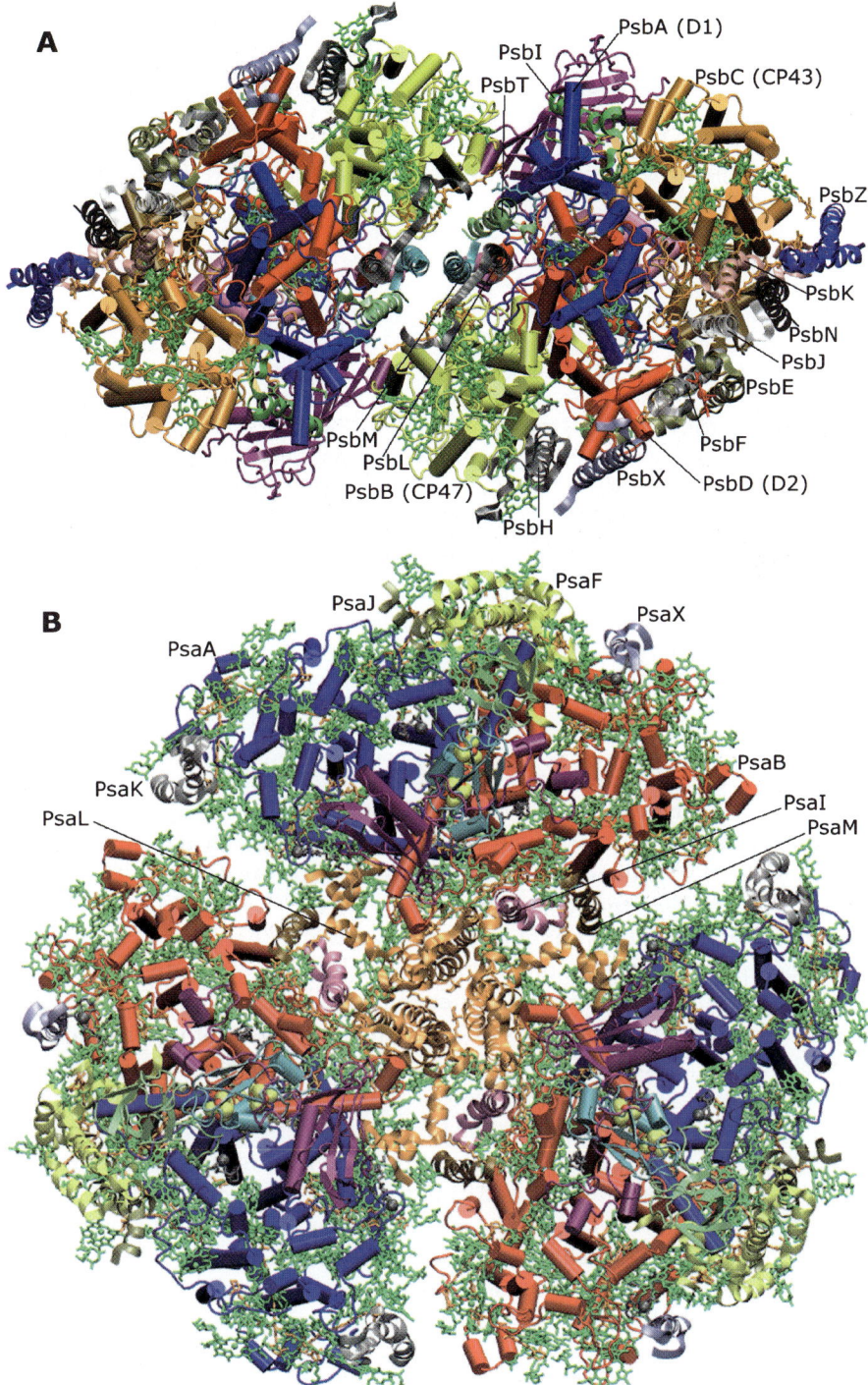

◄ **Fig. 2** Structural models of dimeric PS II and trimeric PS I from cyanobacteria. **A** Structure of PS II. The view direction is from the stromal side onto the membrane plane. The transmembrane helices of the core subunits are shown as columns, while the helices of the small subunits are shown in a ribbon representation. The subunits D1 (*blue*) and D2 (*red*) form the center of the complex. They are flanked on both sides by the antenna proteins CP47 (*yellow*) and CP43 (*orange*). The trimerization domain contains subunits PsbM (*cyan*) PsbL (Brown and Schoch) and PsbT (*green*). The membrane exposed transmembrane helices are assigned to the following subunits (starting from the helix between CP43 and the dimer interface (PsbI), going clockwise around the monomer): psbI (*green*), PsbZ (*blue*), PsbK (*light red*), PsbN (*black*), PsbJ (*grey*), PsbE (*gold*), PsbF (*tan*), Unidentified helix (*grey*), PsbX, (*metallic-blue*), PsbH (*grey*). Partly visible are the lumenal proteins PsbO (*dark-pink*), PsbU (*pink*) and PsbV (Brown and Schoch) (for a better view on these subunits see Fig. 3A). The cofactors are color coded as in Fig. 1. The structural model based on the structure of PS II at 3.5 Å resolution (pdb accession code 1S5L), plus the additional helix of the 3.8 and 3.0 Å structures (2AXT). **B** The trimeric structure of PS I from cyanobacteria. The view direction is from the stromal side onto the membrane plane. The proteins are shown in a backbone representation, with the helices of subunits PsaA (*blue*) and PsaB (*red*), as well as subunits PsaC (*cyan*), PsaD (*dark pink*) and PsaE (*green*) those of the stromal hump, are shown as columns. The small transmembrane helices are shown in ribbon representation and are assigned to the following subunits (starting with PsaM at the monomer/monomer interface and going clockwise around the topmost monomer in the trimer): PsaM (*brown*), PsaI (*pink*), PsaL (*orange*), PsaK (*grey*), PsaJ (*gold*), PsaF (*yellow*) and PsaX, *metallic-grey*). The chlorophylls (depicted in *green*) are represented by their chlorin head groups, their phytyl-tails have been omitted for clarity; the carotenoids are depicted in *orange* and the lipids in *cyan*

evolutionary event (Alfonso et al.), as homo-dimeric reaction centers with homology to PS I can be found in green-sulfur and heliobacteria. PsaA and PsaB consist of 11 transmembrane helices each (see Fig. 3). PsaA and PsaB coordinate the majority of the cofactors of the electron transport chain (P700, A0, A1, and F_X) and 79 of the 90 antenna chlorophylls in PS I. In addition, most of the carotenoids show hydrophobic interactions with either PsaA or PsaB. Both subunits can be structurally divided into a C-terminal domain that surrounds the electron transport chain and a N-terminal domain that flanks the reaction center on both sides and harbors the core antenna system of Photosystem I.

The reaction center domain: The transmembrane part of the C-terminal domain contains five transmembrane α-helices that surround the electron transport chain like a fence. The cofactors of the electron transfer chain, from P700 to F_X, are coordinated by the C-terminal region of PS I. The helices of the C-terminal domain of PsaA and PsaB surround the electron transfer chain like a fence. The outermost helices are the TM helices 7 followed by the TM helices 8, 9 and 11. TM helix 10 forms the innermost part of the C-terminal domain and coordinates P700 and A0. A1 is bond at a hydrophobic pocket formed by the loop between TM helix 10 and 11. The Fx binding domain is the best conserved sequence between PsaA and PsaB and is located in a stromal loop between the TM helices 8 and 9.

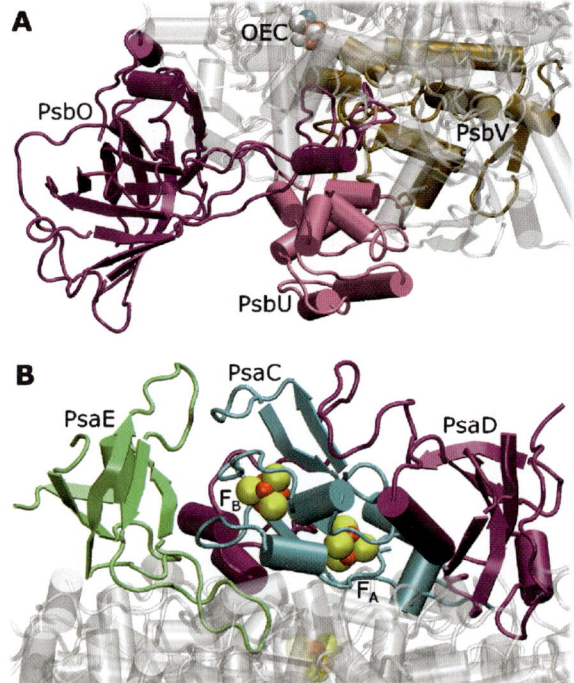

Fig. 3 A The three lumenal membrane-extrinsic subunits of PS II: PsbO (*dark pink*), PsbU (*pink*) and PsbV (*brown*), which stabilize the oxygen-evolving complex. The Mn atoms of the OEC are shown in purple, the Ca atom is depicted in cyan. The picture also shows the loops of the membrane intrinsic subunits that interact with PsbO, PsbU and PsbV in *transparent grey*. **B** The three stromal subunits of PS I, which provide the docking site for ferredoxin/flavodoxin: PsaC (*cyan*) contains the two terminal 4Fe4S clusters F_A and F_B. PsaE (*green*) and PsaD flank PsaC on both sides. Notice the clamp of PsaD that stabilizes PsaC. The picture also shows the loops of the membrane intrinsic subunits that interact with PsaC, PsaD and PsaE in *transparent grey*

The large subunits PsaA and PsaB represent a joint reaction center and core antenna. The C-terminal domain thereby not only coordinates the cofactors of the electron transport chain, but is also coordinating more than one-third of the 90 the antenna chlorophylls (25 out of 90 antenna chlorophylls are coordinated by the C-terminal domain of PsaA and PsaB). The coordination of chlorophylls is not limited to the TM helices but some chlorophylls are even coordinated by the lumenal and stromal loops of PsaA and PsaB. The antenna and reaction center domains are not separated, but about half of the chlorophylls are surrounding the fence of five TM helices surrounding the electron transfer chain.

Whereas the transmembrane helices show a nearly perfect 2-fold symmetry between PsaA and PsaB, the loops show striking differences in sequence, length and secondary structural elements. The loops are inducing an asymmetry into the system, which is functionally important for the docking of the three stromal subunits, PsaC, PsaD and PsaE and the interactions with the small membrane intrinsic subunits of PS I.

Furthermore, PsaA and PsaB build the platform for the docking of the soluble electron donors to PS I. At the lumenal site, PS I contains an indentation, where the docking site for plastocyanin and cytochrome c_6 is located. The major interaction site is formed by two helices in the loop between TM helices 9 and 10. These helices shield P700 from the aqueous surface. They are hydrophobic and may interact with the hydrophobic "north" site of plastocyanin/cytochrome c_6. In cyanobacteria, PsaA and PsaB are exclusively involved in the docking of plastocyanin, but in plants PsaF interact electrostatically with plastocyanin (Hippler et al. 1998, 1996; Ben-Shem et al. 2003).

Core Antenna Domain of PsaA and PsaB

The six N-terminal TM helices of PS I coordinate the peripheral part of the core antenna domain of PS I. The six helices are arranged in pairs (TM helices 1/2, TM helices 3/4, TM helices 5/6) and form an arrangement which can be described as a "trimer of dimers". The helix pair a/b is on the periphery of PS I, whereas the helix pair e/f provides protein–protein interactions with the C-terminal core domain of PSI. This arrangement is very similar to the arrangement of the TM helices of the core antenna proteins PsbB and PsbC in PS II, as shown in Fig. 4. The N-terminal domain of PS I coordinates 40 of the 90 antenna chlorophylls in PS I and has interactions with carotenoids and four lipids. This domain is also involved in the attachment of the small membrane intrinsic subunits: most of these interactions cannot be described by direct protein–protein interactions, but the subunits are attached to the core via protein–cofactor–protein interactions. As an example, PsaK is attached to the PS I core solely by protein cofactor interactions linking it to the TM helices 1 and 2 of PsaA.

2.2.2
Small Transmembrane Subunits in PS I

Seven small membrane intrinsic subunits are surrounding the core of the transmembrane domain of PS I: PsaF, PsaI, PsaJ, PsaK, PsaL, PsaM and PsaX. The names reflect their date of discovery, and are not related to their function or local proximity.

There are two major domains, where some of the small subunits are located. The three subunits PsaI, PsaL and PsaM cluster together and form the trimerization domain at the monomer–monomer interface, whereas the subunits PsaF, PsaJ, PsaK and PsaX that are located at the membrane exposed surface of the trimeric PS I.

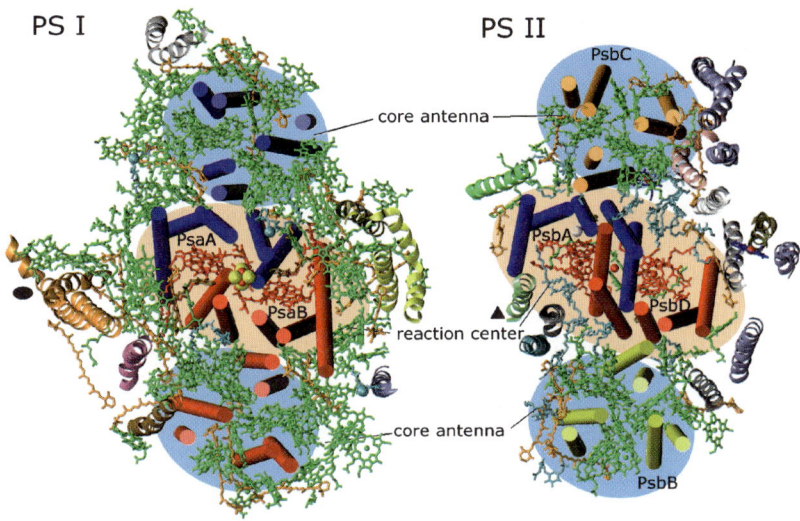

Fig. 4 Comparison of the arrangement of cofactors and transmembrane helices in PS I and II. The general arrangement of the core helices is similar in PS I and II, suggesting that they have evolved from a common ancestor. PS I contains more than twice as many chlorophylls as PS II and PS II lacks the central antenna domain. It is very interesting that the space where massive amounts of lipids are found in PS II is occupied by chlorophylls in PS I. It is also interesting that neither the small subunits nor the position of carotenoids is similar in PS I and II. *Left*: arrangement of the transmembrane helices of protein subunits and cofactors in one monomeric unit of PS I (pdb code 1JB0). The three lumenal subunits and all loops have been omitted for clarity. Transmembrane helices of the membrane intrinsic subunits are represented as cylinders. Color-coding is the same as in Fig. 2B. The C-terminal 5 transmembrane helices of PsaA and PsaB form the reaction center in the core (*orange underlay*), while the N-terminal 6 transmembrane helices represent the core antenna (*blue underlay*). The smaller transmembrane subunits surround the core, stabilize the antenna system, and form the contacts in the trimerization domain (depicted as *black triangle*). The cofactors are color coded as in Fig. 2. *Right*: arrangement of the transmembrane helices of protein subunits and cofactors in one monomeric unit of PS II (from the pdb file 2AXT). The three lumenal subunits and all loops have been omitted for clarity. Transmembrane helices of the membrane intrinsic subunits are represented as cylinders. Color-coding is the same as in Fig. 2A. The reaction center (*orange underlay*) is formed by the core subunits D1 (PsbA) (*blue*) and D2 (PsbD) (*red*). They are flanked on both sides by the core antenna proteins CP43 (PsbC) (*orange*) and CP47 (PsbB) (*yellow*) (*blue underlay*). All small subunits are located outside the inner core. They may modify the properties of the ET chain, stabilize the core antenna system and form the contact at the dimer-interface (twofold axis of the dimer is indicated by a *black oval*)

Monomer–Monomer Interface Subunits: PsaL, PsaI and PsaM

PsaL contains three transmembrane helices. It is located in the center of the trimerization domain of PS I, thereby forming most of the contacts between the monomers. In addition, PsaL coordinates three antenna Chl*a* and forms hydrophobic contacts with carotenoids. These chlorophylls may be important for the excitation energy transfer between the monomers (Sener et al. 2005). The majority of the contact sites between the monomers in the trimerization domain are provided by hydrogen bonds and electrostatic interactions within the loop regions. Three Ca^{2+} binding sites are located at trimer interface with ligands provided by PsaL subunits from neighboring monomers and PsaA, which may further stabilize the trimer.

PsaI contains only one transmembrane helix, located in the vicinity of PsaL. Deletion of PsaI destabilizes the trimer, but not completely hinders the trimer formation (Schluchter et al. 1996). PsaI is not involved in Chl*a* coordination, but forms hydrophobic interactions with carotenoid molecules while forming few contacts with the adjacent monomer. PsaI is located between PsaL and PsaM.

The recent structures of plant PS I showed that close interactions of PsaI and PsaL exist in cyanobacteria and in higher plants, suggesting that the arrangement of these small subunits is a motif that was conserved during evolution (Andersen and Scheller 1993; Janson et al. 1996). This finding is astonishing, taking into account that the interaction partners of PsaL are completely different in cyanobacteria and plants. PsaL forms the trimerization domain in cyanobacteria, thereby interacting with the PsaL subunits of the neighboring monomers in the trimer. In contrast, plant PS I is a monomer, and the region of PsaI and PsaL interacts with PsaH and may form interactions with the Light Harvesting Complex II (LHC II) under favorable light conditions (Andersen and Scheller 1993; Janson et al. 1996; Scheller et al. 2001; Kouril et al. 2005). Despite the different interaction partners, PsaL might have similar functions in plants and cyanobacteria as it is essential for adaptation of the organism to changing light conditions. In cyanobacteria, trimers and monomers coexist in the native membrane. Monomers are the dominant form under high light and, while trimers are formed under low light conditions. In plants, the LHC II moves from PS II to PSI under light conditions that favor PS II (so-called state 2 conditions), thereby increasing the cross section for light capturing in PS I.

PsaM contains only one transmembrane α-helix and is with a MW of 3.4 kDa the smallest subunit of PS I. PsaM is located at the monomer/monomer interface, and forms interactions with PsaI and PsaB. PsaM may be important for the excitation energy transfer between the monomers as it coordinates one chlorophyll that belongs functionally to the next monomer (Sener et al. 2004). Most evidence points in the direction that this subunit is unique to cyanobacterial PS I. Even if some plants may contain an open frame for PsaM, the protein is not be expressed and has not been identified in protein preparations of PSI from plants or green algae.

Membrane Exposed Subunits: PsaF, PsaJ, PsaK and PsaX

The detergent exposed surface of PS I harbors four small hydrophobic protein subunits: PsaF, PsaJ, PsaK and PsaX. These four proteins stabilize the core antenna system of PS I and may play an additional role in forming interactions with the membrane intrinsic peripheral antenna system, the IsiA ring (Bibby et al. 2001; Boekema 2001; Kouril et al. 2005).

While PsaF, PsaJ and PsaK are conserved between cyanobacteria and plants, PsaX is unique to thermophilic cyanobacteria. PsaF and PsaJ form various contacts with PsaA, PsaB and PsaE and are located at the opposite side of the trimerization domain. PsaK is located at the periphery of PsaA. In plants there exists a symmetry related counterpart to PsaK: PsaG. The higher plant subunit PsaG may have evolved from PsaK via gene duplication (Kjaerulff et al. 1993) and forms strong interactions with the peripheral antenna protein Lhca1 (Ben-Shem et al. 2003).

PsaF consists of three domains: the N-terminal lumenal domain, the transmembrane domain and the C-terminal region that is sandwiched between PsaA and the stromal subunit PsaE.

The N-terminal domain contains two α-helices, which are located at the lumenal side of the complex. These helices are hydrophilic and run parallel to the membrane plane. In plants, these helices are extended by 25 amino acids and are actively involved in docking of plastocyanin via electrostatic interactions (Fischer et al. 1998). In *TS elongatus*, PsaF plays no role in docking of plastocyanin as the distance of 15 Å to the putative docking site of cytochrome c6 is too large for an active role of PsaF in the docking process.

The transmembrane region of PsaF comprises one of the most astonishing folds in the PS I structure. It consist of one transmembrane helix followed by two short hydrophobic helices which form a V-shaped structure. The first of the two short helices penetrates the membrane only by 10 Å followed by a sharp turn and the second short helix.

This V-shaped structure is the "outermost part" of the PS I trimer. It is exposed to the membrane even though it contains 2 lysine residues which stick into the membrane. PsaF does not directly coordinate any chlorophylls but is in contact with more than 10 chlorophylls and 6 carotenoids via hydrophobic interactions, thereby stabilizing the core antenna system of PS I. Furthermore, PsaF may be important for the coupling of peripheral antenna systems to PS I both in cyanobacteria and in plants. A *TS elongatus* mutant deficient in PsaF (Muhlenhoff and Chauvat 1996) was not able to grow under low light (unpublished results of our group) and expressed large amounts of allophycocyanin, which may indicate that PsaF is important for docking of the phycobilisomes to PS I.

Plant PsaF directly interacts with the LHC I proteins (Ben-Shem et al. 2003) and can even be isolated as a Chl-protein complex with LHCI proteins (Anandan et al. 1989).

PsaJ is located in close proximity to PsaF and contains one transmembrane α-helix. It is in hydrophobic contact with carotenoids and coordinates three chlorophylls. PsaF and PsaJ are co-expressed in one open reading frame and it was suggested that PsaJ may play an important role in the stabilization of PsaF and the pigment clusters located at the interface between PsaJ/PsaF and the PsaA/PsaB core. PsaJ may interact directly with the IsiA protein under iron deficiency, and the three chlorophylls bound to PsaJ may play an important role in the excitation energy transfer from the chlorophylls in the IsiA ring to the chlorophylls in the PS I core.

PsaK is located at the periphery of the PS I complex, forming only protein contacts with PsaA. It contains two transmembrane α-helices, which are connected in the stroma, with both the C- and N-terminus being located in the lumen. PsaK binds two chlorophylls and forms contacts with carotenoids. As PsaF and J, PsaK may also play an important role in the interaction with the IsiA antenna ring under iron deficiency.

In plants, PsaK interacts with the LHC I proteins. In addition, a role of PsaK in state transitions has been suggested (Varotto et al. 2002).

PsaX is a small subunit that is unique to cyanobacteria and was not identified as an intrinsic subunit of PS I before 2001, when the 2.5 Å structure of PS I revealed that this protein is part of the PS I complex. It is not clear if all cyanobacteria contain PsaX, as it has been so far only identified in PS I from thermophilic cyanobacteria (Koike et al. 1989). It is located at the membrane exposed surface of PS I, coordinates one chlorophyll and forms hydrophobic contacts with several carotenoid molecules and one of the lipids. PsaX may stabilize thermophilic PSI and could play a role in the interaction of PS I with the IsiA antenna ring.

2.2.3
Stromal Subunits of PS I: PsaC, PsaD and PsaE

The subunits PsaC, PsaD and PsaE are located at the stromal side of PS I forming the docking site for ferredoxin.

PsaC is the only small subunit of PSI which binds cofactors of the electron transport chain: the two terminal FeS clusters F_A and F_B. They are coordinated by two CXXCXXCXXXC motifs that provide the signature for 4Fe4S clusters. PsaC has identical structures in cyanobacterial and plant PS I (Ben-Shem et al. 2003; Jolley et al. 2005). The core of PsaC consists of two short α-helices that connect the two FeS clusters, a feature that shows the close evolutionary relationship between PsaC and bacterial ferredoxins. F_A is proximal to the membrane and F_X, while F_B is the terminal FeS cluster that transfers the electron to ferredoxin. The C- and N-terminus of PsaC are unique to PsaC and are very important for the symmetry-breaking docking of PsaC to the PS I core (Antonkine et al. 2003; Jolley et al. 2006).

PsaD is located at the stromal hump, close to the "connecting domain". It is essential for the proper orientation of PsaC and thereby important for the tuning of the electron transfer from PS I to ferredoxin (Setif 2001; Setif et al. 2002). The main part of PsaD consists of a large antiparallel four stranded beta-sheet followed by an α-helix, which forms interactions with PsaC and with PsaA. PsaD locks PsaC on place by forming a clamp wrapping around PsaC. This clamp forms numerous contacts between PsaD, PsaC and PsaE, thereby stabilizing the electron acceptor sites in PS I (Li et al. 1991; Hanley et al. 1996). PsaD may also be directly involved in the docking of ferredoxin. Co-crystals between PSI and ferredoxin have been reported which may serve as a basis for a structure of the PS I ferredoxin complex (Fromme et al. 2002).

PsaE was the first subunit for which a solution structure had been determined by NMR (Falzone et al. 1994). It consists of five anti-parallel stranded β-sheets. Multiple functions have been proposed for PsaE. It is directly involved in anchoring the ferredoxin (Rousseau et al. 1993; Sonoike et al. 1993; Strotmann and Weber 1993), plays a role in cyclic electron transport (Yu and Vermaas 1993) and may interact with FNR. A direct interaction of PsaE with FNR has been shown in barley, where PsaE can be crosslinked to FNR (Andersen et al. 1992). The role of PsaE in the docking of ferredoxin and flavodoxin (Muhlenhoff and Chauvat 1996; Meimberg et al. 1998) was questioned by the finding that PsaE deletion mutants were still able to grow photoautotrophically. The contradiction was finally solved by the discovery that PsaE deletion mutants increased the level of ferredoxin in the cells by orders of magnitude to compensate for deficits caused by the lack of PsaE (van Thor et al. 1999).

2.3
Electron Transfer Chain in PS I

The electron transport chain of PS I consists of six chlorophylls, two phylloquinones and three 4Fe4S clusters. It is located in the center of the PS I complex. The membrane intrinsic part of the electron transfer chain is arranged in two branches, named A and B branches, because most cofactors of the A branch and B branch are coordinated by PsaA and PsaB proteins respectively. From the electronically excited singlet state ^{1}P700* (represented by the first pair of chlorophylls), the electron is transferred via one of the chlorophylls from the second pair chlorophylls to the first stable electron acceptor, A_0, which may be located on one of the chlorophylls located in the middle of the membrane (see Fig. 1). From there, the electron is transferred to the phylloquinone and subsequently to the terminal FeS clusters F_X, F_A and F_B. F_B serves as the terminal FeS cluster and transfers the electron to ferredoxin.

2.3.1
P700: Primary Electron Donor

P700 consists of a pair of chlorophylls, that is located close to the lumenal surface of PS I. The most astonishing feature of PS I is the fact that it consists of two chemically different chlorophyll molecules. The B-branch chlorophyll of P700 is a chlorophyll *a* molecule and thereby chemically identical to all the other 95 chlorophylls in PS I. In contrast, the A-branch chlorophyll of P700 is chlorophyll *a'*, the epimer at the C13 position of the chlorin ring system. The existence of at least one chlorophyll *a'* molecule was first suggested by Watanabe and co-workers (Watanabe et al. 1985) on the basis of chlorophyll extraction experiments.

In addition to the chemical differences, both chlorophylls show strong differences in the hydrogen bonding network. While no hydrogen bonds are formed between the surrounding protein and the chlorin head group of the B-branch Chl a of P700, the A-branch Chl *a'* forms three hydrogen bonds. The central Mg^{2+} ions of the two chlorophylls in P700 are close together with a distance of only 6.3 Å. This distance is shorter than the corresponding distance between the bacteriochlorophylls in the special pair of purple bacterial RC and shorter than the distance of the two chlorophylls of P680 in PS II. Molecular orbital studies of the electronic structure of P700 show that the two chlorophylls are tightly coupled and P700 is a super-molecule (Plato et al. 2003) and ENDOR studies revealed that the spin density in $P700^{+\cdot}$ is asymmetric, with more than 85% of the spin density located on the B-branch Chl*a* of P700 (Kass et al. 2001). It is still an active field of research to prove whether the asymmetry in P700 is essential for the function of PS I. The fact that Chl*a'* is a constituent in cyanobacterial, algal and plant PS I as well as in the homodimeric reaction centers of heliobacteri and green sulfur bacteria may suggest that the Chl*a'* plays a key role for the function of PS I. One possible role might be the "gating" of the electron along the two cofactor branches.

2.3.2
A: Initial Electron Acceptor

The second pair of chlorophylls is located at a very close distance to P700 and serves as the primary electron acceptor. However, it has not yet been detected by spectroscopy, as the first steps of electron transport occur in less than 3 ps. Perhaps these chlorophylls can even act as the primary donor, as there is currently a debate whether the charge separation may start from the electronically excited singlet state of the accessory chlorophyll at the B branch rather than from $^1P700^*$ (Muller et al. 2003). A water molecule provides the fifth ligand to the central Mg^{2+} ion of the second pair of chlorophylls.

2.3.3
A_0: First Stable Electron Acceptor

The third pair of chlorophylls is located in the middle of the membrane at a position that is similar to the position of the two pheophytins in PS II (Ferreira et al. 2004) and the PbRC (Deisenhofer and Michel 1991). These two chlorophylls may form strong interactions with the chlorophylls of the second pair of chloropylls, so that maybe all four of them contribute to the physical-chemical properties of the electron acceptor A_0 (Hastings et al. 1995). The third pair of chlorophylls shows a unusual ligands, as the fifth ligand to the Mg^{2+} ions is provided by sulfur atoms of methionine residues. This ligand binding has been found in heme containing proteins but is unique for chlorophyll coordination, as the concept of hard and soft acids and bases predicts only weak interactions between the hard acid Mg^{2+} and methionine sulfur as a soft base. The weak interaction might be important for the tuning of the redox potential of A0, as recent studies have shown that mutation of the methionine to the strong ligand His slows down the electron transfer chain along the corresponding branches (Fairclough et al. 2003; Ramesh et al. 2004, 2007; Santabarbara et al. 2005).

2.3.4
A_1: Phylloquinone

The phylloquinones QK_A and QK_B (both or one of them) represent the electron acceptor "A_1" and they are located at the stromal side of the membrane. Both phylloquinones are π-stacked with a tryptophane residue. A unique feature of the quinones in PS I is that only one of the two oxygen atoms forms an H-bond to an NH backbone group, whereas the other oxygen atom is not hydrogen bonded. This asymmetry leads to a protein-induced asymmetry in the distribution of the unpaired electron in the radical state A_1-and may be responsible for the extremely negative negative redox potential (–770 mV) of A_1. The electron transfer from A_1 to the FeS cluster, F_X, is the rate limiting step of the electron transfer in PS I. There is experimental evidence that the electron transfer from Q_{KA} and Q_{KB} to F_X proceeds with different rates. In the green alga *Chlamydomonas reinhardtii*, the electron transfer is about a factor of 50 slower on the A- than on the B-branch (Guergova-Kuras et al. 2001), indicating that there might be a higher activation energy barrier on the A- compared to the B-branch. While the protein environments of Q_{KA} and Q_{KB} are very similar, two lipid molecules, which are located close to the phylloquinones, could be responsible for the asymmetry. A negatively charged phospholipid is located on the slower A-branch, which may increase the reorganisation energy, while a neutral galactolipid has replaced the phospholipid on the faster B-branch, thereby lowering the reorganization energy (Ishikita and Knapp 2003). Despite these results, the question as to whether there are one or two active branches in PS I is still a controversially discussed topic. In green algae, mutations of the B-branch have more severe effects on photoautotrophic growth whereas cyanobacteria seem to show more prominent changes in electron transfer upon mutations on the A-branch (Dashdorj et al. 2005), indicating that the branching of the electron along the two chains could differ between different organisms.

2.3.5
F_X: First FeS Cluster

Fx is an essential part of the electron transfer chain and also plays an important role in the stabilization and assembly of the PS I complex. It is coordinated by 2 cysteines from PsaA and 2 cysteines from PsaB, thereby being a rare example of an inter-protein FeS cluster. The Fx binding motif is the best conserved sequence region in PS I and is identical in plants, algae and cyanobacteria. The binding of the stromal subunits PsaC, PsaD and PsaE depends on the functional assembly of F_X and further proteins such as rubredoxin (Andersson et al.) are essentail for the assembly process of Fx (Shen et al. 2002).

2.3.6
F_A and F_B: Terminal FeS Clusters

F_A and F_B are located in the stromal hump of PS I and are bound to the extrinsic subunit PsaC (Ananyev et al.). EPR investigations and mutagenesis studies showed that the cluster in close proximity to F_X represents F_A (Zhao et al. 1992; Mehari et al. 1995), whereas the distal cluster is F_B (Kanervo et al. 1995; Yu et al. 1995). F_B donates electrons to ferredoxin (Fischer et al. 1999; Golbeck 1999; Lakshmi et al. 1999).

2.4
The Antenna System of PS I

Ninety chlorophyll *a* molecules and 22 carotenes form the core antenna system of Photosystem I. The major antenna pigments are the chlorophylls, which capture the light and transfer the excitation energy to P700, where charge separation takes place. PSI is highly efficient in excitation energy transfer with a quantum yield of 99.9% at room temperature. The carotenoids serve in light harvesting and play an important role in photo-protection by quenching of dangerous chlorophyll triplet states.

2.4.1
Chlorophylls

The arrangement of antenna chlorophylls in PS I looks at the first glance chaotic and shows no similarities to the symmetric ring of LHC proteins that surround the RC core in purple bacteria (McDermott et al. 1995). It can be best compared to the arrangement of neurons in the brain, where not a single pathway exists but a network serves as a highly efficient information transfer system. The chlorophyll network in PS I, where each of the chlorophylls has several neighbours at a center to center distance of less than 15 Å, is highly optimized for efficiency and robustness (Sener et al. 2002, 2003). Thereby, energy can be efficiently transferred via multiple pathways to the center of the complex.

It is remarkable that the position and orientation of 85 out of the 90 antenna chlorophylls have been conserved over 1.5 billion years of evolution as they are identical in cyanobacterial and plant PS I (Jordan et al. 2001; Amunts et al. 2007).

The core antenna system of PS I is structurally divided into a central domain, which surrounds the electron transfer chain, and two peripheral domains, flanking the core on both sides. The antenna chlorophylls of the peripheral domains are arranged in two layers, one close to the stromal surface of the membrane and the other close to the lumenal surface of the membrane, while chlorophylls are distributed over the full depths of the membrane in the central domain, thereby facilitating excitation energy exchange between the two layers. Two chlorophylls (named "connecting chlorophylls") form a structural link between the antenna system and the electron transfer chain. Mutagenesis experiments were performed on the ligands of these connecting chlorophylls (Gibasiewicz et al. 2003) showing some alterations in the trapping of the excitation energy.

A special spectral feature that discriminates between PS I and II is the existance of "red" chlorophylls that absorb at wavelengths $\lambda > 700$ nm in PS I. These red chlorophylls have been detected in the 1950s and led to the discovery of the existence of two light reactions (see Allen et al. 1961; Bertsch 1962; Rumberg et al. 1964 and references therein). Dual functions have been proposed for the long-wavelength chlorophylls: they may function in increasing the spectral width of the light absorbed by PS I and/or may be used to funnel the excitation energy to the center of the complex. The exact location of the red chlorophylls is still controversially discussed (Byrdin et al. 2002; Damjanovic et al. 2002; Vasil'ev and Bruce 2004). Several theoretical studies investigated the excitation energy transfer and trapping in PS I, showing that the excitation energy transfer in PS I is probably trap limited and is highly optimized for robustness and efficiency. The results show that orientation of the innermost antenna chlorophylls plays critical role in the efficiency of the antenna system (Gobets et al. 2003).

2.4.2
Carotenoids

The X-ray structure of PS I showed the location of twenty-two carotenoids. They play a structural role, function as additional antenna pigments and prevent the system from damage by over-excitation caused by excess light (photoinhibition). The photo-protective role is critical to the function of PSI, as it allows the carotenoids to quench chlorophyll triplet states ^{3}Chl. Chl triplet states are dangerous and very damaging to the cell, as they can react with oxygen to form singlet oxygen $^1\Delta_g O_2$ that acts as a very potent cell poison. The carotenoids are distributed over the whole antenna system, thereby providing multiple car-chl interactions, that allow transfer of the energy from the triplet chlorophylls to the carotenoids that form the carotenoids triplet state, ^{3}Car, and dissipate the excess energy as heat. It should be noticed that the chemical nature of the carotenoids in PS I has not yet been finally identified. Due to the limitations of the 2.5 Å resolution of the crystal structure, all carotenoids were modeled as beta-carotene; however, biochemical evidence shows that PSI (and the crystals) contain xanthophylls. The resolution of the PS I crystal structure has to be improved to finally identify the chemical nature of the individual carotenoids.

2.4.3
Lipids

The PS I structure at 2.5 Å resolution identified 3 molecules of phosphatidylglycerol (PG) and one molecule of monogalactosyldiacylglycerol (MDGD) in agreement with previous biochemical studies (Kruse and Schmid). Two of the lipids are located close to the electron transfer chain, which may even play an important role in the difference in the rates of electron transfer between the two different branches.

The head groups of these two lipids are not solvent accessible, but are covered by the loops of PsaA and PsaB and the three stromal subunits PsaC, PsaD and PsaE. Thus, it can be assumed that they are incorporated into PS I at a very early stage of the assembly process.

3
Photosystem II

3.1
General Structure of PS II

Photosystem II is one of the most important enzymes on Earth. It has evolved all the oxygen presently found on earth and has thereby changed the atmosphere of our planet from anoxygenic to oxygenic 2.5 billion years ago.

Photosystem II is a large membrane protein complex that catalyzes the light driven electron transfer from water to plastoquinone. The electrons are extracted from water during the process of water splitting, releasing 4 protons into the lumen, and oxygen is evolved. This process provides oxygen for respiration, which is the main energy source for all higher heterotrophic organisms, including humans. Plant, algae and cyanobacteria also have a respiratory chain, but for the process of photosynthesis, oxygen is only a by-product that can even be dangerous for these cells due to its high oxidation potential.

In all photosynthetic organisms, the photosynthetic process is initiated by the process of light-capturing, followed by transfer of the excitation energy to the reaction center. Here, the electron transfer chain is located and the transmembrane charge separation takes place. Both photosystems contain a core antenna, which is smaller in PS II than in PS I. In PS II, the antenna system only consists of 35–36 chlorophylls/P680, which is less than half of the antenna size in PS I (96 Chl/P700). The electron transfer chain of PS II is more complex than the electron transfer chain of PS I and contains the following cofactors (from the lumen to the stroma): the 4MnCa cluster, two redox active tyrosines (Tyr_Z and Tyr_D), the primary electron donor P680, two accessory chlorophylls, two pheophytins, a tightly bound plastoquinone (Q_A) and a mobile plastoquinone (Q_B). After two subsequent electron transfer steps, the doubly reduced Q_B^- binds two protons and leaves the binding pocket as PQH_2. The empty pocket is then refilled by a plastoquinone molecule from the PQ pool. The acceptor side of PS II (from acc Chl to Q_B) shows strong functional and structural similarities to the reaction center of purple bacteria, while the donor site (with Tyr_Z and the oxygen evolving complex), consisting of the 4MnCa cluster, is unique to PS II.

In each round of the photocycle, $P680^+$ is re-reduced by extracting one electron from the 4MnCa cluster. The redox active tyrosine Tyr_Z functions as the intermediate electron carrier between $P680^+$ and the Mn cluster. When four positive charges are accumulated at the Mn cluster after four subsequent electron transfer steps, one molecule of oxygen is released. The proton release pattern is still matter of debate. In complexes that are depleted of the three lumenal subunits, a 1 : 1 : 1 : 1 pattern has been observed, while the proton release pattern of the intact PS II shows a 1 : 0 : 1 : 2 pattern (Schlodder and Witt 1999). Whether these differences are caused by the protein isolation, or represent the buffering of released protons by the protein surrounding protein, is still a matter of further investigation.

One major problem that PS II has had to face since the invention of water splitting 2.5 billion years ago is the severe process of photodamage. This photodamage is caused by $P680^{+\cdot}$, which has the high redox potential of +1.1 V and can easily form a triplet state. The molecular and mechanistic details of the process of photodamage in PS II is a topic of current research. It might occur by direct oxidation of the protein by $P680^{+\cdot}$, or by the formation of the $^3P680$ triplet and subsequently the highly reactive singlet oxygen. During the process of photoinhibition, an irreversible damage of one of the core proteins of PS II, D1, occurs. D1 binds most cofactors of the electron transfer chain, including the Mn cluster, and has to be replaced every 30 min in plants in bright sunlight. The repair mechanism of PS II is a hot topic of current PS II research (Baena-Gonzalez and Aro 2002; Vass et al. 2007; Nowaczyk et al. 2006; Mohanty et al. 2007).

Photosystem II is a dimer in the native membrane of plants, algae and cyanobacteria. It consists of 19 protein subunits in cyanobacteria. In addition to the cofactors of the electron transport chain, it harbors 35–36 antenna chlorophylls and 10–12 carotenoids. In addition, PS II contains a large number of lipids which may play a role in the repair process of PS II. The first crystals of PS II that were able to split water had been grown from the thermophilic cyanobacterium *Thermosynechococcus elongatus* (Zouni et al. 2000) leading to the first X-ray structural model of the intact PS II complex at 3.8 Å resolution. This model provided the first insight into the structure of PS II, and showed for the first time the 3+1 organization of the 4 Mn atoms in the OEC.

All further crystal structures were based on these crystals, with the exception of the 3.7 Å crystal structure, which was determined by Kamyra, Shen and coworkers from *Thermosynechococcus vulcanus* (Kamiya and Shen 2003). In the most recent X-ray structure of PS II at 3.0 Å resolution, 31 antenna chlorophyll molecules, 11 carotenoids, and 14 lipids have been assigned to one monomer of PS II (Loll et al. 2005).

3.2
Arrangement of Protein Subunits in PS II

Twenty-two protein subunits have been assigned in structure of PS II from *Thermosynechococcus elongatus*, PsbA to PsbF, PsbH to PsbO, PsbT to PsbZ, as well as the PsbZ like protein Psb27.

At least 19 proteins of these subunits are present in crystals of PS II, as they have been identified in the PS II crystals by detailed mass spectroscopic analysis (Kern et al. 2005).

Sixteen of these subunits are membrane intrinsic, whereas PsbO, PsbU and PsbV do not contain transmembrane α-helices and are located at the lumenal side of the complex. In the first structural model of the intact PS II at 3.8 Å resolution (Zouni et al. 2001), 36 transmembrane helices were identified. Since this the first structure of PS II, several more structures from PS II have been published at 3.7 to 3.0 Å resolution, revealing more details of the structures. These include assignments of most of the amino acid side chains and identification of the small membrane intrinsic subunits. The initial number and location of the 36 transmembrane helices in PS II were confirmed by the most recent crystal structure of PS II at 3.0 resolution (Loll et al. 2005), whereas one helix is missing in the 3.5 Å structure of PS II (Ferreira et al. 2004). The general arrangement of subunits in PS II is shown in Fig. 2b based on the structural model of PS II at 3.5 Å structure of PS II (Ferreira et al. 2004).

The core of one monomeric unit of PS II, the reaction center (RC), consists of subunits D1 (PsbA) and D2 (PsbD), flanked on both sides by the core antenna proteins CP47 (PsbB) and CP43 (PsbD). This core of PS II shows strong similarities to the arrangement of the two major proteins in PS I (PsaA and PsaB), indicating that PS I and PS II have evolved from a common ancestor, as suggested by Schubert and coworkers (Schubert et al. 1998). A comparison of the membrane intrinsic parts of PS I and II is shown in Fig. 4. None of the extrinsic subunits or small membrane intrinsic subunits of PS I and II show any similarities, which indicates that they have been added to the PS I and II cores after the evolutionary split between the two photosystems. As evident in Fig. 1, the most striking difference between both photosystems is the location of the membrane extrinsic proteins. Whereas PS I contains three extrinsic proteins at the stromal side, which are involved in the docking of ferredoxin/flavodoxin, PS II does not extend more than 10 Å into the stroma. Most of the extrinsic mass of PS II is located at the lumenal site, formed by the large lumenal loops of CP43, CP 47 and the three lumenal proteins PsbO, PsbU and PsbV, which stabilize the oxygen evolving complex in PS II.

3.2.1
Core Subunits D1 and D2 (Baumgartner et al.)

The central core of PS II is often called the reaction center core, as it contains the electron transfer chain. This core is formed by a cluster of 2×5 transmembrane helices, forming an S-type arrangement (see Figs. 2 and 4). These helices are assigned to the protein subunits D1 (PsbA) (depicted in blue) and D2 (PsbD) (red) and resemble the structure of the L and M subunits of the reaction center of purple bacteria (Deisenhofer et al. 1985; Deisenhofer and Michel 1991). The structural and functional comparison of PS II and the PbRC shows that the D1 protein is related to the L subunit, while the D2 protein is related to the M subunit of the PbRC. To a lesser extent, the structure of the D1 and D2 proteins also resembles the C-terminal domains of the large subunits PsaA and PsaB of PS I (Jordan et al. 2001) (see Fig. 4 for a comparison between PS I and PS II). The similarity reveals that *all* actual existing photo-reaction centers might have evolved from a common ancestor as previously proposed (Blankenship and Kindle 1992; Schubert et al. 1998). The D1 protein binds/contains all cofactors of the electron transport chain: The Mn cluster, the redox active tryrosine Tyr_Z, the chlorophyll of the primary donor, $P680_{D1}$, the accessory chlorophyll, Chl_{D1}, the pheophytin $Pheo_{D1}$, and the mobile plastoquione Q_B. The D2 protein contains Tyr_D, a tyrosine which is not directly involved in electron transport but may be important for the photoassembly of the active Mn cluster (Rutherford et al. 2004). Furthermore it coordinates the chlorophylls $P680_{D2}$, the accessory chlorophyll Chl_{D2}, the pheophytin $Pheo_{D2}$ and the tightly bound plastoquione Q_A. In contrast to PS I, where the C-terminal parts of PsaA and PsaB coordinate 28 antenna chlorophylls, the RC core subunits of PS II, D1 and D2, coordinate only two antenna chlorophylls: $ChlZ_{D1}$ and $ChlZ_{D2}$. The most recent structure of PS II revealed that there are two carotenoids associated with the reaction center core in PS II. They form hydrophobic interactions with the outermost transmembrane helix of the reaction center core, which corresponds to the first transmembrane helix of D1 and D2 (helix a). The carotenoids are located in close proximity to $ChlZ_{D1}$ and $ChlZ_{D2}$. It was proposed that the carotenoids associated to D2 may play a role in the non-photochemical quenching process in PS II (Telfer 2002, 2005; Telfer et al. 2003).

The D1 protein has to pay a high price for coordinating most of the carriers of the electron transport chain. It is subject to photodamage, which is most probably caused by the formation of the triplet state of P680, ^{3}P680. D1 has to be replaced constantly: in plants in bright sunlight it has only a half-life of 30 min. The cyanobacterial genome contains three copies of the D1 gene, named *psbA1*, *psbA2* and *psbA3*. While expression of *psbA1* has not been shown so far, psbA2 and psbA3 are both expressed and their expression pattern is adapting under different environmental conditions (Salih and Jansson 1997; Soitamo et al. 1998; Sippola and Aro 2000). The photodamage to the D1 protein leads to a complex degradation and replacement cycle. The complicated process is a very active field of current research and is best studied in plants (see Rokka et al. 2005; Vass et al. 2007 and references therein). It includes phosphorylation of the damaged D1 protein, monomerization of the PS II dimer, de-attachment of the CP43 protein and the three lumenal proteins, proteolytic degradation of D1, synthesis of D1 by membrane-bound ribosomes, cleavage of the N-terminal signal sequence of D1 in the lumen, assembly of the Mn cluster by photo-activation and assembly of all subunits and the re-formation of PS II dimers. Taken the central location and role of the D1 protein into account, this repair cycle could be compared with a heart transplantation taking place every half an hour. The 3.0 Å structure revealed that lipids are essential constituents of PS II. The lipids are bound at the interface between the D1/D2 core, the antenna proteins CP34 and CP47 and two fields of small transmembrane helices (PsaM, L and T at the dimerization domain) and (PsbF, J and K) exposed at the membrane exposed surface of PS II. They may provide the lubricant for the smooth replacement of the D1 protein (Loll et al. 2005, 2007). The lipids may act as a set of "rails" to easily remove the damaged D1 proteins, and facilitate the insertion of the new D1 proteins by sliding into place on these "rails".

3.2.2
Antenna Proteins CP47 and CP43 (PsbB and PsbC)

The two largest subunits of PS II are the antenna proteins CP47 (PsbB) and CP43 (PsbC). They sit on both sides of the central D1/D2 core of PS II, with CP47 (PsbB) flanking the D2 protein and CP43 being located in close proximity to D1. Each of these subunits consists of six transmembrane helices. They form three pairs of dimers, consisting of helices 1/2, 3/4 and 5/6. CP47 is located close to the dimer-dimer interface, while CP43 is located at the periphery of the PS II dimer. This arrangement is in agreement with biochemical studies that have shown that CP43 can be more easily removed from the PS II core than CP47 (Bricker and Frankel 2002). The peripheral location of CP43 allows this subunit to be disassembled from PSII in the process of D1 turnover.

The arrangement of the transmembrane helices in CP47 and CP 43 is very similar to the arrangement of the transmembrane helices in the N-terminal part of the PsaA/PsaB in PS I (see Fig. 4).

CP47 coordinates 16 Chls and CP43 13 (or 14) chlorophylls. While the structures at 3.7 and 3.5 Å reported a 14th chlorophyll in CP43, the 3.0 Å structure identified a lipid (DGDG2 digalactosyldiacylglycerole) at this site.

Both subunits differ from the N-terminal region of the PsaA/B subunits in PS I by the formation of extended loops at the lumenal side. The large loop of CP47 interacts with loops of the D2 protein, the PsbO protein and the PsbU protein, thereby stabilizing the manganese cluster. The long loop of CP43 interacts with loops of the D1 protein and all three extrinsic proteins, PsbO, PsbU and PsbV. Furthermore, amino acid Glu 354 provides a ligand for one of the atoms of the Mn cluster.

3.2.3
Small Membrane Intrinsic Subunits

Fourteen transmembrane helices are located peripherally to the central core and have been assigned to the small membrane intrinsic proteins. They can be structurally divided into subunits located at the dimerization domain and subunits located at the periphery of the PS II dimer. Three helices are located close to the local 2-fold symmetry axis between the dimers, representing PsbM, PsbT and PsbL. One helix is sandwiched between D1 and CP43, which is assigned to PsbI. A field of 10 helices is located at the membrane-exposed periphery of PS II, which constitutes of PsbE, PsbF, PsbH, PsbJ, PsbK and PsbZ. The electron density is much better defined at the dimer interface than at the periphery of PS II, therefore there is still some debate about the subunit-assignment of three of the peripheral helices to the subunits PsbN and PsbX (Ferreira et al. 2004; Loll et al. 2005). However, PsbY but not PsbN, has been identified in the PS II crystals. Therefore it can not be excluded that PsbY may also be one of the three peripheral helices.

Subunits at the Periphery of PS II

Cytochrome b_{559} (PsbE and PsbF)

The membrane intrinsic cytochrome b_{559} consists of two protein subunits: PsbE and PsbF. These are the only small subunits of PS II that directly coordinate a cofactor. The cytochrome is located in close vicinity to helix 1 of D2. The heme is coordinated by His 34 of PsbE and His 24 of PsbF. The cytochrome b_{559} is essential for the function of PS II, as deletion of either PsbE or PsbF are lethal for the function of PS II. The cytochrome can exit in a low potential form with a redox potential of +275 mV or a high potential form with +390 mV. The functional role of this cytochrome and its heme cofactor is under investigation (Bondarava et al. 2003; Lakshmi et al. 2003; Vasil'ev et al. 2003). The structure implies that the heme might be involved in the non-radiative charge recombination between the singly reduced Q_B and P680 as part of a prevention of excessive photodamage, with the help of the carotenoid, Car_{D2} that is placed between the D2 protein and cytochrome b_{559}. A redox active role of carotenoids have been also proposed by (Tracewell and Brudvig 2003).

Subunits Close to Cytochrome b_{559}

Two further helices are located in the close vicinity of PsbE and PsbF. The helix closest to PsbF has been assigned to PsbJ. The helix close to PsaE (named X2 in (Loll et al. 2005)) was present in the 3.8 Å model (Kamiya and Shen 2003) and 3.2 model (Biesiadka et al. 2004) of PS II, but is missing in the 3.5 Å model (Ferreira et al. 2004). The loss may be caused by the higher detergent concentration used for the isolation of PS II in the work of Ferreira et al. (Ferreira et al. 2004). As PsbY is present in the crystals but has not been assigned, one might speculate whether helix X2 may represent PsbY.

The location of the four helices somewhat resembles the arrangement of PsaF and PsaJ in PS I, which are transcribed on one open reading frame. Interestingly, the PS II subunits PsbE, PsbF, PsbL and PsbJ are also expressed in all cyanobacteria on one open reading frame. There are only nine bases between the stop codon of PsaF and the start codon of PsaL, and the Shine Dalgarno sequence of PsaL overlaps with the stop codon for PsaF. Taking into account that the subunits are very likely cotranslationally inserted into the membrane, one might question if the current assignment of PsaL is correct, as the ribosome must move back and forth between the peripheral and dimerization domain in order to insert PsaF, PsaL and PsaJ into the membrane. This problem would be solved if helix X2 would represent PsaL, but this question may have to wait for a higher solution structure to be answered.

Small Subunits Close to CP43

Four transmembrane helices are located at the periphery of PS II in close vicinity to CP43. The helix closest to the interface between CP43 and PsbJ has been assigned to PsbK, while two transmembrane helices that are in the vicinity of the N-terminal region of CP43 have been assigned to PsbZ. One further helix in between PsbZ and PsbK has been assigned to PsbN in the 3.5 Å structure. However, this assignment has not been confirmed at 3.0 Å resolution and PsbN has not been identified in the crystals, so the nature of this helix (named X1 in the 3.0 Å structure) is still unclear. The subunits PsaJ and PsaK shield a field of lipids between the D1 protein and CP43 from the membrane, which shield the cofactors of the D2 branch, including Q_A. The important role of PsbJ has been shown by biochemical studies: A psbJ deletion mutant is impaired in PSII electron flow to plastoquinone (Regel et al. 2001). PsbK may stabilize a cluster of chlorophylls in CP43. A tight interaction of CP43 with PsbK has also been shown by biochemical studies, where PsbK was co-purified with CP43 during ion exchange chromatography (Sugimoto and Takahashi 2003). The stabilization effect may be futher enhanced by the interaction with a carotenoid.

Small Subunits Close to CP47
There is only one helix that is in close contact to chlorophylls of CP47 at the membrane-exposed periphery, and this helix has been assigned to PsbH. It forms hydrophobic contacts with several chlorophylls in CP47 and with one carotenoid, thereby stabilizing the antenna system of CP47.

Small Subunits Close to $ChlZ_{D1}$ and $ChlZ_{D2}$
Two symmetry related single helices are located close to the $ChlZ_{D1}$ and $ChlZ_{D2}$. The helix close to $ChlZ_{D1}$ has been assigned to PsbI, while the assignment of the helix close to $ChlZ_{D2}$ is still under discussion. It has been tentatively assigned to PsbZ in the 3.5 Å structure, but this assignment was not confirmed at 3.0 Å resolution.

Both PsbI and PsbZ are important proteins, as deletion mutants are strongly affected.

PsbI deletion mutants are still able to grow photoautotrophically in dim light, but not in high light, and the amounts of PS II complex and oxygen evolving activity are both reduced to 10–20% of wild-type levels (Kunstner et al. 1995).

Subunits of the Dimerization Domain
Three helices are located in the dimerization domain, which are assigned to the subunits PsbL, PsbM and PsbT. PsbM is closest to the dimerization domain and forms interactions with the PsbM subunit of the neighboring dimer. PsbT is also close to the dimer–dimer interface and is located in close vicinity of the first transmembrane helix of the D1 protein. The helix assigned to PsbL is located between PsbM and PsbT. All three subunits form contacts with several lipids that fill the space between the dimerization domain and the opening between D1 and D2 protein. Thereby all three subunits and the

lipids are in close vicinity to the electron transfer chain and may even stabilize the Q_A binding site. The deletion of each of the three subunits has severe effects on the function of PS II, which are not all in agreement with the proposed location of these subunits. Deletion of PsbM in tobacco leads to high light-sensitivity of PS II and alters the Q_B site properties. The latter finding is difficult to understand with the present assignment of PsbM. Furthermore, the mutant shows reduced phosphorylation of D1 and D2 and may be impaired in PS II repair (Umate et al. 2007). Mutants lacking PsbL are highly sensitive to photoinhibition and it was suggested that PsbL may be important to prevent reduction of PSII by back electron flow from plastoquinol (Ohad et al. 2004). This would also favor a location of PsaL at the periphery close to Q_B and not in the dimerization domain in close vicinity to Q_A. As biochemical and structural evidence do not match, the structural assignment of PsbM and PsbL must be regarded as tentative. In contrast, the assignment of PsbT to one of the helices in the dimerization domain is also strongly supported by biochemical evidence, as results from a deletion mutant of PsbT suggest that PsbT is involved in the stabilization and repair of primary electron acceptor Q_A of PS II during photoinhibition. Furthermore, it may structurally stabilize the Q_A binding site, as half the Q_A was lost from the PSII core complex that lacks PsbT during purification (Ohnishi et al. 2007).

3.2.4
Lumenal Subunits PsbO, PsbV and PsbU

Photosystem II from cyanobacteria contains three extrinsic subunits, which are located at the lumenal side of the core complex: the 33 kDa protein (PsbO), the 12 kDa protein (PsbU) and the cytochrome c_{550} (PsbV) (see Fig. 3B).

PsbU and PsbV are unique to cyanobacteria, whereas PsbO is also present in the PS II complex of higher plants. In the structure at 3.8 Å resolution, the main body of PsbO was identified as a β-barrel structure, which was confirmed in further structures. PsbO forms various contacts with CP43, CP47, D1 and D2 and is thereby important for the stabilization of the Mn cluster, even if it does not directly provide a ligand. The protein subunit PsbV (cyt c_{550}) is located at the side of the lumenal hump that faces away form the dimerization domain and is in close contact to the lumenal loops of CP43. Despite the fact that it contains a heme, its function is unclear, as the reduction of the heme has not been reported. PsbV has strong structural similarity to cytochrome c_6, the electron donor to PS I, and might represent the old electron donor to the non-oxygenic ancestor of PS II, which was trapped during evolution and became an extrinsic subunit of PS II that now stabilizes the oxygen evolving complex (Grotjohann et al. 2004). PsbU is a small protein that is located at the outermost lumenal tip of PS II. It may further stabilize the PS II complex.

3.3
Electron Transport Chain of PS II

The cofactors of the electron transport chain (Svensson et al.) in PS II is shown in Fig. 1B. It consists of four chlorophylls a, two pheophytins, two plastoquinones, one redox active tyrosine Tyr_Z, Tyr_D and the Mn cluster. In contrast to PS I, where the uni/bi-directionality of the ETC is still a matter of debate, the uni-directionality of the ETC in type II reactions centers is well established. Most of the redox-active cofactors of the ETC are located on the D1 site: the 4MnCa cluster, Tyr_Z, $P680_{D1}$, Chl_{D1}, $Pheo_{D1}$ and Q_B. The D2 site is directly involved in the ETC by coordinating $P680_{D2}$ and the tightly bound quinone Q_A. Chl_{D2} and $Pheo_{D2}$ are not directly involved in the forward electron transfer, but might play a role in prevention of photo-damage by facilitating the backward electron transfer from singly reduced $Q_B^{\cdot-}$ to $P680^{\cdot+}$.

3.3.1
Acceptor Side of the Electron Transport Chain in PS II (Alfonso et al.)

The acceptor side of PS II consists of the accessory chlorophyll Chl_{D1}, the pheophytin, $Pheo_{D1}$, the two phylloquionones Q_A and Q_B, and a non-heme iron. The non-heme iron that is located between Q_A and Q_B has only a structural function and is not directly involved in electron transfer from Q_A to Q_B. The arrangement and spectroscopic properties of the cofactors resembles the structural arrangement of the electron transfer chain in reaction centers of purple bacteria. After two subsequent charge separation events, the double-reduced Q_B^{2-} binds two protons and leaves the binding pocket as PQH_2. It is then replaced by a PQ from the PQ-pool. The recent structure at 3.0 Å resolution showed that there may exist two diffusion pathways to the QB site: one that is open to the membrane and one that opens up to the stromal side of PS II (Loll et al. 2005). One might speculate that the pathway that leads to the membrane may be the diffusion pathway for PQ, while the pathway that opens up to the lumen may facilitate the diffusion of protons to the Q_B binding site.

3.3.2
Donor Site of the Electron Transfer Chain of PS II

The donor site of the ETC chain is a is unique feature of PS II as it contains the oxygen evolving complex OEC. It consists of the primary donor $P680^{+\cdot}$, the redox active tyrosine Tyr_Z and a cluster of 4 manganese atoms and one Ca, which catalyzes the water splitting. The tyrosine Tyr_D, located at the D2 site of the ET chain, is not directly involved in electron transfer or water splitting, but may assist in the assembly of the Mn cluster (Sugiura et al. 2004).

In the process of water splitting, one electron is extracted form the OEC in four subsequent charge separation events. The oxidation states of the oxygen evolving center (Ahrling et al.) are named S states. The "clock of water splitting" cycles between S0 (no charge), S1 (+1), S2 (+2), S3 (+3) and S4 (+4). Oxygen is thereby evolved after 4 charge separation events. The two water molecules are already bound in the transition from S4 to S0. The proton release pattern is still a matter of debate. PS II particles depleted of the three lumenal subunits release the protons in a 1 : 1 : 1 : 1 pattern, whereas the intact PS II shows a release of protons in a 1 : 0 : 1 : 2 pattern (Schlodder and Witt 1999).

Primary Electron Donor: P680

The most important difference between the primary donor of PSI (P700) and the primary electron donor of PS II is the redox potential of $P680^{+\cdot}$. The cation radical of the primary donor in PS II, $P680^{+\cdot}$, has a redox potential of 1100 mV, which is one of the highest redox potentials of any cofactor found in biological systems. This provides the redox potential for the unique function of water oxidation. Four Chl molecules are located in the center of the D1/D2 core and they may all contribute to the spectroscopic and electronic properties of the primary electron donor P680*. However, different radical states might be primarily located on individual clorophylls. The four chlorophylls are arranged in two symmetrically related pairs, $P680_{D1}/P680_{D2}$ and Chl_{D1}/Chl_{D2}. The chlorophylls of the first pair, named $P680_{D1}/P680_{D2}$, are oriented perpendicular to the membrane plane. The center-to-center distance varies between the different structures: 10 Å in the 3.8 Å structure (Zouni et al. 2001), 9.56 Å in the 3.7 Å structure (Kamiya and Shen 2003), 8.6 Å in the 3.6 Å structure (Fromme et al. 2002), 8.2 Å in the 3.5 Å structure (Ferreira et al. 2004) and 8.3 Å in the 3.2 Å (Biesiadka et al. 2004) structure and 7.6 Å in the structure at 3.0 Å resolution. The decrease in distance with increase in resolution may be caused by clearer assignment of the 5/6 ring of the chlorin system at higher resolution. Even with the shortest distance reported to be 7.6 Å, the chlorophylls are more separated from each other in PS II than the chlorophylls in P700 or in the special pair of the PbRC, which may indicate a weak excitonic coupling. There is strong evidence from EPR studies that the cation radical of the primary donor, $P680^{+\cdot}$, which extracts electrons from the redox active Tyr_Z and subsequently from the 4Mn–Ca cluster, is located on the chlorophyll $P680_{D1}$. The distance between $P680^{+\cdot}$ and Tyr_Z has been determined to be in the range of 7.9 ± 0.2 Å (Lakshmi et al. 1999) which only matches for the $P680_{D1}$ chlorophyll.

The question can be addressed, why none of the neighboring chlorophylls is oxidized instead of the Tyr_Z (redox potential 1.0 V) by $P680_{D1}$. Taking into account that $Pheo^{-\cdot}$ has a redox potential of 1.4 V, we can assume that all four chlorophylls must have a high redox potential between 1.0 and 1.3 V and may be able to perform the initial charge separation, as suggested in (Barber

2002). Therefore, the excited state P680* may be delocalized – at least at room temperature – among all four chlorophylls.

The plane of the chlorin head group of the second pair of chlorophylls (Chl_{D1}/Chl_{D2}) is tilted at an angle of 30° to the membrane plane. There is strong spectroscopic evidence that the triplet state of the primary donor ^{3}P680 is located on the accessory chlorophyll Chl_{D1}.

Water-Oxidizing Complex

The most interesting feature of PS II is the ability to oxidize water to O_2 and 4 H^+. Photosystem II performs this reaction by a cluster of four Mn and one Ca ion, bound to the protein D1. The first structure of PS II at 3.8 Å resolution unraveled for the first time the location of the manganese cluster and the arrangement of the 4Mn atoms in a 3+1 organization. This principle arrangement was confirmed by the improved crystal structures (Fromme et al. 2002; Kamiya and Shen 2003; Ferreira et al. 2004; Loll et al. 2005) and is also in agreement with EPR and XAFS studies on the Mn cluster (Peloquin and Britt 2001; Britt et al. 2004; Sauer and Yachandra 2004; Haddy 2007; Kern et al. 2007). The 3.5 Å structure elucidated for the first time the location of the Ca in the cluster (Ferreira et al. 2004). It forms a distorted cubane with the Mn atoms 1, 2 and 3. The fourth Mn is more distal to the distorted cubane. The general arrangement of the 4Mn–Ca cluster is in agreement with EPR and XAFS data (for a more detailed discussion see Sauer and Yachandra (2004) and refs. therein).

Another cofactor that plays an important role in the process of water splitting is Cl^-; however, no position for Cl^- ions have been identified in any of the current crystal structures.

Several amino acids of D1 have been identified as potential ligands to the 4Mn–Ca cluster, as Asp 170, Glu 189, His 190, Asp 342, His 332, Glu 333 and the N-terminus of D1 (Ala 344). Furthermore, Glu 354 of CP43 may also provide a ligand to the cluster. Despite the fact that similar amino acids are close to the 4Mn–Ca cluster in all structures, the structures differ significantly in the positioning of the amino acids and the question which amino acid is coordinating which Mn or Ca atom. The assignment of the ligands to the Mn cluster do not agree in all points with mutagenesis studies, either. For example, Glu 189 can be replaced by many other amino acids without affecting oxygen evolution (Clausen et al. 2001), and CP43-E354 has been replaced by other amino acids with only a moderate influence on oxygen evolution.

Neither of the structures could identify the bridging oxygen atoms, nor the substrate water molecules bound to the cluster. The arrangements of the Mn atoms differ significantly in the recent structural models, as shown in Fig. 5. The electron density of the Mn cluster has not been improved in the recent structures, as it is still a papaya-shaped blob (see Fig. 5). The reason for the lack of improvement of the resolution of the Mn cluster is its high sensitivity to photodamage. It has been shown by XAFS studies on single crystals of

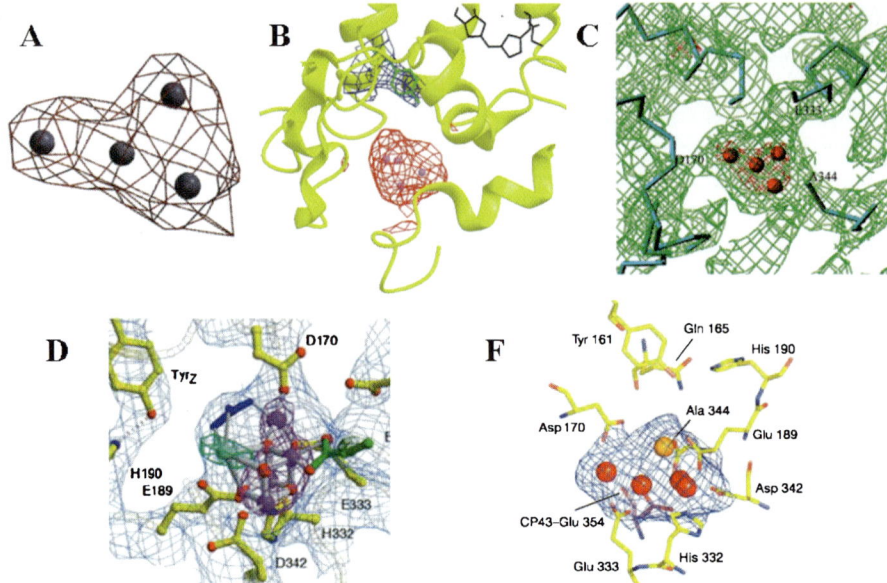

Fig. 5 Electron densities of the Mn cluster in different structures of PS II: **A** First electron density of the Mn cluster from *TS elongatus* at 3.8 Å resolution (Zouni et al. 2001). **B** Electron density of the Mn cluster from *TS elongatus* at 3.6 Å resolution (Fromme et al. 2002). **C** Electron density of the Mn cluster from *TS vulcanus* at 3.7 Å resolution (Kamiya and Shen 2003). **D** Electron density of the Mn cluster from *TS elongatus* at 3.5 Å resolution (Ferreira et al. 2004). **E** Electron density of the Mn cluster from *TS elongatus* at 3.0 Å resolution (Loll et al. 2005)

PS II, that the Mn cluster undergoes severe damage even during short exposure (Yano et al. 2005) and may disassemble. The Mn 4 is very likely the first Mn atom that is lost, as it shows less than 30% occupancy in the 3.0 Å X-ray structure (Loll et al. 2005).

The mechanism of water splitting and the oxidation states of the Mn cluster are still under "hot" discussion, and no final model that can explain all experimental evidence exists. A higher resolution X-ray structure of the undamaged 4Mn–Ca cluster, in combination with spectroscopic results and computational modeling, may be able to solve the secrets of the process of light-driven oxygen evolution. The mechanism of water splitting by PS II is still an open field for future discoveries.

3.4
Antenna System of PS II

In addition to the cofactors of the electron transport chain, 57 additional cofactors have been identified in the structure of PS II at 3.0 Å resolution:

29 antenna chlorophylls, 14 lipids, 11 β-carotene molecules, and 3 detergent molecules (β-dodecylmaltoside).

3.4.1
Chlorophylls

In the 3.0 Å structure, CP47 binds 16 antenna Chl molecules and CP34 binds 13 antenna chlorophylls (Loll et al. 2005). All except three of these chlorophylls have been identified at 3.8 Å resolution (Zouni et al. 2001). All X-ray structural models show very similar locations for the chlorophylls.

The comparison of the antenna systems in PS I and II reveals that PS II lacks the central antenna domain, which harbors more than 50 antenna chlorophylls in PS I. The central PS II-core contains only two chlorophylls ($ChlZ_{D1}$ and $ChlZ_{D2}$). This lack of the central antenna domain may be responsible for the lower efficiency of the excitation energy transfer in PS II compared to PS I and the fact that the fluorescence spectrum of photosynthetic organisms is dominated by fluorescence from PS II. While the excitation energy transfer in PS I has an efficiency of 99.5% the efficiency is decreased to 80% in PS II.

It is very likely that PS II has been stalled in an earlier evolutionary state, where antenna and reaction center were not tightly coupled, while a genefusion in PSI allowed the development of a more efficient joint reaction center and antenna system.

The price PS II has to pay for its ability to use water as an unlimited electron source is the sensitivity to photodamage with the need for repair of the D1 protein, which hindered the development of a closely coupled core antenna system. If PS II contained a central antenna domain, all central chlorophylls would have also to be replaced with the D1 protein, which would be an extremely resource-wasting process. Furthermore, the location of antenna chlorophylls close to the highly oxidizing P680 could have caused severe damage of the antenna pigments.

3.4.2
Carotenoids

Eleven carotenoids have been identified in the 3.0 Å structure of PS II (Loll et al. 2005). They were all modeled as beta-carotenes, as no discrimination between carotenes and xanthophylls can be made at this resolution. In contrast to PS I, all carotenoids are in the all-trans conformation (the cis-conformation proposed for one of the car molecules at 3.5 Å resolution, could be not confirmed in the 3.0 Å structure).

The location of the carotenes shows no striking homologies between PS I and PS II. Furthermore, the arrangement is much more symmetrical in PS II than in PS I.

Only two carotenes are associated with the D1/D2 core, while five carotenoids have been assigned to CP47 and four carotenoids have been assigned to CP43. Three of the carotenoids in CP43 and CP47 show twofold symmetry. They mainly interact with helices 1 and 2 of CP34 and CP47. The main function of the carotenoids is quenching of Chl triplet states. Furthermore, they can act as an additional antenna. Lastly, the carotenoids may also serve as stabilizing elements in PS II. The carotenoids in CP43 bridge CP43 with the cluster of small subunits at the periphery of PS II (PsbZ, PsbN, PsbK and PsbJ) and may be important for the stable binding of the small subunits to the PS II core. The symmetry-related carotenoids in CP47 are located at the dimer interface and may help in the stabilization of the PS II dimer.

The two carotenoids that are located in the RC domain are located at symmetry-related positions but have completely different orientations. Car_{D1}, which is located between helix 1 of the D1 protein and the dimer–dimer interface in close vicinity to $ChlZ_{D1}$, is orientated nearly perpendicular to the membrane plane. In contrast, Car_{D2}, which is located between helix 1 and the cytochrome b_{559} in close vicinity to $ChlZ_{D2}$, is orientated nearly parallel to the membrane plane. The presence of two spectroscopically distinct carotenoids in the D1/D2 core have been shown for plant PS II (Kwa et al. 1992). Loll and coworkers (Loll et al. 2005) proposed that Car_{D1} may represent car489 and Car_{D2} may correspond to car507. Car_{D1} may have the function of quenching triplet states of $ChlZ_{D1}$, whereas Car_{D2} may play an important role in the photoprotective processes of charge recombination between singly reduced Q_B and P680.

3.4.3
Lipids

One important difference between PS I and II is the much higher lipid content of PS II. While only four lipids are present in PS I, 14 lipid molecules have been for the first time identified in the structure of PS II at 3.0 Å resolution (Loll et al. 2005). It has long been known that lipids are essential cofactors in PS II, and mutagenesis studies have shown that they play a role in electron transfer between Q_A and Q_B, as well as in the dimerization of PS II (see Fyfe and Jones 2005; Jones 2007 and refs. therein). Two larger clusters of lipids are located at the interface of the D1/D2 core to CP47 and CP43 (Loll et al. 2007). The cluster that bridges the D1/D2 core with subunits PsbM, PsbL and PsbT might be important for the assembly/disassembly of the dimer, whereas the cluster between the D1/D2 core, CP43 and PsbJ and PsaK may be important for the de-attachment and re-assembly of the CP43 subunit to the PS II core in the process of D1 damage and repair. This would also explain why PS II has so many more lipids than PS I, because PS I does not have to undergo excessive repair in the living cell.

Acknowledgements This work was supported by the National Institute of Health grant number R01GM71619-4 and the National Science Foundation Grant number 0417142. We thank Christopher Vanselow for critical reading of the manuscript.

References

Ahrling KA, Peterson S, Styring S (1997) An oscillating manganese electron paramagnetic resonance signal from the S0 state of the oxygen evolving complex in PS II. Biochemistry 36:13148–13152

Alfonso M, Pueyo JJ, Gaddour K, Etienne AL, Kirilovsky D, Picorel R (1996) Induced new mutation of D1 serine-268 in soybean photosynthetic cell cultures produced atrazine resistance, increased stability of S2QB- and S3QB-states, and increased sensitivity to light stress. Plant Physiol 112:1499–1508

Allen MB, Piette LH, Murchio JC (1961) Observation of two photoreactions in photosynthesis. Biochem Biophys Res Commun 4:271–274

Amunts A, Drory O, Nelson N (2007) The structure of a plant photosystem I supercomplex at 3.4 A resolution. Nature 447:58–63

Anandan S, Vainstein A, Thornber JP (1989) Correlation of some published amino acid sequences for photosystem I polypeptides to a 17 kDa LHCI pigment-protein and to subunits III and IV of the core complex. FEBS Lett 256:150–154

Ananyev GM, Sakiyan I, Diner BA, Dismukes GC (2002) A functional role for tyrosine-D in assembly of the inorganic core of the water oxidase complex of photosystem II and the kinetics of water oxidation. Biochemistry 41:974–980

Andersen B, Scheller HV (eds) (1993) Structure, function and assembly of photosystem I. Academic Press, San Diego

Andersen B, Scheller HV, Moller BL (1992) The PSI-E subunit of photosystem I binds ferredoxin:NADP+ oxidoreductase. FEBS Lett 311:169–173

Andersson J, Wentworth M, Walters RG, Howard CA, Ruban AV, Horton P, Jansson S (2003) Absence of the Lhcb1 and Lhcb2 proteins of the light-harvesting complex of photosystem II – effects on photosynthesis, grana stacking and fitness. Plant J 35:350–361

Antonkine ML, Jordan P, Fromme P, Krauss N, Golbeck JH, Stehlik D (2003) Assembly of protein subunits within the stromal ridge of photosystem I. Structural changes between unbound and sequentially photosystem I-bound polypeptides and correlated changes of the magnetic properties of the terminal iron sulfur clusters. J Mol Biol 327:671–697

Baena-Gonzalez E, Aro EM (2002) Biogenesis, assembly and turnover of photosystem II units. Philos Trans R Soc Lond B Biol Sci 357:1451–1459

Barber J (2002) P680: what is it and where is it? Bioelectrochemistry 55:135–138

Baumgartner BJ, Rapp JC, Mullet JE (1993) Plastid genes encoding the transcription/translation apparatus are aifferentially transcribed early in barley (*Hordeum vulgare*) chloroplast development (evidence for selective stabilization of psbA mRNA). Plant Physiol 101:781–791

Ben-Shem A, Frolow F, Nelson N (2003) Crystal structure of plant photosystem I. Nature 426:630–635

Bertsch WF (1962) Two photoreactions in photosynthesis: evidence from the delayed light emission of *Chlorella*. Proc Natl Acad Sci USA 48:2000–2004

Bialek-Bylka GE, Jazurek B, Dedic R, Hala J, Skrzypczak A (2003) Unique spectroscopic properties of synthetic 15-cis beta-carotene, an important compound in photosynthesis, and a medicine for photoprotective function. Cell Mol Biol Lett 8:689–697

Bibby TS, Nield J, Barber J (2001) Three-dimensional model and characterization of the iron stress-induced CP43'-photosystem I supercomplex isolated from the cyanobacterium *Synechocystis* PCC 6803. J Biol Chem 276:43246–43252

Biesiadka J, Loll B, Kern J, Irrgang KD, Zouni A (2004) Crystal structure of cyanobacterial photosystem II at 3.2 angstrom resolution: a closer look at the Mn-cluster. Phys Chem Chem Phys 6:4733–4736

Blankenship JE, Kindle KL (1992) Expression of chimeric genes by the light-regulated cabII-1 promoter in *Chlamydomonas reinhardtii*: a cabII-1/nit1 gene functions as a dominant selectable marker in a nit1- nit2- strain. Mol Cell Biol 12:5268–5279

Boekema EJ, Hifney A, Yakushevska AE, Piotrowski M, Keegstra W, Berry S, Michel KP, Pistorius EK, Kruip J (2001) A giant chlorophyll–protein complex induced by iron deficiency in cyanobacteria. Nature 412:745–748

Bondarava N, De Pascalis L, Al-Babili S, Goussias C, Golecki JR, Beyer P, Bock R, Krieger-Liszkay A (2003) Evidence that cytochrome b559 mediates the oxidation of reduced plastoquinone in the dark. J Biol Chem 278:13554–13560

Bricker TM, Frankel LK (2002) The structure and function of CP47 and CP43 in Photosystem II. Photosynth Res 72:131–146

Britt RD, Campbell KA, Peloquin JM, Gilchrist ML, Aznar CP, Dicus MM, Robblee J, Messinger J (2004) Recent pulsed EPR studies of the photosystem II oxygen-evolving complex: implications as to water oxidation mechanisms. Biochim Biophys Acta 1655:158–171

Brown JS, Schoch S (1981) Spectral analysis of chlorophyll–protein complexes from higher plant chloroplasts. Biochim Biophys Acta 636:201–209

Byrdin M, Jordan P, Krauss N, Fromme P, Stehlik D, Schlodder E (2002) Light harvesting in photosystem I: modeling based on the 2.5-A structure of photosystem I from *Synechococcus elongatus*. Biophys J 83:433–457

Clausen J, Winkler S, Hays AM, Hundelt M, Debus RJ, Junge W (2001) Photosynthetic water oxidation in *Synechocystis* sp. PCC6803: mutations D1-E189K, R and Q are without influence on electron transfer at the donor side of photosystem II. Biochim Biophys Acta 1506:224–235

Damjanovic A, Vaswani HM, Fleming GR, Fromme P (2002) Chlorophyll excitations in Photosystem I as revealed by semi-empirical ZINDO/CIS calculations. Biophys J 82:293a–293a

Dashdorj N, Xu W, Cohen RO, Golbeck JH, Savikhin S (2005) Asymmetric electron transfer in cyanobacterial Photosystem I: charge separation and secondary electron transfer dynamics of mutations near the primary electron acceptor A0. Biophys J 88:1238–1249

Deisenhofer J, Epp O, Miki K, Huber R, Michel H (1985) Structure of the protein subunits in the photosynthetic reaction center of Rhodopseudomonas-Viridis at 3a resolution. Nature 318:618–624

Deisenhofer J, Michel H (1991) Structures of bacterial photosynthetic reaction centers. Annu Rev Cell Biol 7:1–23

Deisenhofer J, Michel H (1991) Structures of bacterial photosynthetic reaction centers. Ann Rev Cell Biol 7:1–23

Fairclough WV, Forsyth A, Evans MCW, Rigby SEJ, Purton S, Heathcote P (2003) Bidirectional electron transfer in photosystem I: electron transfer on the PsaA side is not essential for phototrophic growth in *Chlamydomonas*. Biochim Biophys Acta-Bioenergetics 1606:43–55

Falzone CJ, Kao YH, Zhao J, Bryant DA, Lecomte JT (1994) Three-dimensional solution structure of PsaE from the cyanobacterium *Synechococcus* sp. strain PCC 7002, a photosystem I protein that shows structural homology with SH3 domains. Biochemistry 33:6052–6062

Ferreira KN, Iverson TM, Maghlaoui K, Barber J, Iwata S (2004) Architecture of the photosynthetic oxygen-evolving center. Science 303:1831–1838

Fischer N, Hippler M, Setif P, Jacquot JP, Rochaix JD (1998) The PsaC subunit of photosystem I provides an essential lysine residue for fast electron transfer to ferredoxin. EMBO J 17:849–858

Fischer N, Setif P, Rochaix JD (1999) Site-directed mutagenesis of the PsaC subunit of photosystem I. F(b) is the cluster interacting with soluble ferredoxin. J Biol Chem 274:23333–23340

Fromme P (1998) Crystallization of Photosystem I for structural analysis. Habilitation. Technical University Berlin, Berlin Germany

Fromme P, Kern J, Loll B, Biesiadka J, Saenger W, Witt HT, Krauss N, Zouni A (2002) Functional implications on the mechanism of the function of photosystem II including water oxidation based on the structure of photosystem II. Philos Trans R Soc Lond B Biol Sci 357:1337–1344

Fyfe PK, Jones MR (2005) Lipids in and around photosynthetic reaction centres. Biochem Soc Trans 33:924–930

Gibasiewicz K, Ramesh VM, Lin S, Redding K, Woodbury NW, Webber AN (2003) Excitonic interactions in wild-type and mutant PSI reaction centers. Biophys J 85:2547–2559

Gobets B, van Stokkum IHM, van Mourik F, Dekker JP, van Grondelle R (2003) Excitation wavelength dependence of the fluorescence kinetics in Photosystem I particles from *Synechocystis* PCC 6803 and *Synechococcus elongatus*. Biophys J 85:3883–3898

Golbeck JH (1999) A comparative analysis of the spin state distribution of in vitro and in vivo mutants of PsaC. Photosynth Res 61:107–144

Grotjohann I, Jolley C, Fromme P (2004) Evolution of photosynthesis and oxygen evolution: Implications from the structural comparison of Photosystems I and II. Phys Chem Chem Phys 6:4743–4753

Guergova-Kuras M, Boudreaux B, Joliot A, Joliot P, Redding K (2001) Evidence for two active branches for electron transfer in Photosystem I. Proc Natl Acad Sci USA 98:4437–4442

Haddy A (2007) EPR spectroscopy of the manganese cluster of photosystem II. Photosynth Res 92:357–368

Hanley J, Setif P, Bottin H, Lagoutte B (1996) Mutagenesis of photosystem I in the region of the ferredoxin cross-linking site: modifications of positively charged amino acids. Biochemistry 35:8563–8571

Hastings G, Hoshina S, Webber AN, Blankenship RE (1995) Universality of energy and electron transfer processes in photosystem I. Biochemistry 34:15512–15522

Hastings G, Reed LJ, Lin S, Blankenship RE (1995) Excited stae dynamics in Photosystem I: effects of detergent and excitation wavelength. Biophys J 69:2044–2055

Hippler M, Drepper F, Haehnel W, Rochaix JD (1998) The N-terminal domain of PsaF: precise recognition site for binding and fast electron transfer from cytochrome c6 and plastocyanin to photosystem I of *Chlamydomonas reinhardtii*. Proc Natl Acad Sci USA 95:7339–7344

Hippler M, Reichert J, Sutter M, Zak E, Altschmied L, Schroer U, Herrmann RG, Haehnel W (1996) The plastocyanin binding domain of photosystem I. EMBO J 15:6374–6384

Humphrey W, Dalke A, Schulten K (1996) VMD: visual molecular dynamics. J Mol Graph 14:33–38

Ishikita H, Knapp EW (2003) Redox potential of quinones in both electron transfer branches of photosystem I. J Biol Chem 278:52002–52011

Janson S, Andersen B, Scheller HV (1996) Nearest-neighbor analysis of higher-plant photosystem I holocomplex. Plant Physiol 112:409–420

Jolley C, Ben-Shem A, Nelson N, Fromme P (2005) Structure of plant photosystem I revealed by theoretical modeling. J Biol Chem 280:33627–33636

Jolley CC, Wells SA, Hespenheide BM, Thorpe MF, Fromme P (2006) Docking of photosystem I subunit C using a constrained geometric simulation. J Am Chem Soc 128:8803–8812

Jones MR (2007) Lipids in photosynthetic reaction centres: structural roles and functional holes. Prog Lipid Res 46:56–87

Jordan P, Fromme P, Klukas O, Witt HT, Saenger W, Krauß N (2001) Three-dimensional structure of cyanobacterial photosystem I at 2.5 Åresolution. Nature 411:909–917

Kamiya N, Shen JR (2003) Crystal structure of oxygen-evolving photosystem II from *Thermosynechococcus vulcanus* at 3.7-A resolution. Proc Natl Acad Sci USA 100:98–103

Kanervo E, Aro EM, Murata N (1995) Low unsaturation level of thylakoid membrane lipids limits turnover of the D1 protein of photosystem II at high irradiance. FEBS Lett 364:239–242

Kass H, Fromme P, Witt HT, Lubitz W (2001) Orientation and electronic structure of the primary donor radical cation P-700· in photosystem I: A single crystals EPR and ENDOR study. J Phys Chem B 105:1225–1239

Kern J, Biesiadka J, Loll B, Saenger W, Zouni A (2007) Structure of the Mn(4)-Ca cluster as derived from X-ray diffraction. Photosynth Res 92:389–405

Kern J, Loll B, Luneberg C, DiFiore D, Biesiadka J, Irrgang KD, Zouni A (2005) Purification, characterisation and crystallisation of photosystem II from Thermosynechococcus elongatus cultivated in a new type of photobioreactor. Biochim Biophys Acta 1706:147–157

Kjaerulff S, Andersen B, Nielsen VS, Moller BL, Okkels JS (1993) The PSI-K subunit of photosystem I from barley (*Hordeum vulgare* L.). Evidence for a gene duplication of an ancestral PSI-G/K gene. J Biol Chem 268:18912–18916

Koike K, Ikeuchi M, Hiyama T, Inoue Y (1989) Identification of photosystem I components from the cyanobacterium *Synechococcus vulcanus* by N-terminal sequencing. FEBS Lett 253:257–263

Kouril R, Arteni AA, Lax J, Yeremenko N, D'Haene S, Rogner M, Matthijs HC, Dekker JP, Boekema EJ (2005) Structure and functional role of supercomplexes of IsiA and Photosystem I in cyanobacterial photosynthesis. FEBS Lett 579:3253–3257

Kouril R, Zygadlo A, Arteni AA, de Wit CD, Dekker JP, Jensen PE, Scheller HV, Boekema EJ (2005) Structural characterization of a complex of photosystem I and lightharvesting complex II of *Arabidopsis thaliana*. Biochemistry 44:10935–10940

Kruse O, Schmid GH (1995) The role of phosphatidylglycerol as a functional effector and membrane anchor of the D1-core peptide from photosystem II-particles of the cyanobacterium *Oscillatoria chalybea*. Z Naturforsch C 50:380–390

Kunstner P, Guardiola A, Takahashi Y, Rochaix JD (1995) A mutant strain of Chlamydomonas reinhardtii lacking the chloroplast photosystem II psbI gene grows photoautotrophically. J Biol Chem 270:9651–9654

Kurisu G, Zhang H, Smith JL, Cramer WA (2003) Structure of the cytochrome b6f complex of oxygenic photosynthesis: tuning the cavity. Science 302:1009–1014

Kwa SLS, Newell WR, Vangrondelle R, Dekker JP (1992) The reaction center of Photosystem-II studied with polarized fluorescence spectroscopy. Biochim Biophys Acta 1099:193–202

Lakshmi KV, Eaton SS, Eaton GR, Brudvig GW (1999) Orientation of the tetranuclear manganese cluster and tyrosine Z in the O(2)-evolving complex of photosystem II: An EPR study of the S(2)Y(Z)(*) state in oriented acetate-inhibited photosystem II membranes. Biochemistry 38:12758–12767

Lakshmi KV, Poluektov OG, Reifler MJ, Wagner AM, Thurnauer MC, Brudvig GW (2003) Pulsed high-frequency EPR study on the location of carotenoid and chlorophyll cation radicals in photosystem II. J Am Chem Soc 125:5005–5014

Li N, Zhao JD, Warren PV, Warden JT, Bryant DA, Golbeck JH (1991) PsaD is required for the stable binding of PsaC to the photosystem I core protein of Synechococcus sp. PCC 6301. Biochemistry 30:7863–7872

Loll B, Kern J, Saenger W, Zouni A, Biesiadka J (2005) Towards complete cofactor arrangement in the 3.0 Å resolution structure of photosystem II. Nature 438:1040–1044

Loll B, Kern J, Saenger W, Zouni A, Biesiadka J (2007) Lipids in photosystem II: interactions with protein and cofactors. Biochim Biophys Acta 1767:509–519

McDermott G, Prince A, Freeer A, Hawthornwhite-Lawless M, Papiz M, Cogdell R, Asaacs N (1995) Crystal Structure of an integral membrane light-harvesting complex from photosynthetic bacteria. Nature 374:517–521

Mehari T, Qiao F, Scott MP, Nellis DF, Zhao J, Bryant DA, Golbeck JH (1995) Modified ligands to FA and FB in photosystem I. I. Structural constraints for the formation of iron-sulfur clusters in free and rebound PsaC. J Biol Chem 270:28108–28117

Meimberg K, Lagoutte B, Bottin H, Muhlenhoff U (1998) The PsaE subunit is required for complex formation between photosystem I and flavodoxin from the cyanobacterium *Synechocystis* sp. PCC 6803. Biochemistry 37:9759–9767

Mohanty P, Allakhverdiev SI, Murata N (2007) Application of low temperatures during photoinhibition allows characterization of individual steps in photodamage and the repair of photosystem II. Photosynth Res. Published online June 2007. DOI 10.1007/s11120-007-9184-y

Muhlenhoff U, Chauvat F (1996) Gene transfer and manipulation in the thermophilic cyanobacterium *Synechococcus elongatus*. Mol Gen Genet 252:93–100

Muller MG, Niklas J, Lubitz W, Holzwarth AR (2003) Ultrafast transient absorption studies on Photosystem I reaction centers from *Chlamydomonas reinhardtii*. 1. A new interpretation of the energy trapping and early electron transfer steps in Photosystem I. Biophys J 85:3899–3922

Nowaczyk MM, Hebeler R, Schlodder E, Meyer HE, Warscheid B, Rogner M (2006) Psb27, a cyanobacterial lipoprotein, is involved in the repair cycle of photosystem II. Plant Cell 18:3121–3131

Ohad I, Dal Bosco C, Herrmann RG, Meurer J (2004) Photosystem II proteins PsbL and PsbJ regulate electron flow to the plastoquinone pool. Biochemistry 43:2297–2308

Ohnishi N, Kashino Y, Satoh K, Ozawa S, Takahashi Y (2007) Chloroplast-encoded polypeptide PsbT is involved in the repair of primary electron acceptor QA of photosystem II during photoinhibition in *Chlamydomonas reinhardtii*. J Biol Chem 282:7107–7115

Peer W, Silverthorne J, Peters JL (1996) Developmental and light-regulated expression of individual members of the light-harvesting complex b gene family in *Pinus palustris*. Plant Physiol 111:627–634

Peloquin JM, Britt RD (2001) EPR/ENDOR characterization of the physical and electronic structure of the OEC Mn cluster. Biochim Biophys Acta 1503:96–111

Plato M, Krauss N, Fromme P, Lubitz W (2003) Molecular orbital study of the primary electron donor P700 of photosystem I based on a recent X-ray single crystal structure analysis. Chem Phys 294:483–499

Ramesh VM, Gibasiewicz K, Lin S, Bingham SE, Webber AN (2004) Bidirectional electron transfer in photosystem I: accumulation of A0- in A-side or B-side mutants of the axial ligand to chlorophyll A0. Biochemistry 43:1369–1375

Ramesh VM, Gibasiewicz K, Lin S, Bingham SE, Webber AN (2007) Replacement of the methionine axial ligand to the primary electron acceptor A0 slows the A0-reoxidation dynamics in photosystem I. Biochim Biophys Acta 1767:151–160

Regel RE, Ivleva NB, Zer H, Meurer J, Shestakov SV, Herrmann RG, Pakrasi HB, Ohad I (2001) Deregulation of electron flow within photosystem II in the absence of the PsbJ protein. J Biol Chem 276:41473–41478

Rokka A, Suorsa M, Saleem A, Battchikova N, Aro EM (2005) Synthesis and assembly of thylakoid protein complexes: multiple assembly steps of photosystem II. Biochem J 388:159–168

Rousseau F, Setif P, Lagoutte B (1993) Evidence for the involvement of PSI-E subunit in the reduction of ferredoxin by photosystem I. EMBO J 12:1755–1765

Rumberg B, Schmidt-Mende P, Witt HT (1964) Different demonstrations of the coupling of two light reactions in photosynthesis. Nature 201:466–468

Rutherford AW, Boussac A, Faller P (2004) The stable tyrosyl radical in photosystem II: why D? Biochim Biophys Acta 1655:222–230

Salih GF, Jansson C (1997) Activation of the silent psbA1 gene in the cyanobacterium *Synechocystis* sp strain 6803 produces a novel and functional D1 protein. Plant Cell 9:869–878

Santabarbara S, Kuprov I, Fairclough WV, Purton S, Hore PJ, Heathcote P, Evans MCW (2005) Bidirectional electron transfer in photosystem I: Determination of two distances between P-700(+) and A(1)(–) in spin-correlated radical pairs. Biochemistry 44:2119–2128

Sauer K, Yachandra VK (2004) The water-oxidation complex in photosynthesis. Biochim Biophys Acta 1655:140–148

Scheller HV, Jensen PE, Haldrup A, Lunde C, Knoetzel J (2001) Role of subunits in eukaryotic Photosystem I. Biochim Biophys Acta 1507:41–60

Schlodder E, Witt HT (1999) Stoichiometry of proton release from the catalytic center in photosynthetic water oxidation. Reexamination by a glass electrode study at ph 5.5–7.2. J Biol Chem 274:30387–30392

Schluchter WM, Shen G, Zhao J, Bryant DA (1996) Characterization of psaI and psaL mutants of *Synechococcus* sp. strain PCC 7002: a new model for state transitions in cyanobacteria. Photochem Photobiol 64:53–66

Schubert WD, Klukas O, Saenger W, Witt HT, Fromme P, Krauss N (1998) A common ancestor for oxygenic and anoxygenic photosynthetic systems: a comparison based on the structural model of photosystem I. J Mol Biol 280:297–314

Sener M, Park S, Lu DY, Damjanovic A, Ritz T, Fromme P, Schulten K (2003) Excitation transfer dynamics in monomeric and trimeric forms of cyanobacterial photosystem I. Biophys J 84:274a

Sener MK, Jolley C, Ben-Shem A, Fromme P, Nelson N, Croce R, Schulten K (2005) Comparison of the light-harvesting networks of plant and cyanobacterial photosystem I. Biophys J 89:1630–1642

Sener MK, Lu DY, Ritz T, Park S, Fromme P, Schulten K (2002) Robustness and optimality of light harvesting in cyanobacterial photosystem I. J Phys Chem B 106:7948–7960

Sener MK, Park S, Lu D, Damjanovic A, Ritz T, Fromme P, Schulten K (2004) Excitation migration in trimeric cyanobacterial photosystem I. J Chem Phys 120:11183–11195

Setif P (2001) Ferredoxin and flavodoxin reduction by photosystem I. Biochim Biophys Acta 1507:161–179

Setif P, Fischer N, Lagoutte B, Bottin H, Rochaix JD (2002) The ferredoxin docking site of photosystem I. Biochim Biophys Acta 1555:204–209

Shen G, Antonkine ML, van der Est A, Vassiliev IR, Brettel K, Bittl R, Zech SG, Zhao J, Stehlik D, Bryant DA, Golbeck JH (2002) Assembly of photosystem I. II. Rubredoxin is required for the in vivo assembly of F(X) in Synechococcus sp. PCC 7002 as shown by optical and EPR spectroscopy. J Biol Chem 277:20355–20366

Sippola K, Aro EM (2000) Expression of psbA genes is regulated at multiple levels in the cyanobacterium Synechococcus sp. PCC 7942. Photochem Photobiol 71:706–714

Soitamo AJ, Sippola K, Aro EM (1998) Expression of psbA genes produces prominent 5' psbA mRNA fragments in Synechococcus sp. PCC 7942. Plant Mol Biol 37:1023–1033

Sonoike K, Hatanaka H, Katoh S (1993) Small subunits of Photosystem I reaction center complexes from Synechococcus elongatus. II. The psaE gene product has a role to promote interaction between the terminal electron acceptor and ferredoxin. Biochim Biophys Acta 1141:52–57

Stroebel D, Choquet Y, Popot JL, Picot D (2003) An atypical haem in the cytochrome b(6)f complex. Nature 426:413–418

Strotmann H, Weber N (1993) On the function of PsaE in chloroplast Photosystem I. Biochim Biophys Acta 1143:204–210

Sugimoto I, Takahashi Y (2003) Evidence that the PsbK polypeptide is associated with the photosystem II core antenna complex CP43. J Biol Chem 278:45004–45010

Sugiura M, Rappaport F, Brettel K, Noguchi T, Rutherford AW, Boussac A (2004) Site-directed mutagenesis of Thermosynechococcus elongatus photosystem II: the O2-evolving enzyme lacking the redox-active tyrosine D. Biochemistry 43:13549–13563

Svensson B, Etchebest C, Tuffery P, van Kan P, Smith J, Styring S (1996) A model for the photosystem II reaction center core including the structure of the primary donor P680. Biochemistry 35:14486–14502

Telfer A (2002) What is beta-carotene doing in the photosystem II reaction centre? Philos Trans R Soc Lond B Biol Sci 357:1431–1439

Telfer A (2005) Too much light? How beta-carotene protects the photosystem II reaction centre. Photochem Photobiol Sci 4:950–956

Telfer A, Frolov D, Barber J, Robert B, Pascal A (2003) Oxidation of the two beta-carotene molecules in the photosystem II reaction center. Biochemistry 42:1008–1015

Tracewell CA, Brudvig GW (2003) Two redox-active beta-carotene molecules in photosystem II. Biochemistry 42:9127–9136

Umate P, Schwenkert S, Karbat I, Dal Bosco C, Mlcochova L, Volz S, Zer H, Herrmann RG, Ohad I, Meurer J (2007) Deletion of PsbM in tobacco alters the QB site properties and the electron flow within photosystem II. J Biol Chem 282:9758–9767

van Thor JJ, Geerlings TH, Matthijs HC, Hellingwerf KJ (1999) Kinetic evidence for the PsaE-dependent transient ternary complex photosystem I/Ferredoxin/Ferredoxin: NADP(+) reductase in a cyanobacterium. Biochemistry 38:12735–12746

Varotto C, Maiwald D, Pesaresi P, Jahns P, Salamini F, Leister D (2002) The metal ion transporter IRT1 is necessary for iron homeostasis and efficient photosynthesis in Arabidopsis thaliana. Plant J 31:589–599

Vasil'ev S, Bruce D (2004) Optimization and evolution of light harvesting in photosynthesis: the role of antenna chlorophyll conserved between photosystem II and photosystem I. Plant Cell 16:3059–3068

Vasil'ev S, Brudvig GW, Bruce D (2003) The X-ray structure of photosystem II reveals a novel electron transport pathway between P680, cytochrome b559 and the energy-quenching cation, ChlZ+. FEBS Lett 543:159–163

Vass I, Cser K, Cheregi O (2007) Molecular mechanisms of light stress of photosynthesis. Ann NY Acad Sci 1113:114–122

Watanabe T, Kobayashi M, Hongu A, Nakazato M, Hiyama T (1985) Evidence, that a chlorophyll a' dimer constitutes the photochemical reaction centre 1 (P700) in photosynthetic apparatus. FEBS Lett 235:252–256

Yamashita E, Zhang H, Cramer WA (2007) Structure of the cytochrome b6f complex: quinone analogue inhibitors as ligands of heme cn. J Mol Biol 370:39–52

Yano J, Kern J, Irrgang KD, Latimer MJ, Bergmann U, Glatzel P, Pushkar Y, Biesiadka J, Loll B, Sauer K, Messinger J, Zouni A, Yachandra VK (2005) X-ray damage to the Mn4Ca complex in single crystals of photosystem II: a case study for metalloprotein crystallography. Proc Natl Acad Sci USA 102:12047–12052

Yu J, Vermaas WF (1993) Synthesis and turnover of photosystem II reaction center polypeptides in cyanobacterial D2 mutants. J Biol Chem 268:7407–7413

Yu L, Bryant DA, Golbeck JH (1995) Evidence for a mixed-ligand [4Fe-4S] cluster in the C14D mutant of PsaC. Altered reduction potentials and EPR spectral properties of the FA and FB clusters on rebinding to the P700-FX core. Biochemistry 34:7861–7868

Yu L, Vassiliev IR, Jung YS, Bryant DA, Golbeck JH (1995) Modified ligands to FA and FB in photosystem I. II. Characterization of a mixed ligand [4Fe-4S] cluster in the C51D mutant of PsaC upon rebinding to P700-Fx cores. J Biol Chem 270:28118–28125

Zhao J, Li N, Warren PV, Golbeck JH, Bryant DA (1992) Site-directed conversion of a cysteine to aspartate leads to the assembly of a [3Fe-4S] cluster in PsaC of photosystem I. The photoreduction of F_A is independent of F_B. Biochemistry 31:5093–5099

Zouni A, Jordan R, Schlodder E, Fromme P, Witt HT (2000) First photosystem II crystals capable of water oxidation. Biochim Biophys Acta 1457:103–105

Zouni A, Witt HT, Kern J, Fromme P, Krauß N, Saenger W, Orth P (2001) Crystal structure of photosystem II from *Synechococcus elongatus* at 3.8 Å resolution. Nature 409:739–743

Microbial Rhodopsins:
Scaffolds for Ion Pumps, Channels, and Sensors

Johann P. Klare[1] · Igor Chizhov[2] · Martin Engelhard[3] (✉)

[1]Fachbereich Physik, University Osnabrück,
Barbarastraße 7, 49069 Osnabrück, Germany

[2]Medizinische Hochschule Hanover,
Carl-Neuberg-Straße 1, 30623 Hannover, Germany

[3]Max-Planck-Institute for Molecular Physiology,
Otto Hahn Str. 11, 44227 Dortmund, Germany
martin.engelhard@mpi-dortmund.mpg.de

Abstract Microbial rhodopsins have been intensively researched for the last three decades. Since the discovery of bacteriorhodopsin, the scope of microbial rhodopsins has been considerably extended, not only in view of the large number of family members, but also their functional properties as pumps, sensors, and channels. In this review, we give a short overview of old and newly discovered microbial rhodopsins, the mechanism of signal transfer and ion transfer, and we discuss structural and mechanistic aspects of phototaxis.

1
Introduction

The discovery of bacteriorhodopsin (BR) almost forty years ago (Oesterhelt and Stoeckenius 1971) caused an intense interest in the mechanism and function of the light-activated proton pump. This seven helix membrane protein carrying a retinal chromophore was originally thought to be a bacterial rhodopsin – hence its name – providing a model system for vision in vertebrates. However, it soon became evident that, despite structural similarities, the functions were quite different, with rhodopsin being a photosensor and BR being a photoreceptor, converting light energy into a proton gradient. Further research on *Halobacterium salinarum* the archaeal strain, from which BR was first isolated, demonstrated other rhodopsin-like pigments with functions other than proton pumping. The first evidence was provided by action spectra of halobacterial phototaxis (Hildebrand and Dencher 1975). In this work, two photosystems were recognized, enabling the bacteria to avoid harmful blue light, while seeking optimal conditions for the function of BR. Subsequently, two responsible pigments were identified. Sensory rhodopsin I, SRI, (Bogomolni and Spudich 1982; Spudich and Bogomolni 1984; Tsuda et al. 1982) enables *H. salinarum* to seek favourable light (above 500 nm), and, in a two photon process, to flee conditions of UV light. A second sensor, sensory rhodopsin II, SRII, (also named phoborhodopsin), is responsible for avoidance of blue light, which is especially important for the bacteria to escape conditions of oxidative stress (Spudich et al. 1986; Takahashi et al. 1985; Wolff et al. 1986). Further investigations showed ion pump activity different to that of BR, for which the name halorhodopsin (HR) was coined (Matsuno-Yagi and Mukohata 1977; Mukohata et al. 1980). Later, it was proven that HR functions as a chloride pump (Schobert and Lanyi 1982). Thus, in the early 1980s, four different pigments had been identified in *H. salinarum*. These can be grouped into two sub-families: BR and HR, which function as ion pumps, generating an ion gradient across the membrane, and SRI and SRII, which serve the bacteria as photoreceptors to find favourable light conditions at low oxygen concentration. Originally thought to be confined to Archaea, they have been detected in all three kingdoms of life (reviewed in Spudich et al. 2000).

A new chapter in the elucidation of microbial rhodopsins was opened when a rhodopsin-like protein was described in Clamydomonas (Foster et al. 1984). This membrane protein belongs to a family of retinylidene proteins, with properties different from those assigned for archaeal rhodopsins (Harz and Hegemann 1991). Although still quite elusive, with respect to their in vitro accessibility, it is now evident, mainly from electrophysiological experiments, that this class of proteins constitute light-activated ion channels (reviewed in Nagel et al. 2005a).

Since the discovery of BR, the scope of microbial rhodopsins has been considerably extended, not only in view of the large number of family members, but also with respect to their functional properties as pumps, sensors, and channels. Best understood are the bacterial rhodopsins, because of a wealth of spectroscopic, physiological, and structural data. New advances in vibrational spectroscopy (mainly developed because of the experimental demands in the BR field; reviewed in Siebert 1990) turned out to be especially fruitful for an understanding of the mechanism of proton transfer and signal transfer (Rothschild et al. 1981; Dollinger et al. 1986; Siebert et al. 1982). Concomitantly, site-directed mutagenesis (Mogi et al. 1988; Gerwert et al. 1989; Ni et al. 1990) and amino acid specific isotope labelling (Engelhard et al. 1985; Eisenstein et al. 1987) became available, providing tools for unequivocal assignment of amino acids involved in proton transfer. BR was one of the first membrane proteins whose structure could be determined by electron cryo-microscopy (Henderson et al. 1990). The introduction of lipid cubic phases for membrane protein crystallisation (Landau and Rosenbusch 1996; Pebay-Peyroula et al. 1997) even enabled the determination of the structure of intermediates (Sass et al. 1997).

The present knowledge allows a detailed understanding of proton transfer in BR. This data, together with data obtained for the other microbial rhodopsins, revealed general mechanistic principles, despite their different function. Apparently, nature has chosen common structures and initial events, which are individually utilized by the proteins to fulfil their particular function. Light activation triggers a *trans–cis* isomerisation of retinal and photoreaction cycle, whose turnover rate is specific for sensors (second range) and ion pumps (millisecond range). Concomitantly, conformational changes seem to be similar, especially in view of helix movements. In this review, we will describe three subgroups of microbial rhodopsins, and we will focus on two general questions related to the mechanism of proton transfer and the mechanism of transmembrane signal transfer. These aspects are certainly of general interest, not only to scientists working in the field of photoreceptor research, but also to those interested in challenges of proton transfer in proteins and in transmembrane signal transduction.

2
Microbial (Type 1) Rhodopsins

2.1
Archaeal Rhodopsins

2.1.1
Ion Pumps

Light-driven ion transport is accomplished by two different types of bacterial rhodopsins: Bacteriorhodopsin (BR) and halorhodopsin (HR). Up until now, 14 archaeal BR proteins and 11 HR proteins have been functionally identified (Tables 1 (BR) and 2 (HR)). BR, first identified in the extreme halophilic archaeon *Halobacterium salinarum* (*H. salinarium*), transports protons from the cytoplasm to the extracellular space (Blaurock and Stoeckenius 1971; Oesterhelt and Stoeckenius 1971).

BR was the first microbial rhodopsin identified, and its structure and function have been studied extensively during the past 30 years. The apoprotein, bacterioopsin, binds all-*trans* retinal covalently, via a protonated Schiff base to the ε-amino group of Lys216. BR shows a broad absorption band, with a maximum at 570 nm, and undergoes a so-called light-dark adaptation. In the dark-adapted form, (λ_{max} = 558 nm) the ratio between 13-*cis* retinal and all-*trans* retinal is 2 : 1, which shifts to 100% all-*trans* retinal in the light-adapted state (Scherrer et al. 1989). Upon light excitation, BR undergoes a photocycle, which is initiated by retinal isomerization from all-*trans* to 13-*cis*. Relaxation back to the original ground state occurs in about 10 ms by passing through several distinct intermediate states. A more detailed description is given in Sect. 3.1.

Structural information for BR was already available in 1975, when Henderson and Unwin reported a three-dimensional model of the purple membrane at 7 Å resolution obtained by electron microscopy (Henderson and Unwin

Table 1 The bacteriorhodopsin family

BR-Family (proton pumps)		
Strain	Name	Refs.
1. *Hb. salinarum*	bacteriorhodopsin (BR)	(Oesterhelt and Stoeckenius 1971b)
2. strain mex	mex-bacteriorhodopsin (mex-BR)	(Otomo et al. 1992a)
3. strain port	port-bacteriorhodopsin (port-BR)	(Mukohata et al. 1999c)
4. strain shark	shark-bacteriorhodopsin (shark-BR)	(Otomo and Muramatsu 1995)
5. *Hr.* spez. aus1 (= strain SG-1)	archaerhodopsin-1 (aR-1, SGbR)	(Mukohata et al. 1988)
6. *Hr.* spez. aus2	archaerhodopsin-2 (aR-2)	(Mukohata et al. 1999b)
7. *Hr. sodomense*	archaerhodopsin-3 (aR-3)	(Mukohata et al. 1999a)
8. *Ha. argentinensis* (sp. arg-1)	cruxrhodopsin-1 (cR-1)	(Tateno et al. 1994b)
9. *Ha. mukohataei* (sp. arg-2)	cruxrhodopsin-2 (cR-2)	(Tateno et al. 1994a; Sugiyama et al. 1994)
10. *Ha. vallismortis*	cruxrhodopsin-3 (cR-3)	(Kitajima et al. 1996)
11. *Ha. japonica*	*Ha. japonica* cruxrhodopsin (*Hj*cR) (cruxrhodopsin-4 (cR-4))[1]	(Yatsunami et al. 2000)
12. *Ha. marismortui*	cruxrhodopsin-5 (cR-5)[1]	(Baliga et al. 2004d)
13. *Ht.* arg-4	deltarhodopsin-1 (dR-1)	(Ihara et al. 1999)
14. *Ht. turkmenica*	*Ht. turkmenica* deltarhodopsin (*Ht*dR) (deltarhodopsin-2 (dR-2))[1]	(Kamo et al. 2006)

[1] Name suggested in this review

1975). During the following years, several electron microscopy studies finally enhanced the resolution to 2.8 Å for a projection structure (Baldwin et al. 1988) and 3.5 Å for 3-D model (Grigorieff et al. 1996). In 1997, the first X-ray structure from a three-dimensional crystal grown in the lipidic cubic phase was reported (Pebay-Peyroula et al. 1997) with resolution of 2.5 Å. Today, a large number of BR crystal structures are available, even for different intermediate states of bacteriorhodopsin, providing a detailed picture of BR structure and function on the atomic level (reviewed, for example, in Heberle et al. 2000 and Edmonds and Luecke 2004).

In 2006, the crystal structures of archaerhodopsin-1 and archaerhodopsin-2 (aR-1 and aR-2, respectively), two light-driven proton pumps detected in *Halorubrum* sp. aus-1 and aus-2, became available (Enami et al. 2006). They share 55–58% sequence identity with bacteriorhodopsin (BR), but exhibit some remarkable structural differences (see Sect. 2.4). Both aR-1 and aR-2 show an absorption maximum at ∼550 nm, and they undergo dark/light adaptation in a similar way as observed for BR (Ihara et al. 1994). Meanwhile, several BR-like proteins have been identified in different organisms (see Table 1).

Table 2 The halorhodopsin family

HR-Family (chloride ion pumps)		
Strain	Name	Refs.
1. Hb. salinarum	Halorhodopsin (HR)	(Matsuno-Yagi and Mukohata 1977b; Mukohata and Kaji 1981b)
2. strain mex	mex-halorhodopsin (mex-HR)	(Otomo et al. 1992b)
3. strain port	port-halorhodopsin (port-HR)	(Mukohata et al. 1999e)
4. strain shark	shark-halorhodopsin (shark-HR)	(Mukohata et al. 1999f)
5. Hr. spez. aus1 (= strain SG-1)	archaehalorhodopsin-1 (aHR-1, SGHR)	(Soppa et al. 1993b)
6. Hr. sodomense	archaehalorhodopsin-3 (aHR-3)	(Ihara et al. 1999)
7. Ha. Argentinensis (sp. arg-1)	cruxhalorhodopsin-1 (cHR-1)	(Tateno et al. 1994c)
8. Ha. vallismortis	cruxhalorhodopsin-3 (cHR-3)	(Kitajima et al. 1996)
9. Ha. marismortui	cruxhalorhodopsin-5 (cHR-5)[1]	(Baliga et al. 2004c)
10. Ht. arg-4	deltahalorhodopsin-1 (dHR-1)	(Ihara et al. 1999)
11. N. pharaonis	pharaonis halorhodopsin (pHR)	(Bivin and Stoeckenius 1986b)

[1] Name suggested in this review

The second type is the chloride ion pump, HR, which was also first discovered in *H. salinarum* (Matsuno-Yagi and Mukohata 1977, 1980; Mukohata and Kaji 1981). HR was initially thought to function as a sodium ion pump, but it was later recognized as an inward-directed chloride pump (Schobert and Lanyi 1982). Its amino acid sequence was reported in 1987 (Blanck and Oesterhelt 1987). *H. salinarum* HR (*Hs*HR) also binds all-*trans* retinal at Lys 242, and it exhibits an absorption maximum at 578 nm (reviewed in Dencher 1983). The photocycle of *Hs*HR and the photocycle of *Natronomonas pharaonis*, *Np*HR, (Bivin and Stoeckenius 1986) have been studied in detail, and they display similar properties (summarized in (Chizhov and Engelhard 2001)). Like BR, after light-excitation *Np*HR passes a linear sequence of intermediates, which is generally similar to that of *Hs*HR, although, in the latter case, one intermediate (O-intermediate) has not yet been proven to exist (see below). Analogous to BR, *Hs*HR also undergoes a dark/light adaptation. Dark-adapted *Hs*HR contains ~45% of all-*trans* retinal, which is shifted to ~75% all-*trans* chromophore under irradiation. In contrast, a dark/light adaptation has not been found in *Np*HR. Both the dark-adapted and the light-adapted states contain ~85% all-*trans* retinal (Zimányi and Lanyi 1997).

A two-dimensional projection structure of *Hs*HR at 6 Å resolution (Havelka et al. 1993) and a three-dimensional structure of *Hs*HR based on electron microscopy data at 7 Å (Havelka et al. 1995) have been published.

X-ray crystal structures of *Hs*HR are available at resolutions of 1.8 Å (Kolbe et al. 2000) for its mutant, T203V, at 1.6 Å for the ground state, and at 1.9 Å for the L1 intermediate (Gmelin et al. 2006), providing an atomic picture of the halide pump HR.

BR and HR carry out light-driven ion-translocation, leading to a hyperpolarization of the cell membrane. The proton gradient, resulting from BR, enables the bacterium to produce ATP through the ATP- synthase machinery, supplying the organism with energy under anaerobic conditions, where the respiratory chain is non-functional (Danon and Stoeckenius 1974). At the same time, HR maintains the osmotic balance between the cytoplasm and the extracellular medium, by importing chloride ions against the membrane potential (Oesterhelt 1998).

2.1.2
Sensory Receptors

Phototactic behaviour of *H. salinarum* was first reported by Hildebrand and Dencher in 1975. A few years later, retinal proteins were identified as the responsible receptors (Dencher 1978; Dencher and Hildebrand 1979; Spudich and Stoeckenius 1979; Sperling and Schimz 1980). In 1982, this conclusion was substantiated by Spudich and co-workers, who were able to show that retinal pigments other than BR and HR are indeed responsible for phototaxis (Spudich and Spudich 1982; Bogomolni and Spudich 1982). The authors identified a retinal protein absorbing between 580 nm and 590 nm, which was originally named slow rhodopsin-like pigment (later renamed sensory rhodopsin I (SRI)) because of its slow photocycle turnover, determined to be about two orders of magnitude slower than that of BR (800 ms vs. 10 ms). SRI turned out to be responsible for both the attractant response on light with $\lambda > 500$ nm, as well as the repellent response towards visible light at $\lambda \approx 370$ nm, thereby having the ability of colour discrimination (Spudich and Bogomolni 1984). The model proposed by Spudich and Bogomolni suggested the involvement of one photon and two photon processes. The first excitation of SRI with a photon of $\lambda > 500$ nm triggers the photocycle, and it triggers the attractant response. However, a second photon can be absorbed by a long-lived photointermediate ($\lambda_{max} = 373$ nm), thereby activating the photophobic response of the bacteria.

In 1986, a fourth retinal protein, solely responsible for the repellent response to blue-green light, was identified in *H. salinarum*. It was named phoborhodopsin (PR) (Takahashi et al. 1985) or sensory rhodopsin II (SRII) (Wolff et al. 1986). The amino acid sequences of the two sensory rhodopsins from *H. salinarum* have been determined (Blanck et al. 1989; Zhang et al. 1996). Analogous proteins have been identified in *Natronomonas pharaonis* (*Np*SRII) and *Haloarcular vallismorts* (crux-sensory rhodopsin-3, csR-3) (Seidel et al. 1995), as well as in *Halorubrum* spez. aus-1 and *Halorubrum*

Table 3 Sensory rhodopsins

Sensory rhodopsins Strain	Name	Refs.
Sensor I		
1. Hb. salinarum	sensory rhodopsin 1 (SRI)	(Bogomolni and Spudich 1982b)
2. Hr. spez. aus1 (= strain SG-1)	archae-sensory rhod. 1 (aSR-1, SGSR)	(Soppa et al. 1993a)
3. Hr. sodomense	archae-sensory rhod. 3 (aSR-3)	(Ihara et al. 1999)
4. Ha. vallismortis	crux-sensory rhodopsin 3 (cSR-3)	(Kitajima et al. 1996)
5. Ha. marismortui	crux-sensory rhodopsin 5 (cSR-5)[1]	(Baliga et al. 2004b)
Sensor II		
1. Hb. salinarum	sensory rhodopsin 2 (SRII) phoborhodopsin (pR)	(Takahashi et al. 1985b)
2. Ha. vallismortis	vallismortis sensory rhodopsin 2 (vSRII) cruxphoborhodopsin 3 (cpR-3 = valpR)	(Seidel et al. 1995b)
3. Ha. marismortui	cruxphoborhodopsin 5 (cpR-5)[1]	(Baliga et al. 2004a)
4. N. pharaonis	N. pharaonis sensory rhodopsin 2 (NpSRII) pharaonis phoborhodopsin (ppR)	(Seidel et al. 1995b)

[1] Name suggested in this review

sodomense, where they have been named archeae-sensory rhodopsins 1 (asR-1) and archeae-sensory rhodopsins 3 (asR-3), respectively, according to the BR analogues (aR's) found in the genus *Halorubrum* (see Table 3).

Structural information on sensory rhodopsins is presently only available for *Np*SRII in the ground state (Luecke et al. 2001; Royant et al. 2001) and in the K-state (Edman et al. 2002). Further structural data was obtained for *Np*SRII in complex with N-terminal fragment of its cognate transducer, *Np*HtrII, in the ground state (Gordeliy et al. 2002), as well as in the intermediate states K and M (Moukhametzianov et al. 2006).

2.2
Eubacterial Rhodopsins

2.2.1
Proteorhodopsin

In 2000, a eubacterial retinal protein exhibiting BR-like proton pump activity was discovered through genomic analyses of naturally occurring marine bacterioplankton. Beja and co-workers cloned the gene of proteorhodopsin (PR) from a member of the γ-proteobacteria of the SAR86 group (Beja et al. 2000). Since then, several PR genes have been reported in marine plankton from the Red Sea, the Mediterranean Sea (Man et al. 2003; Sabehi et al. 2003), and several other regions (de la Torre et al. 2003; Man-Aharonovich et al. 2004; Sabehi et al. 2004). The vast abundance of rhodopsin-like photoreceptors in marine organisms was impressively demonstrated by Venter and co-workers, using a "whole-genome shotgun sequencing" approach on microbial populations collected from seawater samples of the Sargasso Sea near Bermuda, revealing the presence of about 800 PR genes (Venter et al. 2004). In 2005, Beja and co-workers calculated the fraction of PR-utilizing organisms to be 13% of the microorganisms in the photic zone.

Sequence analysis of the PR genes revealed that almost all of them are predicted to encode for light-driven proton pumps. In the case of PR, this has been experimentally proven by electrophysiological measurements (Friedrich et al. 2002). Additionally, it has been shown that, for the "prototype" PR (Beja et al. 2000) and for the green-absorbing and blue-absorbing PRs (GPR and BPR, respectively) (Sineshchekov and Spudich 2004; Wang et al. 2003), they efficiently pump protons when expressed in *E. coli*. Comparing their amino acid sequence, there is a strong evidence that function as a transmembrane proton pump PR is more closely related to the sensory rhodopsins than to BR. In particular, residues responsible for the release of protons to the extracellular medium are found to be significantly different from that of BR (Beja et al. 2000). In a recent publication, Spudich proposed the possible existence of "sensory proteorhodopsins", if a unified mechanism of sensors of pumps applies (Spudich 2006). His argument was based on the carboxylate residue motif near the Schiff base and the lack of a second carboxylate on helix C, typical for sensor proteins. Twenty-two closely related rhodopsins were found to fit this criteria, indicating that not only pumps but also sensors have evolved in marine organisms. A further argument in favour of this assumption is the finding that 19 of the putative sensory PRs are found to be under the control of the same promoter as a gene encoding a 659-residue protein with sequence homology similar to that of the halobacterial transducers. Although there is no physiological evidence for sensory PRs at the moment, there are at least strong indications for their existence.

2.2.2
Anabaena Sensory Rhodopsin

The first sensory rhodopsin found in the eubacterial domain is a green-light activated photoreceptor in the freshwater cyanobacterium *Anabaena (Nostoc)* sp. PCC7120 (Jung et al. 2003). In 2004, Spudich, Luecke, and co-workers reported the crystal structure of *Anabaena* sensory rhodopsin (ASR) at resolution of 2.0 Å (Vogeley et al. 2004). ASR binds all-*trans* retinal and exhibits an absorption maximum at 543 nm. Like BR, ASR undergoes light–dark adaptation. However, it exhibits unique features. Firstly, the retinal configuration is all-*trans* in the dark-adapted state and 13-*cis* in the light-adapted state with a blue-shifted and lower extinction absorption spectrum. Secondly, ASR shows an efficient reversible light-induced interconversion between the all-*trans* and the 13-*cis* unphotolyzed states. As the relative amounts of the two chromophore states depend on the wavelength of the ambient light, ASR might be able to differentiate between distinct qualities of light (Sineshchekov et al. 2005b).

Another difference between *Anabaena* sensory rhodopsin and the archaeal rhodopsins turned out to be the highly hydrophilic character of its cytoplasmic half-channel, networked by water molecules. This property gives rise for the interaction with the soluble transducer protein, which is encoded in the same operon as the receptor. The 14.7 kDa protein, which was named ASR transducer (ASRT), seems to be complexed in a 4 : 1 ratio with ASR (Vogeley and Luecke 2006), and it has been shown to accelerate the ASR photocycle when expressed in *E. coli* (Jung et al. 2003).

2.2.3
Xanthorhodopsin

In 2005, another eubacterial retinal protein was reported in the extreme halophile *Salinibacter ruber*, isolated from salt-crystallizer ponds that was named Xanthorhodopsin (XR) (Balashov et al. 2005). This rhodopsin is quite unusual, as it is a light-driven ion pump similar to BR, AR, or PR, but it has two chromophores. In addition to retinal, the carotenoid salinixanthin is used to harvest light energy for the transmembrane proton transport. The two chromophores are present in a 1 : 1 ratio, and they strongly interact. Light energy absorbed by salinixanthin is transferred to retinal with a quantum efficiency of about 40%. As a result, the spectral region, which can be used to drive the pump, is much broader than in the other type 1 rhodopsins, making XR much more flexible and efficient in production of energy for the cell. Due to the dual chromophore system, the absorption spectrum of XR shows two maxima, at 487 nm and at 519 nm, with an additional shoulder around 460 nm, accounting for salinixanthin, and a broad shoulder at around 560 nm, due to retinal chromophore (Balashov et al. 2006).

2.2.4
Other Eubacterial Rhodopsins

Recently, rhodopsin genes have been identified in two other eubacterial organisms, namely in the radiation-resistant actinobacteria *Kineococcus radiotolerans* and *Rubrobacter xylanophilus* during genome sequencing of these organisms at the Joint Genome Institute (http://www.jgi.doe.gov/). Up to now, these proteins have neither been isolated nor characterized, but phylogenetic analyses (Sharma et al. 2006) revealed clear differences between them (fur-

Table 4 Eubacterial rhodopsins

Eubacterial rhodopsins Organism	Name	Refs.
1. γ-proteobacteria (SAR86)	Proteorhodopsin (PR)	(Beja et al. 2000)
2. *Anabaena (Nostoc)* sp. PCC7120	*Anabaena* sensory rhodopsin (*A*SR)	(Jung et al. 2003e)
3. *Salinibacter ruber*	Xanthorhodopsin (XR)	(Balashov et al. 2005a)
4. *Kineococcus radiotolerans*	n.n.	(Sharma et al. 2006b)
5. *Rubrobacter xylanophilus*	n.n.	(Sharma et al. 2006a)

ther discussed in Sect. 2.5). Table 4 summarises the properties and origin of eubacterial rhodopsins currently known.

2.3
Eukaryotic Type 1 Rhodopsins

2.3.1
Proton Pumps and Channels in Algae

As early as 1968, Schilde suggested that a rhodopsin-like protein is responsible for the light-induced response of transmembrane potential in the giant unicellular marine algae *Acetabularia acetabulum* (*A. mediterranea*) (Schilde 1968). This was the first time it was indicated that rhodopsin-like pigments existed outside the animal kingdom. Almost 40 years later, Hegemann and co-workers successfully cloned the full-length *opsin*-cDNA, expressed it in *Xenopus* oocytes, and they characterized *Acetabularia* rhodopsin (AR), confirming its proton-pump activity (Tsunoda et al. 2006). AR turned out to have a light-driven H^+ pump activity, similar to BR. However, in contrast to the latter, its sensitivity to the external pH was high, whereas the internal pH had only minor influence on the protein kinetics. Also, in contrast to BR, AR exhibited a clear dependence on the transmembrane voltage. Up to now, AR has been the sole example of ion-pumping rhodopsin in a photosynthetic eukaryote.

The presence of rhodopsin-like retinal proteins responsible for the phototactic behaviour of green algae was first demonstrated for unicellular algae *Chlamydomonas reinhardtii* in 1984 (Foster et al. 1984). Almost ten years later, Hegemann and co-workers were able to identify a retinal binding protein in eyespot membranes of *C. reinhardtii*, which was named chlamyopsin or chlamyrhodopsin in the retinal-bound form (Deininger et al. 1995). Chlamyrhodopsin turned out not to be a typical seven-helix receptor. It showed no significant homology to the archaeal Type 1 rhodopsins, but it exhibited a conserved retinal binding site, and it seemed to be more related to animal opsins. At that time, it was proposed that chlamyrhodopsin might be a light-gated ion channel, based on the high abundance of polar and charged residues. Subsequent work revealed that chlamyrhodopsin is not responsible for the phototactic and photophobic behaviour of the organism (Fuhrmann et al. 2001). In 2001, Spudich and co-workers were able to identify two more rhodopsins from cDNA sequences, which function as low-light and high-light intensity phototaxis receptors in *C. reinhardtii* (Sineshchekov and Govorunova 2001). They were named *Chlamydomonas* sensory rhodopsins A and B (CSRA and CSRB), and their function as phototaxis receptors was shown in vivo. The receptors consist of two distinct parts, a N-terminal (~300 residues) Type 1 rhodopsin-like domain, followed by a membrane-associated domain of about 400 residues. It was shown that photoexcitation leads to photoreceptor currents, a fast one at high light intensity, in case of CSRA, and a slow one at low light intensity, for CSRB. Nevertheless, the mechanism of current generation by the two rhodopsins was not clear at that point, although it was suggested that at least one of them might function as a light-gated ion channel. This light-gated channel activity was subsequently proven by electrophysiological measurements on CSRA-expressing and CSRB-expressing *Xenopus* oocytes (Nagel et al. 2002, 2003). CSRA was shown to conduct protons, whereas CSRB is a channel also for other cations. Interestingly, these functions are solely carried out by their rhodopsin domain. Based on these findings, the authors of the studies named the proteins channelrhodopsin-1 (ChR1), corresponding to CSRA, and channelrhodopsin-2 (ChR2), corresponding to CSRB. It remains unclear if the activity of these rhodopsins is the same in *Chlamydomonas*. The role of cytoplasmic domain needs to be elucidated. The light-gated ion channel activity of the channelrhodopsins has recently become a valuable tool for neuroscience, in terms of light-dependent activation of neurons. In 2005, Boyden et al. reported the photostimulation of mammalian neurons, by using lentiviral gene delivery of channelrhodopsin-2 to cultivated rat hippocampal neurons (Boyden et al. 2005). Remarkably, ChR2 has also been used to stimulate neurons in vivo, as reported for the light-induced triggering of behavioural responses in the nematode *Caenorhabditis elegans* (Nagel et al. 2005b), for triggering of appetitive or aversive learning in *Drosophila* larvae (Schroll et al. 2006), and, very recently, for transgenic mice, where it has been used for mapping of synaptic connectivity (Arenkiel et al. 2007; Wang et al. 2007).

Another example of a rhodopsin-like gene found in algae is the one in the cryptomonad algae *Guillardia theta* (Ruiz-Gonzalez and Marin 2004), which exhibits positive phototaxis with a maximum sensitivity at 450 nm, and an additional band at 500 nm. Moreover, in 2005, Sineshchekov et al. (Sineshchekov et al. 2005a) identified two cDNA sequences from *G. theta* and an additional one from *Cryptomonas* sp. The authors succeeded in expressing one of the *G. theta* rhodopsins in *E. coli*, and they analysed the photochemical reaction cycle. *G. theta* rhodopsin 1 (*Gt*R1) exhibits K-like, M-like, and O-like intermediates with a turnover time of about 80 ms. There is no indication for proton pump activity, although the proton acceptor and proton donor residues are conserved as in BR. Therefore, the authors suggest that *Gt*R1 is a sensory rhodopsin. Also *G. theta* rhodopsin 2 (*Gt*R2) as well as the *Cryptomonas* rhodopsin are thought to function as sensors, as they lack carboxylated residues at the Schiff base donor positions.

2.3.2
Fungal Rhodopsins

Although indications for rhodopsin-like proteins in eukarya (but outside the animal kingdom) have existed since the 1960s (see Sect. 2.3.1), the first rhodopsin identified in such an organism was that of the fungus *Allomyces reticulatus* (Saranak and Foster 1997). Its zoospores (reproductive cells) are phototactic. Sarank and Foster were able to show that phototaxis is dependent on retinal, and that, indeed, a protein from the rhodopsin-type 1 family is responsible for the phototactic behavior of the cell. The gene coding for this opsin has not yet been investigated further. Another eukaryotic BR-like protein identified was that from the filamentous fungus *Neurospora crassa* (Bieszke et al. 1999a,b). Although *Neurospora* rhodopsin (NR) possesses most of the residues important for proton translocation, it does not exhibit a proton pump activity, and it seems to have a sensory function as well (summarized in Brown and Jung 2006). Light-activated proton transport in a eukaryote was first observed in *Leptosphaeria maculans*, with Leptosphaeria rhodopsin (LR) being the responsible pigment (Waschuk et al. 2005).

For quite a number of fungal rhodopsins, (or, more precisely, opsin-like genes), identified so far, their in vivo function remains to be elucidated. One example is the *carO* gene from *Fusarium fujikuroi* (*Gibberella fujikuroi* mating group C) (Prado et al. 2004), which codes for protein similar to Type 1 rhodopsins and also putative fungal heat-shock chaperones. The presence of the conserved lysine residue indicates that retinal can be bound, and it suggest its function as a photoreceptor. However, it remains unclear if it has a sensory or ion-translocating activity. Other examples recently reported and phylogenetically analysed in studies by Ruiz-Gonzalea and Marin (Ruiz-Gonzalez and Marin 2004b) and Sharma et al. (Sharma et al. 2006) are found in *Saccharomyces cerevisiae* (Acc. nos. CAA86397, CAA92370 and CAA42313 for a putative rhodopsin genes), *Sclerotinia sclerotiorum* (within whole genome shotgun sequence AAGT01000049), and *Botryotinia fuckeliana* (whole genome shotgun sequence AAID01000494).

2.3.3
Other Eukaryotic Type 1 Rhodopsins

For the first time, in 2004, a rhodopsin-related sequence was found in a marine dinoflagellate, *Pyrocystis lunula* (Ruiz-Gonzalez and Marin 2004). This rhodopsin-like protein is not yet characterized, and, therefore, its identity

Table 5 Eukaryotic type 1 rhodopsins

Eukaryotic type 1 rhodopsins Organism	Name	Refs.
Algae		
1. *Acetabularia acetabulum*	*Acetabularia* rhodopsin (AR)	(Tsunoda et al. 2006a)
2. *Chlamydomonas reinhardtii*	Chlamyrhodopsin	(Deininger et al. 1995)
3. *Chlamydomonas reinhardtii*	*Chlamydomonas* sensory rhodopsin A (CSRA)	(Sineshchekov and Govorunova 2001c)
4. *Chlamydomonas reinhardtii*	*Chlamydomonas* sensory rhodopsin B (CSRB)	(Sineshchekov and Govorunova 2001b)
5. *Guillardia theta*	*Guillardia theta* rhodopsin 1 (*Gt*R1)	(Sineshchekov et al. 2005c)
6. *Guillardia theta*	*Guillardia theta* rhodopsin 2 (*Gt*R2)	(Sineshchekov et al. 2005b)
7. *Cryptomonas* sp.	*Cryptomonas* rhodopsin	(Sineshchekov et al. 2005a)
Fungi		
1. *Neurospora crassa*	*Neurospora* rhodopsin (NR)	(Bieszke et al. 1999a)
2. *Fusarium fujikuroi*	n.n. (*CarO* gene product)	(Prado et al. 2004b)
3. *Saccharomyces cerevisiae*	n.n. (three putative rhodopsin genes, see text)	
4. *Sclerotinia sclerotiorum*	n.n.	
5. *Botryotinia fuckeliana*	n.n.	
other		
1. *Pyrocystis lunula*	n.n.	(Ruiz-Gonzalez and Marin 2004f)

as a Type 1 rhodopsin and its putative function remain unknown. However, rhodopsin-like protein led to speculations about possible lateral gene transfer events leading to its appearance in an alveolate (see Sect. 2.5). Table 5 summarises properties and origin of eukarytic rhodopsins presently known.

2.4
Structure of the Rhodopsins

Generally, Type 1 rhodopsins are characterized by seven membrane-spanning helices, usually named helices A–G (Fig. 1). These seven helices are arranged in a bundle-like structure, forming a channel, where a retinal molecule close to the membrane centre is bound via a Schiff base to the ε-amino group of a conserved lysine residue located on helix G (Table 6). A comparison of the available structural data reveals a strong structural conservation between the different type 1 rhodopsins. The overall arrangement of the transmembrane

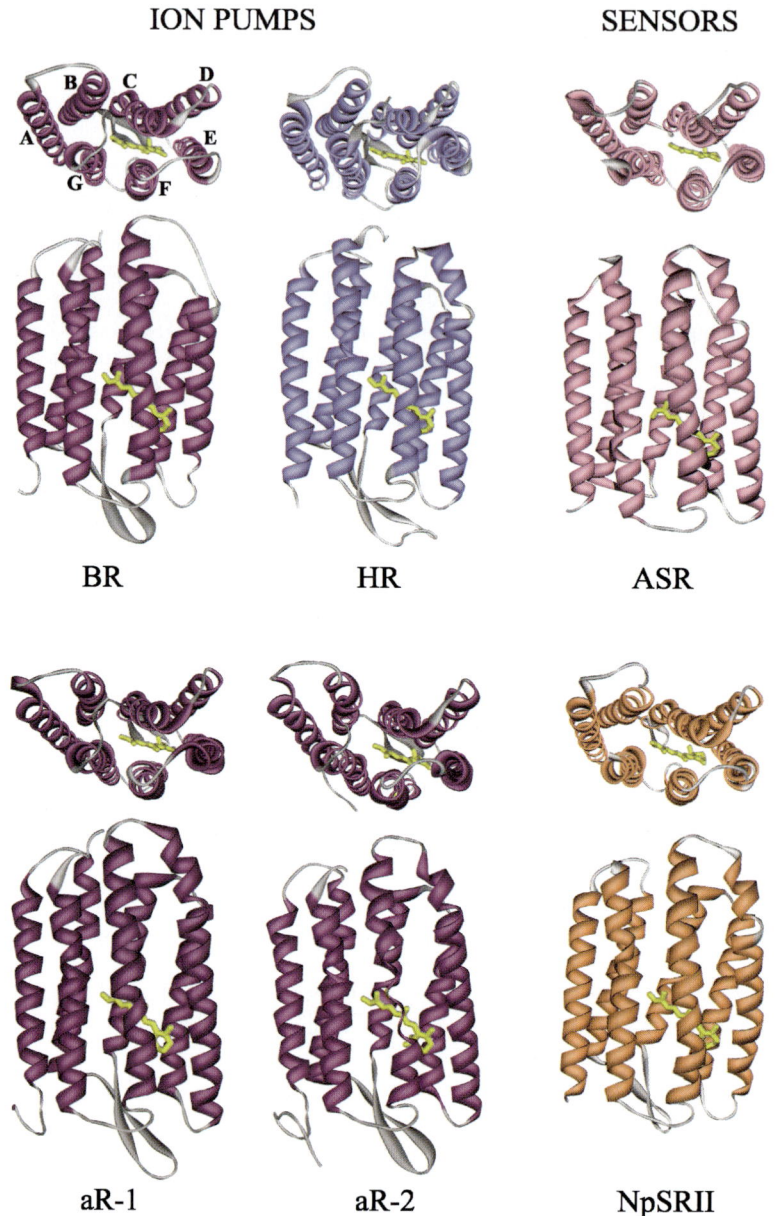

Fig. 1 Type 1 Rhodopsin Crystal Structures: Crystal structures of type 1 rhodopsins, for which structural information is available. The proteins are shown in ribbon representation, viewed from the cytoplasm (*upper structure*) and along the membrane plane (*lower structure*). The colour used for representation of helices corresponds to the appearance of proteins according to their absorption maxima. Helix names are given for BR. Protein data bank identifiers for the respective structures are: 1C3W (BR), 1E12 (*p*HR), 1VGO (aR-1), 1UAZ (aR-2), 1XIO (*A*SR), and 1JGJ (*Np*SRII)

Table 6 Conserved Lys residue for retinal binding and Schiff base donor/acceptor groups in type 1 rhodopsins

Protein	BR	aR-1	aR-2	HsHR	NpHR	PR	HsSRI	HsSRII	NpSRII
Ret.-binding (Lys)	216	216	216	242	256	231	205	202	205
Acceptor	Asp (85)	Asp (90)	Asp (91)	Thr (111)	Thr (126)	Asp (97)	Asp (76)	Asp (73)	Asp (75)
Donor	Asp (96)	Asp (101)	Asp (102)	Ala (122)	Ala (137)	Glu (108)	Tyr (87)	Tyr (84)	Phe (86)

helices is very similar, although the tilting angles appear to be slightly different (Fig. 2).

More expressed differences are found for the loop regions, especially for the loop connecting helices E and F on the cytoplasmic side. Whereas in bacteriorhodopsins a long and flexible loop is present in the archaerhodopsins, a loop–helix–loop motif is present in the halide pump HR and the sensor proteins NpSRII and ASR. A structural feature unique for the *Anabaena* sensory

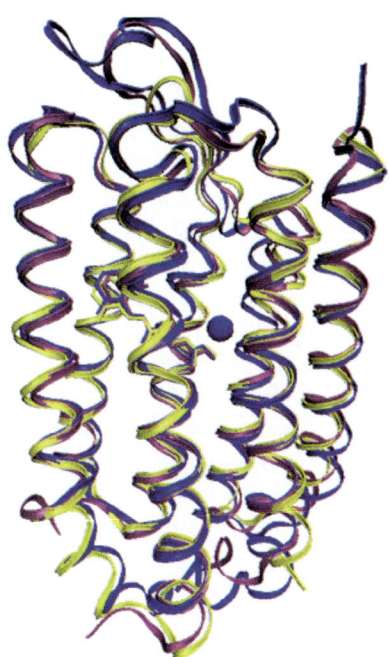

Fig. 2 Backbone overlay of BR, HR (*blue*), and NpSRII. The *blue ball* represents the position of chloride in HR. The PDB structures (Protein Data Bank: 1KGB.pdb, 1E12.pdb, and 1JGJ.pdb, respectively) were superimposed using SAP program

rhodopsin is the absence of the antiparallel beta-sheet connecting helices B and C on the extracellular side in the other rhodopsins. Remarkably, ASR also has the shortest, and thereby the least flexible E–F loop. Specific features are recognized if the recently solved crystal structures of aR-1 and aR2 (Enami et al. 2006) with that of bacteriorhodopsin. Among the three proton pumps, the region surrounded by helices B, C, and G, as well as the motif of aromatic residues forming the retinal binding pocket, are highly conserved. In contrast, the sequence of helices E and F is poorly conserved. Additionally, a so-called omega loop is found in both archaerhodopsins near the N terminus, in which the carboxyl group of Asp1 is hydrogen bonded to the backbone amides of residues 4, 5, and 6. Furthermore, the proton release channel on the extracellular side is found to be more opened in aR-2 compared to aR-1 or BR.

Beside the conserved lysine residue responsible for the covalent attachment of retinal and those residues characteristic for sensors or pumps, a number of additional residues are well-conserved within the respective subgroups. The two key residues, which determine if the protein acts as an ion pump or as a sensor protein, are the acceptor-/donor-groups for the Schiff base de-/re-protonation during the photocycle. In the proton pumps, bacteriorhodopsin, aR-1, and aR-2, the initial proton acceptor is Asp85, and the proton donor is Asp96. These residues are fully conserved among these proteins. A comparison of these positions between the different type 1 rhodopsins (see Table 6) shows that the two carboxyl groups in BR and the aR's are optimized for fast proton transport, whereas the presence of a hydrophobic side chain at the donor position in the sensory rhodopsins assures a significantly longer lifetime of the signalling state by retardation of the Schiff base reprotonation (see Sect. 4.3.2). In HR, the aspartate at the "donor"-position is replaced by Ala and at the "donor"-position a threonine residue is present, creating a halide binding site (Kolbe et al. 2000). Although the detailed mechanism of halorhodopsin function is still under debate, this combination of residues seems to facilitate the chloride ion transport towards the cytoplasm in an optimal way. Indeed, it has been shown that replacement of Asp85 by Thr (D85T) in bacteriorhodopsin converts the proton pump into a chloride ion pump (Sasaki et al. 1995), demonstrating that the ion specificity of the transport resides in the side-chain of residue 85.

The retinal chromophore resides in a binding pocket build-up by the transmembrane helices, which is responsible for the so-called opsin shift (This is the shift in the absorption maxima of protonated retinylidene Schiff bases when bound to rhodopsins from λ_{max} around 460 nm to 495–600 nm). For most of the type 1 rhodopsins characterized so far, retinal is bound to the protein in its all-*trans* configuration (in contrast to type 2 rhodopsins (see Sect. 2.5)). Upon light excitation, the molecule isomerizes across one of the double bonds in the polyene chain to form the 13-*cis* isomer. Although the presence of the 13-*cis* configuration was confirmed in a number of crystal structures of trapped photocycle intermediates of BR (Luecke et al. 1999; Sass et al. 2000; Lanyi 2000; Lanyi and Schobert 2003, 2004, 2006, 2007; Schobert et al. 2003; Edman et al. 1999; Royant et al. 2001b; Neutze et al. 2002; Kouyama et al. 2004; Takeda et al. 2004; Nishikawa et al. 2005) and *Np*SRII (Edman et al. 2002; Moukhametzianov et al. 2006), the retinal does not exhibit the expected bent shape of this configuration. Moreover, strong distortions of the retinal bond angles along the polyene chain were found to keep the chromophore in an almost straight configuration, most likely induced by constraints caused by the retinal binding pocket.

2.5
Evolution of Type I Rhodopsins

Type 1 rhodopsins, originally named "archaeal rhodopsins" because of their first discovery in halophilic archaea, are present in all three kingdoms of life: Archaea, Eubacteria, and Eukaryotes. They have been detected in a limited number of halophilic archaea, in γ-Proteobacteria (recently also in the α-Proteobacterium *Magnetospirillum magnetotacticum*), in a single cyanobacterium, in fungi, in a green algae, and also in a marine dinoflagellate. Additionally, several rhodopsin-related sequences had been described, although lacking canonical Lys residues, which enables the protein to bind retinal (Sharma et al. 2006). This observation would exclude their function as a photoactive pigment. However, one has to keep in mind that non-covalently bound retinal can also bind into a retinal binding pocket (Schweiger et al. 1994).

In general, however, despite their presence in all three domains of life and their high abundance in marine organisms, they have been found only in a few species within each domain. For example, they have been found only in a limited number of closely related halophilic archaea or in a single cyanobacterium. This fact and the recent discovery of numerous new Type 1 rhodopsins, motivated to analyse the evolution of Type 1 rhodopsins. Two evolutionary scenarios are discussed in the literature, which could explain the broad but patchy distribution of type 1 rhodopsins. (1) The last universal common ancestor had a type 1 rhodopsin gene. Today's distribution would then be explained by losses of the gene in most of the lineages independently from each other. (2) Lateral gene transfer across large evolutionary distances could also be responsible for the distribution observed today. A number of studies favour scenario (1) based on the high sequence similarities, and the fact that evolutionary distances between the organisms containing type 1 rhodopsins are remarkably large (Ruiz-Gonzalez and Marin 2004; Zhai et al. 2001). However, at least in the recent past, additional horizontal gene transfer events could not be ruled out. Despite the arguments favouring scenario (1), nowadays most of the phylogenetic analyses are interpreted to support that solely lateral gene transfer events caused the widespread type 1 rhodopsin distribution (scenario (2)). Several instances of such interdomain horizontal gene transfer events are now documented (for example, between *S. ruber* and the halophilic archaea or vice versa (Mongodin et al. 2005), between marine planctonic bacteria and archaea in the case of proteorhodopsin, (Frigaard et al. 2006) or between fungi and haloarchaea (Sharma et al. 2006)). The discovery of a rhodopsin-related sequence in the dinoflagellate *Pyrocystis lunula* (see Sect. 2.3.3) gave rise for another example of interdomain horizontal gene transfer, but its direction is still under debate. Ruiz-Gonzales and Marin (Ruiz-Gonzalez and Marin 2004) argued that a transfer event most probably took place from a dinoflagellate to proteobacteria, thus from eukaryote to prokaryote. Contrarily, Sharma and co-workers derived from their phylogenetic analyses that this protein originated in bacteria and was subsequently passed on to dinoflagellate.

In summary, a complete picture of the evolution of type 1 rhodopsins is not yet available, but the recent discovery of numerous "new" rhodopsin-like sequences and phylogenetic analyses carried out so far give at least strong indications that these proteins did not evolve from an universal common ancestor, but that their wide, though sparse, distribution arose from several lateral gene transfer events and losses of genes in multiple lineage.

Another question concerns the evolution of sensors and ion pumps. Are ion pumps derived from sensors, or did sensors follow the ion pump? Two reports from 1999 address this issue in detail for the subgroup of archaeal rhodopsins (Ihara et al. 1999; Mukohata et al. 1999). Both reports state that the four so-called rhodopsin clusters (bacteriorhodopsin, halorhodopsin, sensors rhodopsin I, and sensory rhodopsin II) have occurred by gene duplications, before the generic speciation of halophilic archaea. Based on the branching topology and relative evolution rates determined in these studies, the following picture of rhodopsin speciation seems to be most likely: Starting from a single gene encoding, the halobacterial rhodopsin ancestor, which is speculated to be closely related to the proton pump bacteriorhodopsin, the first gene duplication event resulted in a proto-ion pump and a proto-sensor gene. From the proto-ion pump, in the second gene duplication, the two ion pump genes corresponding to bacteriorhodopsin and halorhodopsin evolved. On the other hand, at this stage, the proto-sensor gene possibly captured a proto-transducer gene, putatively by acquisition of a chemoreceptor gene, to build a functional sensor-transducer unit. From the second duplication event in the sensor branch, the two sensory systems, SRI/HtrI and SRII/HtrII, arose. Based on the fact that the evolution rate of pre-SRI was found to be significantly faster than that of pre-SRII, Ihara and co-workers furthermore speculate that the original function of the ancestral sensor was similar to that of SRII, and that SRI evolved from the proto-sensor by the accumulation of mutations.

A possible relation between type 1 (microbial) rhodopsins and type 2 rhodopsins found in the eyes of mammalian species has been frequently discussed. Although structural similarities, such as the presence of seven transmembrane helices and the covalent attachment of a similar chromophore

(11-*cis* instead of all-*trans* retinal), seem to point to an evolutionary connection between the microbial rhodopsins and the mammalian GPCR, a closer inspection of the structural and functional features and phylogenetic analyses revealed that they evolved independently. First, the domain organization is different between type 1 and type 2 rhodopsins. Whereas the microbial family is characterised by short, non-functional loop regions connecting the transmembrane helices, mammalian rhodopsins exhibit large hydrophilic loops interacting with the heterotrimeric G-proteins, receptor kinases, and other signalling proteins (Khorana 1993; Helmreich and Hofmann 1996; Sakmar 1998). Moreover, an additional helix (helix VIII) in type 2 rhodopsins oriented along the membrane surface perpendicular to the other helices is most likely involved in the signalling process and represents a clear difference from type 1 rhodopsins. Closely related to the above argument, the primary structures of the two rhodopsins classes show significant differences, indicating that an evolutionary relationship is very unlikely (reviewed, e.g. in Spudich et al. 2000). It should be mentioned that, although both protein families use retinal as a photo-excitable chromophore, type 1 rhodopsins rely on all-*trans* retinal, whereas type 2 rhodopsins bind 11-*cis* retinal. In the first case, photoexcitation leads to a 13-*cis* configuration, which thermally relaxes back to all-*trans* retinal. In contrast, in type 2 rhodopsins all-*trans* retinal is formed on irradiation, which has to be removed from the protein in order to re-isomerise to the original 11-*cis* retinal. This complicated reaction scheme is probably due to the function of vertebrate rhodopsin, which should not be able to thermally isomerize in order to avoid activation in the dark (a comparison of type 1 and type 2 rhodopsins is given by Engelhard and Hofmann (2006)).

3
Ion Transfer and Signal Transfer Mechanisms

3.1
Photocycle

The mechanism of ion transfer in microbial rhodopsins has been the subject of intensive research for almost four decades. The most detailed picture has been obtained from studying BR, which, as a prototype of light-activated ion pumps, directs the analysis of other proteins from the family. Much like haemoglobin does, it also serves as a model for the elucidation of fundamental questions, such as mechanism of proton transfer, water structure in proteins, or time spacing of conformational changes leading to vectoriality. From the onset, the investigations were guided by the observation of a reaction cycle, during which proton is transferred across the membrane from the cytoplasm to the extra cellular space.

As early as 1975, the first model of the BR photocycle was presented by Lozier et al., which is still valid in its main features (Lozier et al. 1975). The authors proposed a model, which is depicted in Fig. 3. After light excitation, a K-intermediate is formed in the ps range. The further course of the photocycle includes at least four more intermediates (L-, M-, N-, and O-states) before BR relaxes back to the ground state in about 10 ms.

Important features of this model are its cyclic nature, the assumption of irreversible sequence of first-order reactions, and distinct intermediates, kinetically and spectroscopically defined. The M-intermediate (M_{412} in the model) is characterized by a deprotonated Schiff base, which represents a key element in the photoreaction cycle. Although experiments, to clarify the mechanism of the photocycle, are plentiful using a wide variety of methods, the general scheme has not been changed substantially (reviewed, e.g. in Lanyi and Váró 1995). Generally, the photocycle kinetics turned out to be multiexponential, which did not allow design of a scheme to describe a true picture of the reaction cycle (Nagle et al. 1995). As a consequence, several models rely on different assumptions, like multiple photocycles originating from BR substates (e.g. Eisfeld et al. 1995) or branched photocycles (e.g. Korenstein et al. 1978). However, in most publications, sequential reactions (e.g. Chizhov et al. 1996; Váró and Lanyi 1991b) are presumed, leading to a generally accepted overall model, closely related to the original proposal of Lozier and Stoeckenius (Lozier et al. 1975). It is quite clear that fast equilibria between intermediates are accompanied by irreversible reactions. For example, studying

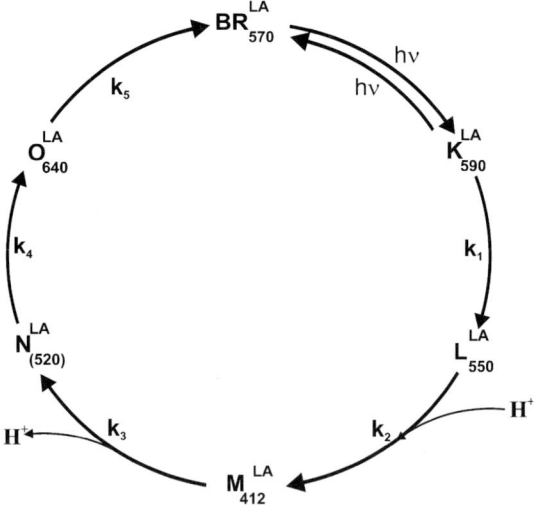

Fig. 3 Photocycle model proposed by Lozier et al. Figure adapted from Lozier et al. (1975). It should be noted that, in that early photocycle model, the release and uptake of a proton was placed between N → O and O → BR, respectively

pressure dependence of the photocycle kinetics, Klink et al. (2002) were able to unravel the L to M transitions as a sequence of different L ↔ M equilibria, gradually shifting the balance to higher M concentrations (Fig. 4). Commonly, a description of the photocycle becomes more accurate if more than one parameter is changed (e.g. temperature, measuring wavelength, infrared, ionic strength), and the data is fitted globally (Müller and Plesser 1991; Müller et al. 1991).

One key observation made in these analyses concerns the proof of two M-intermediates (M1 and M2). M2 is formed in an irreversible reaction from M1 (Váró and Lanyi 1990, 1991a), with a time constant of about 100 µs (Druckmann et al. 1992). This step is of significance, since it represents the reprotonation switch that changes access of the Schiff base from the extracellular side to the cytoplasmic side (Fig. 4). From a conceptual point of view, such a transition is a prerequisite to vectorial proton transfer. In numerous earlier experiments, it has been shown that, at physiological conditions, one proton is released to the extracellular side during M transitions, and it is picked up from the cytoplasm with the formation of O (reviewed in Lanyi

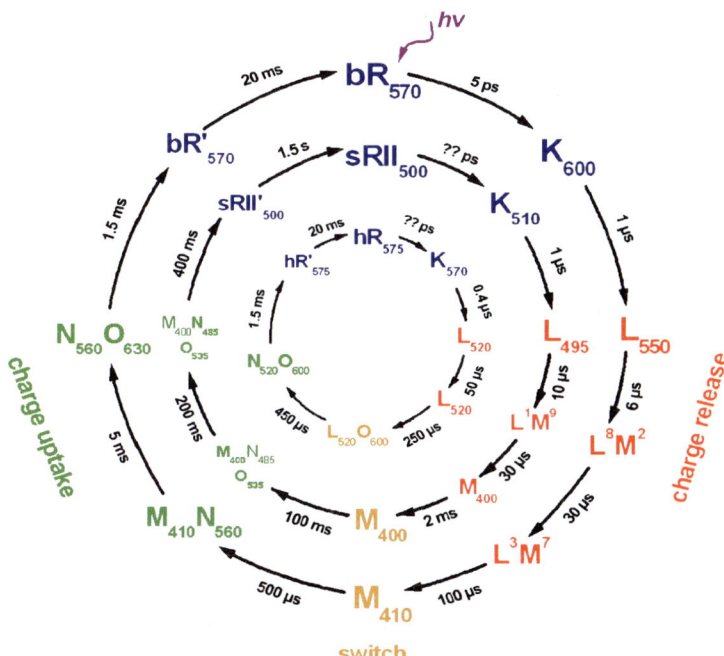

Fig. 4 Photocycles for BR (*inner circle*), NpSRII (*middle circle*), and NpHR (*outer circle*) composed of the canonical intermediates K, L, M, N, and O. The *lower index numbers* indicate λ_{max} (nm). Fast equilibria between adjacent intermediates are represented by the corresponding abbreviations, with the upper index numbers denoting the concentration in per cent/10

2004a). It is interesting to note that the uptake and release of protons is sensitively dependent on external conditions. This observation bears importance when comparing the coupling of protein conformational changes to ion transfer steps in BR to those of the other bacterial rhodopsins HR and SR (see below).

Two more general models have been proposed, which combine photocycle and proton transfer data. Lanyi and co-workers explained proton release and uptake in the frame of the alternating access model of ion pumps, stating that the connectivity of the Schiff base changes from one membrane surface to the other at different times of the photocycle (Kataoka et al. 1994). In the Isomerisation-Switch-Transfer (IST) model, Oesterhelt and co-workers propose a kinetic independence of ion transfer (T) and switch (S), a time-dependent process, which changes the accessibility of the Schiff base by isomerisation (Haupts et al. 1997). One central question in any model – although not really addressed in the literature – concerns the problem of rate-limiting steps: Are the conformational changes of the opsin, those of the ion transfer, or are they both rate limiting? The first order rate constants of the BR photocycle are surprisingly independent on the pH over a wide range (Xie et al. 1987). At high pH, the proton concentration of the bulk becomes rate-limiting, thus slowing down the reprotonation of Asp96 (Chizhov et al. 1992). Similar effects are observed for other microbial rhodopsins, such as proteorhodopsin (Friedrich et al. 2002) and NpSRII (Chizhov et al. 1998). In the former example, almost no dependence of the kinetic constants on pH was observed, even though at acidic pH no M-intermediate was detected. Similar results were obtained for the D75N mutant of NpSRII, which lacks the acceptor group for the Schiff base proton (Schmies et al. 2000). Interestingly, this mutation does not interfere with the normal phototaxis behaviour, if expressed in *H. salinarum* (Inoue et al. 2007). Also, in the case of the chloride pump, HR independency of the first order rate constants on chloride concentration is found (Chizhov and Engelhard 2001). Although these observations have to be analysed in further detail, it seems that protein conformational changes are rate-limiting, rather than ion transfer steps, which is a notion in line with the IST model.

The photocycles of sensors and ion pumps can be described by quite similar schemes, although the turnover of sensors is characteristically two orders of magnitude slower. This observation has been taken as evidence to differentiate between the two of them (Jung et al. 2003). It is striking that the crucial M1 → M2 type conformational switch has also been identified in the photoreaction cycle of sensory rhodopsin II and halorhodopsin from *N. pharaonis*. In the former example, the functionally important outward movement of helix F, concomitantly with the rotation of TM2 of the transducer, occurs with a rate constant of about 3 ms (Klare et al. 2004b; Wegener et al. 2000). This time had already been determined for the M1 → M2 step in earlier experiments (Chizhov et al. 1998). The photocycle of halorhodopsin has also been subject of intensive investigations (reviewed in (Oesterhelt 1995; Váró et al. 1995)). These original studies on HR were done from *H. salinarum*. In more recent years, HR from *N. pharaonis* became available (Bivin and Stoeckenius 1986a; Lanyi et al. 1990; Duschl et al. 1990; Scharf and Engelhard 1994) with the benefit of an easier biosynthetic access (Hohenfeld et al. 1999; Sato et al. 2002) and a better stability, which allowed an investigation of its photocycle under various conditions. The turnover time is similar to that of BR, but the M-intermediate is missing due to the Schiff base counterion, which is a halide and not a carboxyl group. Instead of the M1 → M2 switch, a similar O1 → O2 transition is observed. The functionally important properties of the M1 → M2 switch are taken over by an O1 → O2 transition, during which the monovalent anion accessibility changes (Chizhov and Engelhard 2001; Hackmann et al. 2001), contrary to BR, which transports a cation from the cytoplasmic to the extracellular side.

Apparently, the switch is the key event in the mechanism of ion pumps and the mechanism of sensors. What is known about the conformational changes leading to this switch? Most data is again from BR, whose structure has been determined at high resolution. Furthermore, it has also been possible to obtain structural data on most of the intermediates (for a review see Lanyi 2004b). Although substantial effort has been put into the structural analysis of other microbial rhodopsins, only a few have been obtained so far, including HsHR (Kolbe et al. 2000a), NpSRII (Luecke et al. 2001; Royant et al. 2001a), and the NpSRII/NpHtrII complex (Gordeliy et al. 2002). As already mentioned, the protein backbone of three microbial rhodopsins can be superimposed almost congruently(Fig. 2).

Evidently, only subtle changes are necessary to convert an ion pump into a sensor. An impressive corroboration of this conclusion has been described by Spudich and Sudo (2006). Only three amino acids had to be modified in BR to alter the proton pump into a sensory receptor (see below). Recently, the first structure of a eubacterial rhodopsin became available, showing distinct differences to the archaeal rhodopsins (Vogeley et al. 2004).

3.2
Mechanism of Signal and Ion Transfer

The high resolution structures of BR, HR, and SRII, together with other biophysical studies – most notably FTIR-spectroscopy – have considerably contributed to our understanding of the mechanism of ion transfer and signal transfer. The active site near the middle of the bilayer comprises the retinylidene Schiff base, with its counterion Asp85 (numbering according to the BR sequence) and specific water molecules. It separates the cytoplasmic channel from the extracellular channel (abbreviated CP-channel and EC-channel, respectively). The EC-channel (depicted in Fig. 5), formed by charged and polar amino acids and specific water molecules, is quite hydrophilic with access to bulk water (Harbison et al. 1988). Three trapped water molecules are observed in the neighbourhood of the Schiff base, forming a pentagonal like arrangement with Asp85 and Asp212. This structural element is also found in NpSRII (Royant et al. 2001a; Luecke et al. 2001; Gordeliy et al. 2002). Further structural water molecules connect a conserved Arg residue (Arg82) to the pentagonal by a water molecule coordinated to Tyr57 and Thr205. Adjacent to this region and closer to the bilayer surface, diffusive water molecules (average residence time 95 ps, (Grudinin et al. 2005)) have been recognized in molecular dynamics simulations. These water molecules and specific polar residues in the extracellular channel play a decisive role for the proton transfer from the Schiff base to the membrane surface during the L → M transitions. FTIR data (Garczarek and Gerwert 2006) and molecular dynamics calculations encompassing this proton release pocket (Mathias and Marx 2007) provided evidence for a one-dimensional wire-like proton network, which is different from the classic Zundel or Eigen motives. The highly polar region encountered in BR is not as distinct in NpSRII or AR1 and HR, perhaps reflecting their particular function. Networks and dynamics of water within the interior of proteins are certainly of fundamental importance,

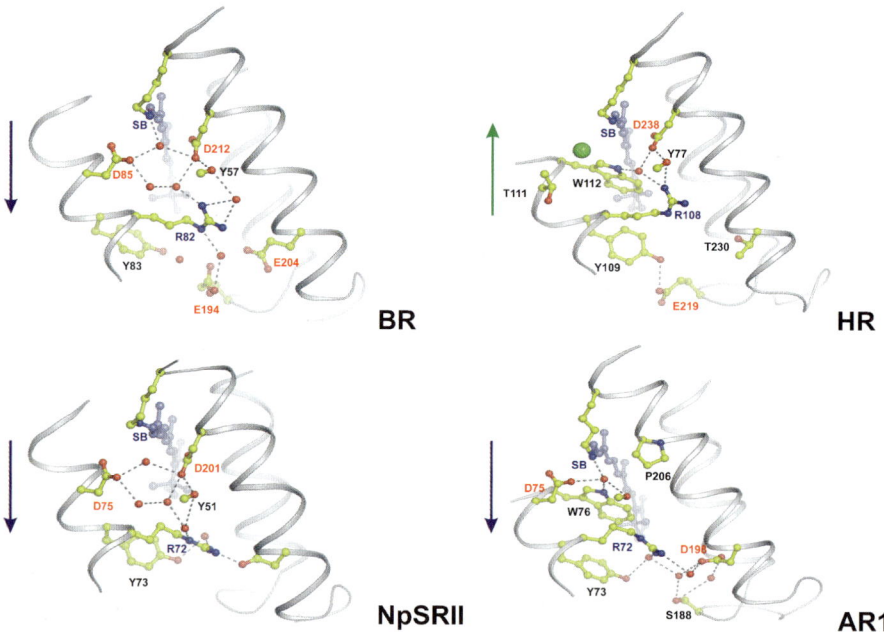

Fig. 5 Comparison of the Schiff base region for BR, HR, NpSRII, and AR (AR1 denotes the structure containing all-trans retinal analog (Vogeley et al. 2004a)). The pictures were generated using pdb files 1C3W (BR), 1E12 (HR), 1H2S (NpSRII), and 1XIO (AR1). The *green ball* in the HR structure represents chloride. The direction of proton transfer is indicated by *blue arrows*, and direction of chloride by the *green arrow*

not only for bacterial rhodopsins, but also for other proton translocating membrane proteins. For example, molecular dynamics calculations of water molecules in cytochrome oxidase revealed dynamical and structural properties tuned to the specific function of this protein as proton pump and oxygen reduction (Olkhova et al. 2004). Further experiments and calculations are needed to work out common principles, which govern the structure and dynamics of intrinsic water molecules.

The organisation of the extracellular channel of HR and AR1 is quite distinct from that of BR and NpSRII (Fig. 5). HR obviously has to harbour the chloride ion, which has been identified in the pocket close to the protonated Schiff base, replacing the negative charge of the Asp_{85}^{BR}-carboxylate (Kolbe et al. 2000). This residue is replaced by a Thr, whose methyl group faces the chloride ion. The cavity accommodating chloride is formed by a cluster of water molecules and OH groups from Ser residues. Interestingly, a hydrogen bond from Trp 112-N^1H connects via a water molecule to the Asp238 and Arg108, as a guardian to the extracellular channel. A similar pattern is observed for AR1, where Trp76 also participates in the configuration of the half channel (Vogeley et al. 2004). A further dramatic difference concerns the corresponding position of Asp_{212}^{BR}, whose place is taken by Pro_{206}^{AR} (Jung et al. 2003), rendering this part of helix G more flexible.

The cytoplasmic channel is generally less hydrophilic, with no direct access to the Schiff base (Fig. 6), but again with characteristic differences between pumps and sensors. In BR, Asp_{96} protonated in the ground state is separated by about 10 Å from the Schiff base in a relatively hydrophobic environment (Pebay-Peyroula et al. 1997). EPR experiments revealed the aprotic character of this half-channel (Steinhoff et al. 2000). Also in HR, a chloride-conducting pathway between Schiff base and the cytoplasmic surface cannot be recognized (Kolbe et al. 2000a). The hydrophobic nature of this part is further increased in sensors, where the position of Asp_{96}^{BR} is occupied by Phe (NpSRII) (Seidel et al. 1995) or Tyr (SRI) (Blanck and Oesterhelt 1987). Contrary to this generally observed water-excluding cytoplasmic channel, AR possess a continuous hydrogen-bonded network, providing a connection between the active site around the protonated Schiff base and the cytoplasm, probably the binding site of the soluble transducer (Vogeley et al. 2007).

Irrespective of these structural differences, it is noteworthy that the intrinsic function of a particular archaeal rhodopsin can easily be converted into that of another pigment. Adding azide to natronobacterial HR (NpHR) converts the chloride pump into an efficient proton translocating protein (Kulcsár et al. 2000). Conversely, BR can be modelled into a chloride pump by replacing Asp85 by a Thr residue (Tittor et al. 1997; Sasaki et al. 1995). More recently, Spudich and Sudo replaced three amino acids in BR with those from NpSRII (Ala215 → Thr; Pro200 → Thr; Val210 → Tyr) to generate a sensory rhodopsin (Sudo and Spudich 2006). Thr200 and Tyr210 are involved in binding NpHtrII to the receptor, whereas Thr215 alters the structure of the

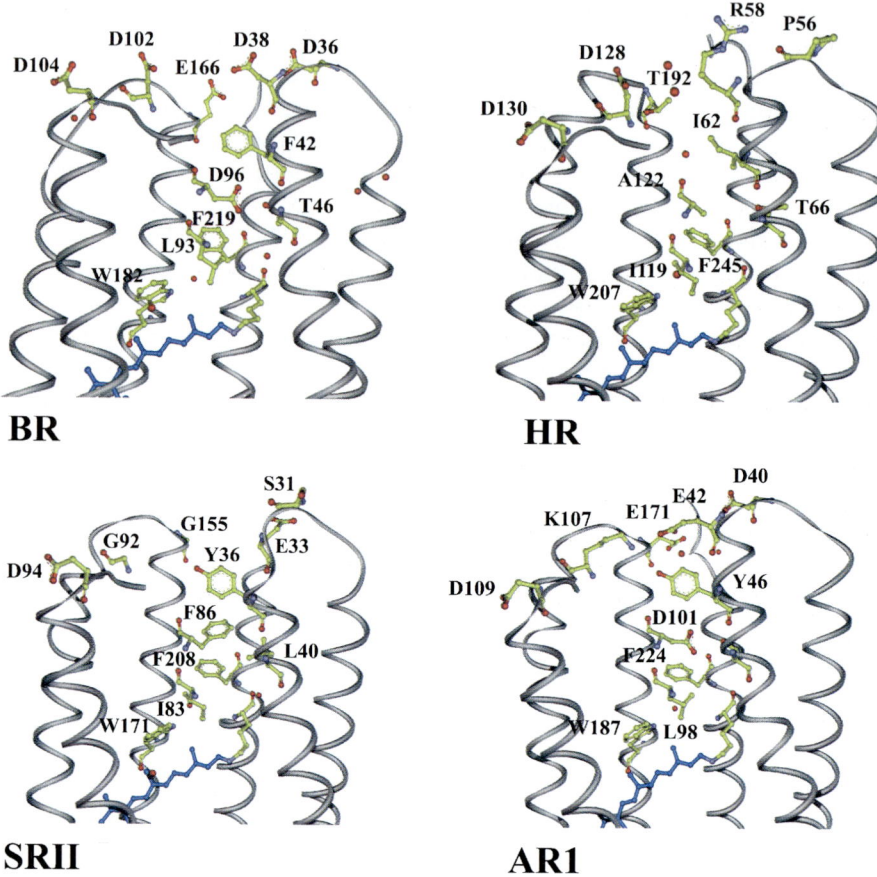

Fig. 6 Comparison of amino acids comprising the extracellular channel of BR, HR, NpSRII, and AR. The pictures were generated using pdb files 1C3W (BR), 1E12 (HR), 1H2S (NpSRII), and 1XIO (AR1)

π-bulge. In earlier experiments, it has already been demonstrated that a mutation in the latter position can confer a fast BR-like photocycle turnover in NpSRII (Klare et al. 2002). These examples show the close inter-relationship of archaeal rhodopsins, indicating common principles of function.

Due to a wealth of structural, spectroscopic, and theoretical data, a comprehensive picture of key steps of signal transduction or ion-translocation have emerged (for BR this has been detailed in (Subramaniam et al. 2002) and (Lanyi and Schobert 2004)). The first step, light activated *trans* → 13-*cis* isomerisation of retinal, changes the local environment of the protonated Schiff base from a favourable to a less favourable configuration, with the N–H bond now pointing towards the cytoplasm. Consequently, the pK of Asp85 and Schiff base is altered, such that a proton transfer to Asp85 can occur during the L → M transition. Because the β-ionone ring of the retinal chromophore is blocked in its pocket by steric interactions, other parts have to respond to the isomerisation. One observation concerns the movement of the Schiff base nitrogen about 0.7 Å towards the cytoplasmic side, concomitantly with a displacement of the distal C-atom of Lys216 (Subramaniam and Henderson 2000; Subramaniam et al. 1999; Luecke et al. 1999, 2000; Sass et al. 2000). These structural rearrangements can be correlated with the M1–M2 transitions, in which the accessibility of the Schiff base is changed to the extracellular side, ready to pick up a proton from Asp96. Concomitantly, the C9 and C13 methyl groups of retinal, which are in direct Van der Waals contact with Trp182 (helix F) (Fig. 7) pushes helix F in an outward movement by about 6°, thereby opening the extracellular channel to water molecules, which enables a proton to reprotonate the Schiff base via Asp96 (Grudinin et al. 2005). The helix F movement is facilitated because of an interruption of the hydrogen bond to water 501, which itself is hydrogen bonded to Ala215 on helix G, resulting in a disconnection of both helices. The helix F movement has been verified earlier by a number of experiments, including X-ray crystallography, electron microscopy, and EPR spectroscopy (Koch et al. 1991; Kamikubo et al. 1996; Subramaniam et al. 1993; Vonck 1996), which was later confirmed by higher resolution data (Sass et al. 2000; Subramaniam and Henderson 2000; Luecke et al. 1999; Radzwill et al. 2001) (see Fig. 8).

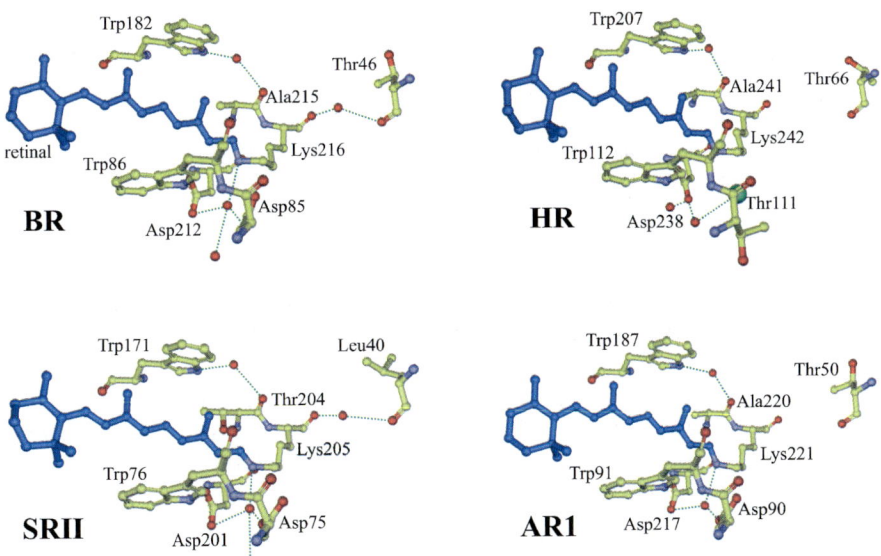

Fig. 7 Structural characteristics connecting Schiff base with Helix F and Helix G. The pictures were generated using pdb files 1C3W (BR), 1E12 (HR), 1H2S (NpSRII), and 1XIO (AR1)

The question arises, if this helix F movement is of general significance, not only for ion transport, but also for signal transfer. From a structural point of view, the four structures available show all the necessary structural elements to connect the retinal binding pocket with the helix F and helix G region (Fig. 7). Trp182 finds its counterpart also in HR, NpSRII, and AR. A water molecule (water 501 in BR) links this Trp with the π-bulge via the main chain carbonyl group of an Ala residue. An exception is found in the sensor NpSRII where a Thr replaces Ala, a signature obviously decisive for being a sensor or ion pump (Sudo et al. 2007; Klare et al. 2002). AR containing an Ala at this site seems to be an exception to the rule. However, its sequence in this region is quite different, in the first place reflecting its role of a sensor interacting with a soluble transducer probably on the cytoplasmic surface (Vogeley et al. 2004, 2007). The helix F movement was not only experimentally observed in BR, but also in NpSRII (Klare et al. 2004a; Wegener et al. 2000). Also, in bovine rhodopsin a similar movement of corresponding helix IV has been detected (Farrens et al. 1996). Obviously, the conformational change connected to helix F movement is of functional importance not only for microbial rhodopsins, but also for the other seven helix membrane proteins, like GPCRs. It remains to be elucidated from case to case, how these subtle modifications of a confined structural element are translated into a particu-

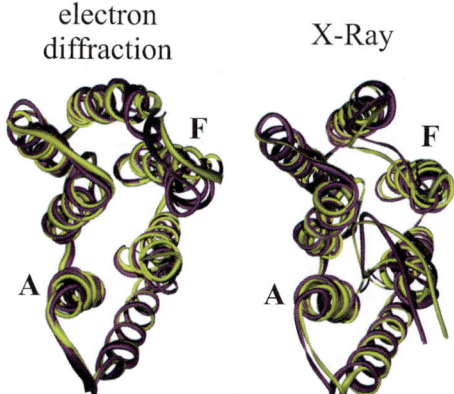

Fig. 8 Comparison of light-induced conformational changes in bacteriorhodopsin observed with electron diffraction (*left panel*) and X-ray crystallography (*right panel*), respectively. The structures are rendered in ribbon representation. They are viewed from the cytoplasm. Ground state structures are coloured in *purple*, and the M-trapped structures are shown in *yellow*. The electron diffraction data clearly shows a significant outward-bending motion of the cytoplasmic part of helix F, whereas conformational changes in the crystal structure seem to be strongly damped by packing effects. PDB identifiers for the respective structures are: 1FBB (electron diffraction, ground state), 1FBK (el. diff., M state) (Subramaniam and Henderson 2000b), and 1CWQ (X-ray, Ground+M state) (Sass et al. 2000b)

lar function. NpSRII serves as a first example where the signal transfer to its cognate transducer has been elucidated in detail (see below).

4
Phototaxis

Phototactic responses of a microbial organism by means of retinal proteins were first reported for the archaeon *H. salinarum* (Spudich and Stoeckenius 1979). It was observed, that besides chemical gradients, different light conditions also alter the intervals of forward swimming, stopping, and backward swimming, by increasing or decreasing the probability of this switching event. Without an external stimulus, the swimming pattern of the cells is like a random walk. In case of a positive signal, the frequency of reversals is reduced, thereby prolonging the swimming periods, whereas a repellent signal leads to shorter runs, and, consequently, more frequent angular changes of the swimming direction, caused by Brownian motion or mechanical obstacles. The combination of these responses then results in a net movement towards the more favourable conditions. The phototactic system used by this organism to seek light conditions, which are optimal for the functioning of the light-driven ion pumps, BR and HR, and for avoidance of harmful oxidative stress, consists of the two sensory rhodopsins, SRI and SRII, which form a 2 : 2 complex with their cognate transducer proteins, HtrI and HtrII, respectively (for recent reviews on archaebacterial phototaxis see, e.g. Hoff et al. 1997; Engelhard et al. 2003; Spudich et al. 2000). Under aerobic conditions, when the cell is using the respiratory chain to gain energy for its function, the two ion pumps are not expressed and SRII is the only photoreceptor present in the cell membrane. This repellent receptor covers the blue-green region of the spectrum (λ_{max} = 495 nm), which represents the intensity maximum of daylight, thereby directing the cell away from bright sunlight, which would otherwise cause oxidative stress under the condition of high oxygen tension. If the oxygen supply is not sufficient to maintain the energy needs of the cells via the respiratory chain, BR and HR together with SRI are expressed, while the synthesis of SRII is suppressed. SRI has its absorption maximum at 587 nm, thereby sensing the region of the spectrum which drives the function of the ion pumps. In the two-photon reaction described above, SRI also mediates a photophobic response by means of a photoactive long-lived M-intermediate with a fine-structured absorption band at 373 nm. The mechanism of this dual functionality is explained by a model brought up by Spudich and Bogomolni (Spudich and Bogomolni 1984). Absorption of just one photon with $\lambda > 500$ nm triggers the photocycle, and it results in a photophilic answer of the cell. If light in the UV region is present as well, the photoactive intermediate is excited and the repellent signalling cascade is activated.

4.1
The Transducer Proteins

4.1.1
Archaeal MCP-Type Transducers

The light signal sensed by the archaeal sensory rhodopsins I and II is firstly transferred to their cognate transducer proteins HtrI and HtrII, which are members of the family of halobacterial transducer proteins (Htp's) (Rudolph et al. 1996). The transducers relay the signal to a cytoplasmic signal transduction cascade, a two-component system homologous to the chemotactic system of enteric bacteria like *E. coli* (see Sect. 4.2). Their primary sequences, of which about 20 have been reported so far, reveal a consensus secondary structure dominated by α-helices. The N-terminal transmembrane domain consists of two helices. It is followed directly by two so-called HAMP domains, which have been identified bioinformatically in histidine kinases, adenylyl cyclases, methyl-accepting chemotaxis proteins, and phosphatases (Aravind and Ponting 1999). A striking difference between the halobacterial transducer proteins and the chemotaxis receptors from *E. coli* or *B. subtilis* concerns the number of HAMP domains. Whereas eubacterial proteins contain only one domain, the archaeal counterparts possess two consecutive HAMP domains, which seem to be a unique property of the Htr's (Koch 2005). Quite recently, the structural features of such HAMP domains were investigated. In 2006, Bordignon et al. reported on the structural properties of the first HAMP domain of *Np*HtrII (Bordignon et al. 2006), revealing remarkable high dynamics for this domain. In the same year, Hulko and co-workers published the NMR structure of a HAMP domain from the thermophile *Archaeoglobus fulgidus* (Hulko et al. 2006). Contrary to the finding of Bordignon and co-workers, this HAMP consists of a four-helical, parallel coiled-coil structure. Hulko et al. proposed a model, explaining the switch between two states by a rotation of the helices in a cogwheel- like manner, thereby altering the helix–helix packing. More structural information and data on dynamics and kinetics are needed to understand the function of the HAMP domain as mediator of transmembrane signalling.

Further downsteam of the HAMP domains, methylation and signalling domains follow, which are highly homologous to that of chemotaxis receptors. Most of the cytoplasmic domain of the Htr's is characterized by a seven-residue repeat (a-b-c-d-e-f-g) with hydrophobic side chains in positions a and d, indicative for a coiled-coil arrangement of the helices (Le Moual and Koshland 1996). This structural element was confirmed by solving the crystal structure of the cytoplasmic domains of *E. coli* serine and *T. maritima* MCP_{1143} chemoreceptors (Kim et al. 1999; Park et al. 2006) (see Fig. 9).

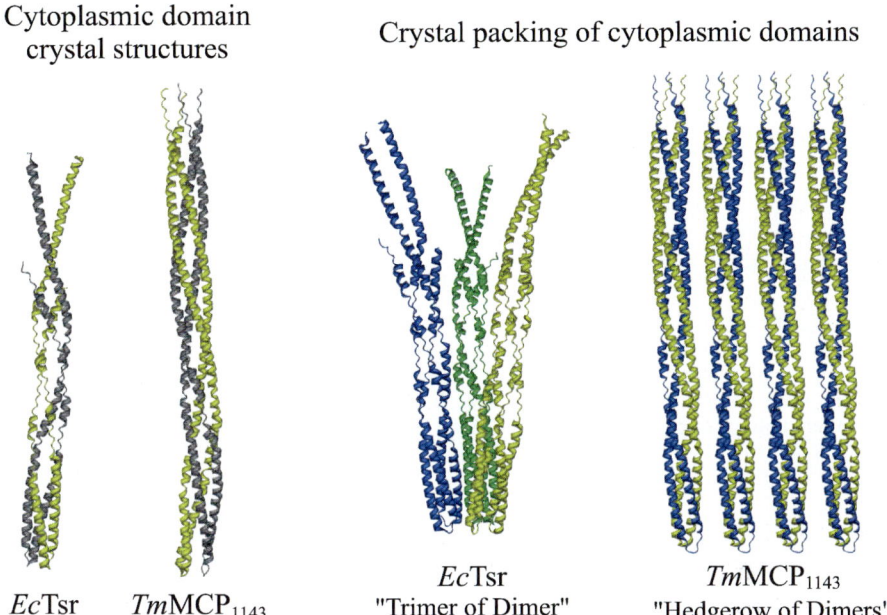

Fig. 9 *Left panel*: Crystal structures of the cytoplasmic domains of EcTsr (pdb: 1QU7) and *T. maritima* MCP$_{1143}$ (pdb: 2CH7). The structures are shown in ribbon representation. *Right panel*: Crystal packing of the EcTsr and TmMCP$_{1143}$. The *E. coli* chemoreceptor is arranged as a trimer of dimmers, whereas the *T. maritima* protein is packed like a "hedgerow" of receptor dimers

4.1.2
The *Anabaena* Soluble Transducer

The apoprotein of *Anabaena* sensory rhodopsin is expressed from an operon, containing an additional gene coding for a 125-residue (14 kDa) soluble protein (Jung et al. 2003), which was found to interact with ASR. Based on these findings, Jung and co-workers concluded that this small soluble protein acts as a putative transducer, and, consequently, they named it ASR transducer (ASRT). Up to now, ASRT has been the only example for a soluble transducer protein transmitting the signal from a sensory rhodopsin type receptor. In 2007, the crystal structure of ASRT was reported (Vogeley et al. 2007). ASRT forms a planar tetramer (see Fig. 10), which is also consistent with biochemical data. These experiments revealed a 4 : 1 stoichiometry with a dissociation constant in the micromolar range for the binding of the transducer to the receptor. The mechanism of signal transfer to and via ASRT is still unknown, although ASRT is expected to bind at the cytoplasmic surface of ASR. The signalling state, defined either by the long-lived M intermediate or by the all-*trans* or 13-*cis* configuration of the retinal chromophore, might alter the

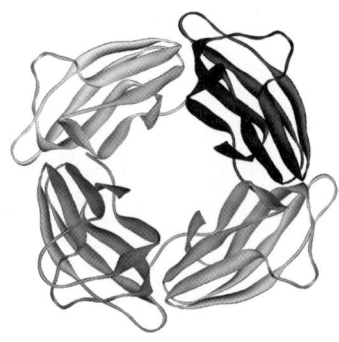

*A*SRT tetramer

Fig. 10 Crystal structure of the Anabaena soluble transducer (pdb: 2I18), revealing its tetrameric quaternary structure, which is thought to represent the functional unit (see text)

cytoplasmic surface, as observed for vertebrate rhodopsin. Obviously, a signal transfer via a membrane embedded binding domain, like in the case of SRs and corresponding Htrs, does not happen. Similarly, *A*SRT also lacks the signalling and methylation domains typical for the MCP-Type sensory transducers, indicating a signalling cascade significantly different from the component system of phototaxis and chemotaxis.

4.2
Two-Component Systems

The swimming behaviour of chemotactic and phototactic microbial organisms like *H. salinarum* is determined by the rotational direction of the polarly inserted motor-driven helical flagella (Hildebrand and Dencher 1975; Spudich and Stoeckenius 1979; Marwan and Oesterhelt 1990). Because the flagella helices are right-handed, clockwise rotation results in forward swimming, whereas switching the motor to a counter-clockwise rotation pulls the bacterium backwards (unipolarly flagellated bacteria like *E. coli* tumble rather than swim backwards because the flagellar bundles unwind in the counter clockwise mode). The probability of the switching event is controlled by a cytoplasmic response regulator, CheY, which is part of a two-component signal transduction chain, identified and thoroughly investigated in a number of bacteria, e.g. in *Bacillus subtilis*, enteric bacteria like *E. coli* (reviewed, e.g. in Stock et al. 2000), and in the halophilic archaebacteria *H. salinarum* (Rudolph and Oesterhelt 1995, 1996) (see Fig. 11). Further members of this signalling chain are: CheB, another response regulator involved in adaptation, CheA, a histidine kinase, and CheR, a constitutively active methyltransferase. The phospatase CheZ is not found in *H. salinarum*. CheW is necessary for binding of CheA to the archaeal transducer proteins and eubacterial methyl-accepting

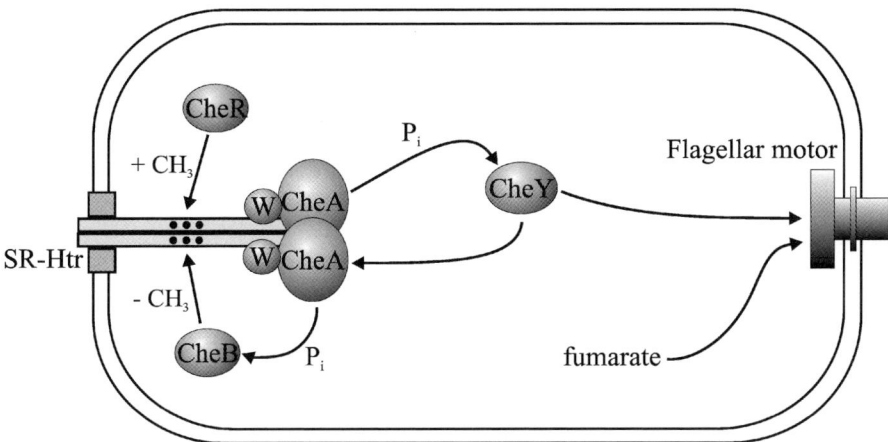

Fig. 11 Schematic representation of the two-component system found in *E. coli* and *H. salinarum*. For a detailed description of the components and interactions, see text

proteins. Additionally, the *che* gene clusters analyzed so far include additional proteins like CheC in *B. subtilis* (CheJ in *H. salinarum*) or CheD, but the function of these proteins is still unknown, although deletions of these proteins produce a clear phenotype, concerning the chemotactic behaviour of the cells (Rudolph and Oesterhelt 1996). It was also found that some of the Che proteins are present in several different forms in the same organism, such as the presence of two CheW and three CheJ forms in *Halobacterium* NRC-1 (CheW1,2 and CheJ1,2,3, respectively).

The signal transduction chain appears to be similar in most of the systems. The histidine kinase, CheA, is bound to the cytoplasmic tip of the signalling domain of the receptor/transducer via the adapter molecule, CheW. The exact mode of binding is still unknown, although a recent study, combining X-ray crystallography and EPR spectroscopy, revealed first insights into the possible organization of the ternary complex (Park et al. 2006). The signalling state of the receptor/transducer, dependent on the presence of attractant or repellent external stimuli, decreases or increases the autophosphorylation activity of CheA. In the next step, one of the two response regulators, CheY or CheB, is phosphorylated by CheA. Phosphorylated CheY, (CheY $\sim$ P), is a switch factor of the flagellar motor. High concentrations of CheY $\sim$ P increase the reversal frequency, whereas low levels lead to prolonged swimming periods. On the other hand, activation of the methylesterase CheB by phosphorylation leads to demethylation of Asp or Glu residues, located nearby the signalling domain of the receptor/transducer molecule, which are constitutively methylated by the methyltransferase, CheR. Consequently, the methylation level is regulated in dependence of the external attractant or repellent concentrations. The level of methylation also modulates the signalling state of the molecule in an inverse manner. A high degree of methylation leads to an attractant signalling state and vice versa. The interplay of these components enables the organism to adapt to constant stimuli, thereby increasing the dynamic range of sensing by orders of magnitude. Also in the presence of high levels of attractant or repellent substances, the bacteria retain the ability to sense altering gradients of stimulants.

4.3
Mechanisms of Signal Transduction

4.3.1
Receptor Transducer Interaction

The receptor/transducer complex exhibits a 2 : 2 stoichiometry in all cases studied so far. This finding parallels with the oligomeric state of chemoreceptors, for which a dimeric structural unit was observed (Chervitz and Falke 1996 and literature therein). Detailed structural information was obtained for the NpSRII-NpHtrII complex which shed insight into the mechanism of signal transfer from the receptor to the transducer (Wegener et al. 2001; Moukhametzianov et al. 2006; Gordeliy et al. 2002). In these studies, a C-terminal transducer fragment comprising residues 1–114 NpsRII was complexed to NpSRII and used either for EPR investigations or for crystallisation. Both data sets revealed topologies of the complex (see Fig. 12) with the four helices of the transducer dimer sandwiched between two sensor molecules.

The main interactions in the complex are Van der Waals contacts with a few hydrogen bonds (Fig. 13). The tightest packing is observed between helices TM2 and G. The interaction between TM2 and F is not as pronounced. The transducer helices in the transmembrane four-helical bundle also interact strongly by intercalation of bulky hydrophobic side chains. The four hydrogen bonds are formed connecting helices F and TM2 (Tyr-199^{NpSRII} to Asn-74^{NpHtrII}) and the F–G loop region, with both transducer helices on the extracellular side (Thr-189^{NpSRII} to Ser-62^{NpHtrII} (two H-bonds) and Thr-189^{NpSRII} to Glu-43^{NpHtrII}). Calorimetric studies have shown that the hydrogen bond between Tyr and Asn is not essential for the thermodynamics of the complex formation (Hippler-Mreyen et al. 2003). However, this bond seems to be important for the receptor-transducer signal transfer (Sudo et al. 2002). This latter finding is corroborated by the sensor properties of a triple mutant BR, which included those amino acids involved in hydrogen bonds (Sudo and Spudich 2006). Convers results were obtained for the role of the Tyr199^{NpSRII} – Asn74^{NpHtrII} hydrogen bond in receptor-transducer signal transfer. The crystal structure of the M-intermediate of the complex (Moukhametzianov et al. 2006) revealed an unperturbed hydrogen bond in the activated state. In contrast to that, FTIR data obtained by two different groups gave indications for either an increased strength of this bond in

Microbial Rhodopsins: Scaffolds for Ion Pumps, Channels, and Sensors   107

RECEPTOR/TRANSDUCER COMPLEX

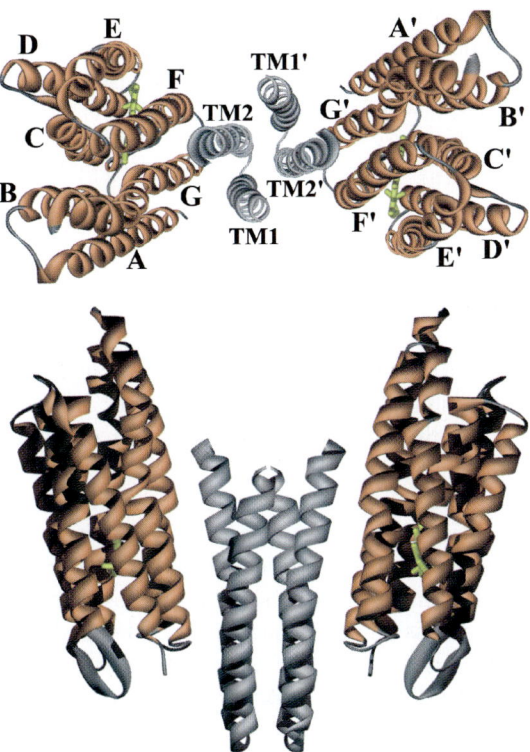

Fig. 12 Crystal structure of the *Np*SRII-*Np*HtrII complex in ribbon representation. *Upper panel*: View from the cytoplasm, *lower panel*: Structure viewed along the membrane. The helices are annotated. (PDB identifier: 1H2S)

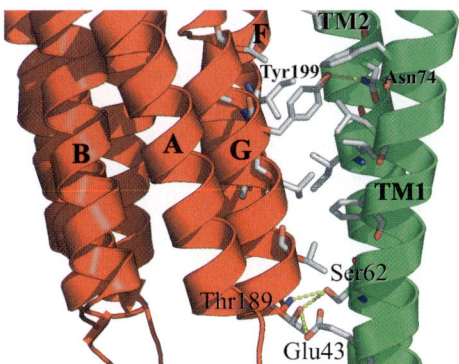

Fig. 13 Receptor–Transducer binding interface, comprising the interface helices. The transducer is depicted in *green*, and the receptor is coloured *red*

the M-intermediate (Furutani et al. 2005), or its disruption and the formation of other stronger H-bonds to Tyr199 and Asn74. The crystal structure of the complex did not reveal electron densities of the cytoplasmic peptide, indicating a highly dynamic structure, which was also proposed from EPR experiments. One possibility to obtain structural information about this part would be to apply solid state NMR techniques, as was already demonstrated for NpSRII (Etzkorn et al. 2007).

4.3.2
The Molecular Mechanism of Signal Transfer

The mechanism of signal transfer was most extensively investigated in the photophobic receptor-transducer complex *Np*SRII–*Np*HtrII from *N. pharaonis*. Based on the results obtained for this system, the sequence of events leading to the signalling state of the complex can be described as follows (reviewed in Klare et al. 2004; Spudich and Luecke 2002). As outlined in Sect. 3.2, the absorption of a photon leads to a transition in which helix F is pushed in a tilt-like motion to the outside. This reaction occurs during the $M_1 \rightarrow M_2$ transition with a time constant of about 3 ms (Chizhov et al. 1998). The formation of the signalling state seems to occur with this spectrally silent transition, which is mainly characterized by protein conformational changes, rather than alterations of the chromophore, as shown by Fourier transform infrared (FTIR) measurements (Hein et al. 2003). From physiological experiments, it has been shown that the signalling state is correlated with formation of M and the decay of O (Yan et al. 1991). The nature of the conformational changes, occurring during formation of the active signalling state, was characterized in detail, applying electron paramagnetic resonance (EPR) spectroscopy in combination with site-directed spin labelling (SDSL). The studies revealed a transient mobilisation of helix F during the $M_1 \rightarrow M_2$ transition (Klare et al. 2004b), which becomes immobilized again during the reformation of the receptor ground state (Wegener et al. 2000). In contrast to that, the neighbouring helix G did not show significant changes in its mobility during the photocycle. These findings were interpreted as an outward-tilt motion of the the cytoplasmic half of helix F similar to the observations made for BR (see Sect. 3.2). Subsequently, a more detailed investigation using EPR confirmed the earlier conclusions. In this SDSL work, the label positions were located on helices B, C, F, and G, and clear evidence was provided for a distance increase between helices C and F, which occurred upon receptor activation (Bordignon et al. 2007). Further support of the helix-tilt model came from biochemical (Yoshida et al. 2004) and computational studies (Sato et al. 2005).

Having established a helix F movement, the question arose if this motion is of functional importance, and if it can trigger the activation of the transducer. The answer was given by Wegener et al., who proposed a model that connected helix F tilt with a clockwise rotation of the transducer helix TM2 (Wegener et al. 2001). TM2 is in close contact with helix F, and the authors suggested that the outwardly tilting F helix would collide tangentially with TM2, thereby inducing its rotary motion as depicted in Fig. 14. Additional experiments on transducer mutants where cysteines positioned in the interior of the four-helix bundle showed an increase in the cross-linking efficiency of TM2 upon light activation supported the transducer helix rotation (Yang and Spudich 2001). Similar results were obtained by using a dye (Yoshida et al. 2004). It should be noted that a small piston-type movement (the favourite model of the chemotaxis research community, for example (Chervitz and Falke 1996)) cannot be excluded and would result in an overall screw-like motion of TM2.

How signal transfer to the cytoplasmic methylation/signalling domain occurs remains still unclear. Upon activation, conformational changes have not yet been experimentally observed, neither in the chemoreceptors nor in the phototransducers. Data revealed by EPR showed the presence of at least two conformational states in the first HAMP domain of the NpSRII–NpHtrII complex being in a temperature- dependent equilibrium. Additionally, the structural characterisation of this domain revealed it to be very dynamic, at least in the inactive state (Bordignon et al. 2005). This supports a model proposed by Kim and co-workers in 1994 (Kim 1994) for the signalling mechanism of chemoreceptors. The so-called "frozen dynamic dimer model" suggests that the dynamic properties of the whole cytoplasmic four-helix bundle are modulated by ligand binding to the periplasmic sensor domain. A more dynamic state would therein reflect the "attractant" signalling state, wheras a more

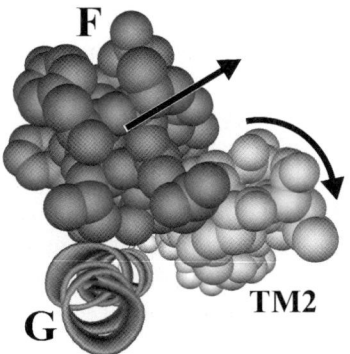

view from the cytoplasm

Fig. 14 Light-induced conformational changes of the receptor, helix F, and the transducer, helix TM2. The spacefill representation reveals the tight interaction between F and TM2. The outward movement of helix F (*straight arrow*) ejects TM2 in the way of a clockwise rotation (*bent arrow*)

rigid structure of the cytoplasmic domain would correspond to the "repellent" signalling state.

4.4
Comparison Between Phototaxis and Chemotaxis

Several lines of evidence support that the molecular mechanisms of chemotaxis and phototaxis are closely related. In 1989, Maqsudul and co-workers reported the presence of methyl-accepting taxis proteins in *H. halobium* (*H. salinarum*) being responsible for the chemotactic, as well as for the phototactic responses of this archaeon (Alam et al. 1989). The high amino acid sequence conservation between the *E. coli* chemoreceptors and the haloarchaeal phototransducer proteins was recognized with the first primary sequence of an Htr available (Yao and Spudich 1992). These sequence similarities were identified in all membrane-bound phototransducer proteins documented so far. Also, the haloarchaeal two-component system shows a close relationship to that of enteric bacteria. Using chimera Htr-chemoreceptor proteins phototactic active *E. coli* were constructed, demonstrating the evolutionary proximity of the two signalling chains (Jung et al. 2001). In these constructs, the cytoplasmic domain of the chemoreceptors was fused to the transmembrane part of the transducer, which, in turn, was fused to *Np*SRII to maintain the 1 : 1 stoichiometry in cells. In 2003, the same group reported measurements on the CheA-autophosphorylation and the phosphotransfer to CheY, modulated by one of the previously reported chimeras (*Np*SRII–*Np*HtrII–*Ec*Tsr-fusion), thereby showing that the light-response reported previously is indeed mediated by activation of the two-component, signalling cascade of the host *E. coli* (Trivedi and Spudich 2003). This data emphasizes the close relationship between chemotaxis and phototaxis, and it seems probable to assume a common mechanism of signal transfer across the membrane and ion transport. More structural data of the signalling state, especially of the membrane domain of chemoreceptors, are needed to clarify the controversy between piston-like movement and a screw-like rotation of TM2. The photophobic receptor (HsSRII) from *H. salinarum* contains an extracellular serine-binding domain (Zhang et al. 1996; Hou et al. 1998). Its dual functionality might serve as connective link, not only between the mechanism of chemotaxis and phototaxis, but also between structural equivalents of negative and positive extracellular inputs. One striking difference between eubacterial chemoreceptors and archaeal transducers concerns the number of HAMP domains, which is doubled in the latter proteins. If this observation has an impact on the primary signal transfer, mechanism and the function of HAMP domains in general still has to be elucidated.

5
Outlook

Since their first discovery in the early 1970s, research on microbial rhodopsins has steadily increased, and is now a field reaching into diverse areas of biochemistry and cell biology. Representatives are now found in all three domains of life, and their function is only proven for a minute number of proteins. It will be of great interest to complement in vitro data with in vivo investigations, which will provide not only a better understanding of underlying molecular mechanism, but also about the coupling of receptor activation with downstream signalling cascades. The discovery of marine phototrophic Eubacteria will have an impact on the discussion on energy conversion beyond classical photosynthesis. A completely new field of research has been opened by the expression of channelopsins and halorhodopsin in eukaryotic cells and in whole organisms (Zhang et al. 2007). Both pigments function in mammals without exogenous cofactors, and they can be used to analyse neural circuits by all-optical means. Research will certainly not stop with these novel developments. On the contrary, it is foreseeable that elucidating structure and function of microbial rhodopsins will lead to new insights and unexpected applications.

Acknowledgements Our own work cited in this review was funded by the Deutsche Forschungsgemeinschaft and the Max Planck Society. We would like to thank Antje Schulte and Wulf Blankenfeldt for help designing the structural figures.

References

Alam M, Lebert M, Oesterhelt D, Hazelbauer GL (1989) Methyl-accepting taxis proteins in *Halobacterium halobium*. Eur Mol Biol Org J 8:631–639

Aravind L, Ponting CP (1999) The cytoplasmic helical linker domain of receptor histidine kinase and methyl-accepting proteins is common to many prokaryotic signaling proteins. FEMS Microbiol Lett 176:111–116

Arenkiel BR, Peca J, Davison IG, Feliciano C, Deisseroth K, Augustine GJ, Ehlers MD, Feng G (2007) In Vivo Light-Induced Activation of Neural Circuitry in Transgenic Mice Expressing Channelrhodopsin-2. Neuron 54:205–218

Balashov SP, Imasheva ES, Boichenko VA, Anton J, Wang JM, Lanyi JK (2005) Xanthorhodopsin: A Proton Pump with a Light-Harvesting Carotenoid Antenna. Science 309:2061–2064

Balashov SP, Imasheva ES, Lanyi JK (2006) Induced Chirality of the Light-Harvesting Carotenoid Salinixanthin and Its Interaction with the Retinal of Xanthorhodopsin. Biochemistry 45:10998–11004

Baldwin JM, Henderson R, Beckman E, Zemlin F (1988) Images of purple membrane at 2.8 A resolution obtained by cryo-electron microscopy. J Mol Biol 202:585–591

Beja O, Aravind L, Koonin EV, Suzuki MT, Hadd A, Nguyen LT, Jovanovich SB, Gates CM, Feldman RA, Spudich JL, Spudich EM, DeLong EF (2000) Bacterial Rhodopsin: Evidence for a New Type of Phototrophy in the Sea. Science 289:1902–1906

Bieszke JA, Braun EL, Bean LE, Kang S, Natvig DO, Borkovich KA (1999a) The *nop*-1 gene of *Neurospora crassa* encodes a seven transmembrane helix retinal-binding protein homologous to archeal rhodopsins. Proc Natl Acad Sci USA 96:8034–8039

Bieszke JA, Spudich EM, Scott KL, Borkovich KA, Spudich JL (1999b) A Eukaryotic Protein, NOP-1, Binds Retinal To Form an Archaeal Rhodopsin-like Photochemically Reactive Pigment. Biochemistry 38:14138–14145

Bivin DB, Stoeckenius W (1986) Photoactive Retinal Pigments in Haloalkaliphilic Bacteria. J Gen Microbiol 132:2167–2177

Blanck A, Oesterhelt D (1987) The halo-opsin gene. II. Sequence, primary structure of halorhodopsin and comparison with bacteriorhodopsin. EMBO J 6:265–273

Blanck A, Oesterhelt D, Ferrando E, Schegk ES, Lottspeich F (1989) Primary structure of sensory rhodopsin I, a prokaryotic photoreceptor. EMBO J 8:3963–3971

Blaurock AE, Stoeckenius W (1971) Structure of the Purple Membrane. Nat New Biol 233:152–155

Bogomolni RA, Spudich JL (1982) Identification of a third rhodopsin-like pigment in phototactic *Halobacterium halobium*. Proc Natl Acad Sci USA 79:6250–6254

Bordignon E, Klare JP, Döbber MA, Wegener AA, Martell S, Engelhard M, Steinhoff H-J (2005) Structural Analysis of a HAMP Domain: The Linker Region of the Phototransducer in Complex with Sensory Rhodopsin II. J Biol Chem 280:38767–38775

Bordignon E, Klare JP, Holterhues J, Martell S, Krasnaberski A, Engelhard M, Steinhoff H-J (2007) Analysis of Light-Induced Conformational Changes of Natronomonas pharaonis Sensory Rhodopsin II by Time Resolved Electron Paramagnetic Resonance Spectroscopy+. Photochem Photobiol 83:2–272

Boyden ES, Zhang F, Bamberg E, Nagel G, Deisseroth K (2005) Millisecond-timescale, genetically targeted optical control of neural activity. Nat Neurosci 8:1263–1268

Brown LS, Jung K-H (2006) Bacteriorhodopsin-like proteins of eubacteria and fungi: the extent of conservation of the haloarchaeal proton-pumping mechanism. Photochem Photobiol Sci 5:538–546

Chervitz SA, Falke JJ (1996) Molecular mechanism of transmembrane signaling by the aspartate receptor – A model. Proc Natl Acad Sci USA 93:2545–2550

Chizhov I, Chernavskii DS, Engelhard M, Müller KH, Zubov BV, Hess B (1996) Spectrally silent transitions in the bacteriorhodopsin photocycle. Biophys J 71:2329–2345

Chizhov I, Engelhard M (2001) Temperature and halide dependence of the photocycle of halorhodopsin from *Natronobacterium pharaonis*. Biophys J 81:1600–1612

Chizhov I, Engelhard M, Chernavskii DS, Zubov B, Hess B (1992) Temperature and pH sensitivity of the O_{640} intermediate of the bacteriorhodopsin photocycle. Biophys J 61:1001–1006

Chizhov I, Schmies G, Seidel R, Sydor JR, Lüttenberg B, Engelhard M (1998) The photophobic receptor from *Natronobacterium pharaonis* – temperature and pH dependencies of the photocycle of sensory rhodopsin II. Biophys J 75:999–1009

Danon A, Stoeckenius W (1974) Photophosphorylation in Halobacterium halobium. Proc Natl Acad Sci USA 71:1234–1238

de la Torre JR, Christianson LM, Beja O, Suzuki MT, Karl DM, Heidelberg J, DeLong EF (2003) Proteorhodopsin genes are distributed among divergent marine bacterial taxa. Proc Natl Acad Sci USA 100:12830–12835

Deininger W, Kroger P, Hegemann U, Lottspeich F, Hegemann P (1995) Chlamyrhodopsin represents a new type of sensory photoreceptor. EMBO J 14:5849–5858

Dencher NA (1983) The Five Retinal -Protein Pigments of Halobacteria: Bacteriorhodopsin, Halorhodopsin, P 565, P 370 and Slow-Cycling Rhodopsin. Photochem Photobiol 38:753–756

Dencher NA (1978) Light-induced behavioral reactions in Halobacterium halobium: Evidence for two rhodopsins acting as photopigments. In: Energetics and Structure of Halophilic Microorganisms. Elsevier/North Holland Biomedical Press, Amsterdam, pp 67-88

Dencher NA, Hildebrand E (1979) Sensory transduction in Halobacterium halobium: retinal protein pigment controls UV-induced behavioral response. Z Naturforsch C 34:841-847

Dollinger G, Eisenstein L, Lin S-L, Nakanishi K, Odashima K, Termini J (1986) Bacteriorhodopsin: Fourier Transform Infrared Methods for Studies of Protonation of Carboxyl Groups. Meth Enzym 127:649-662

Druckmann S, Friedman N, Lanyi JK, Needleman R, Ottolenghi M, Sheves M (1992) The back photoreaction of the M intermediate in the photocycle of bacteriorhodopsin: Mechanism and evidence for two M species. Photochem Photobiol 56:1041-1047

Duschl A, Lanyi JK, Zimányi L (1990) Properties and photochemistry of a halorhodopsin from the haloalkalophile, *Natronobacterium pharaonis*. J Biol Chem 265:1261-1267

Edman K, Nollert P, Royant A, Belrhali H, Pebay-Peyroula E, Hajdu J, Neutze R, Landau EM (1999) High-resolution X-ray structure of an early intermediate in the bacteriorhodopsin photocycle. Nature 401:822-826

Edman K, Royant A, Nollert P, Maxwell CA, Pebay-Peyroula E, Navarro J, Neutze R, Landau EM (2002) Early Structural Rearrangements in the Photocycle of an Integral Membrane Sensory Receptor. Structure 10:473-482

Edmonds BW, Luecke H (2004) Atomic resolution structures and the mechanism of ion pumping in bacteriorhodopsin. Front Biosci 9:1556-1566

Eisenstein L, Lin SL, Dollinger G, Odashima K, Ding WD, Nakanishi K (1987) FTIR difference studies on apoproteins; protonation states of aspartic- and glutamic acid residues during the photocycle of bacteriorhodopsin. J Am Chem Soc 109:6860-6862

Eisfeld W, Althaus T, Stockburger M (1995) Evidence for parallel photocycles and implications for the proton pump in bacteriorhodopsin. Biophys Chem 56:105-112

Enami N, Yoshimura K, Murakami M, Okumura H, Ihara K, Kouyama T (2006) Crystal Structures of Archaerhodopsin-1 and -2: Common Structural Motif in Archaeal Light-driven Proton Pumps. J Mol Biol 358:675-685

Engelhard M, Gerwert K, Hess B, Kreutz W, Siebert F (1985) Light-driven protonation changes of internal aspartic acids of bacteriorhodopsin: An investigation by static and time-resolved infrared difference spectroscopy using [4-13C] aspartic acid labeling purple membrane. Biochemistry 24:400-407

Engelhard M, Hofmann KP (2006) Photoreceptors. In: Rückpaul R, Ganten D (eds) Encyclopedic Reference of Genomics and Proteomics in Molecular Medicine. Springer Verlag, Berlin, pp 1407-1413

Engelhard M, Schmies G, Wegener AA (2003) Archaebacterial Phototaxis. In: Batschauer A (ed) Photoreceptors and Light Signaling. Royal Society of Chemistry, Cambridge, pp 1-39

Etzkorn M, Martell S, Andronesi OC, Seidel K, Engelhard M, Baldus M (2007) Secondary Structure, Dynamics, and Topology of a Seven-Helix Receptor in Native Membranes, Studied by Solid-State NMR Spectroscopy. Angew Chem Int Ed Engl 46:459-462

Farrens DL, Altenbach C, Yang K, Hubbell WL, Khorana HG (1996) Requirement of rigid-body motion of transmembrane helices for light activation of rhodopsin. Science 274:768-770

Foster KW, Saranak J, Patel N, Zarilli G, Okabe M, Kline T, Nakanishi K (1984) A rhodopsin is the functional photoreceptor for phototaxis in the unicellular eukaryote Chlamydomonas. Nature 311:756-759

Friedrich T, Geibel S, Kalmbach R, Chizhov I, Ataka K, Heberle J, Engelhard M, Bamberg E (2002) Proteorhodopsin is a Light-driven Proton Pump with Variable Vectoriality. J Mol Biol 321:821–838

Frigaard NU, Martinez A, Mincer TJ, DeLong EF (2006) Proteorhodopsin lateral gene transfer between marine planktonic Bacteria and Archaea. Nature 439:847–850

Fuhrmann M, Stahlberg A, Govorunova E, Rank S, Hegemann P (2001) The abundant retinal protein of the Chlamydomonas eye is not the photoreceptor for phototaxis and photophobic responses. J Cell Sci 114:3857–3863

Furutani Y, Kamada K, Sudo Y, Shimono K, Kamo N, Kandori H (2005) Structural Changes of the Complex between pharaonis Phoborhodopsin and Its Cognate Transducer upon Formation of the M Photointermediate. Biochemistry 44:2909–2915

Garczarek F, Gerwert K (2006) Functional waters in intraprotein proton transfer monitored by FTIR difference spectroscopy. Nature 439:109–112

Gerwert K, Hess B, Soppa J, Oesterhelt D (1989) Role of aspartate-96 in proton translocation by bacteriorhodopsin. Proc Natl Acad Sci USA 86:4943–4947

Gmelin W, Zeth K, Efremov R, Heberle J, Tittor J, Oesterhelt D (2007) The Crystal Structure of the L1 Intermediate of Halorhodopsin at 1.9 Å Resolution. Photochem Photobiol 83:369–372

Gordeliy VI, Labahn J, Moukhametzianov R, Efremov R, Granzin J, Schlesinger R, Büldt G, Savopol T, Scheidig AJ, Klare JP, Engelhard M (2002) Molecular basis of transmembrane signaling by sensory rhodopsin II-transducer complex. Nature 419:484–487

Grigorieff N, Ceska TA, Downing KH, Baldwin JM, Henderson R (1996) Electron-crystallographic Refinement of the Structure of Bacteriorhodopsin. J Mol Biol 259:393–421

Grudinin S, Büldt G, Gordeliy V, Baumgaertner A (2005) Water Molecules and Hydrogen-Bonded Networks in Bacteriorhodopsin–Molecular Dynamics Simulations of the Ground State and the M-Intermediate. Biophys J 88:3252–3261

Hackmann C, Guijarro J, Chizhov I, Engelhard M, Rödig C, Siebert F (2001) Static and time-resolved step-scan fourier transform infrared investigations of the photoreaction of halorhodopsin from *Natronobacterium pharaonis*: consequences for models of the anion translocation mechanism. Biophys J 81:394–406

Harbison GS, Roberts JD, Herzfeld J, Griffin RG (1988) Solid state NMR detection of proton exchange between the bacteriorhodopsin Schiff base and bulk water. J Am Chem Soc 110:7221–7223

Harz H, Hegemann P (1991) Rhodopsin-regulated calcium currents in *Chlamydomonas*. Nature 351:489–491

Haupts U, Tittor J, Bamberg E, Oesterhelt D (1997) General concept for ion translocation by halobacterial retinal proteins – the isomerization/switch/transfer (IST) model. Biochemistry 36:2–7

Havelka WA, Henderson R, Heymann JAW, Oesterhelt D (1993) Projection Structure of Halorhodopsin from Halobacterium halobium at 6 A Resolution Obtained by Electron Cryo-microscopy. J Mol Biol 234:837–846

Havelka WA, Henderson R, Oesterhelt D (1995) Three-dimensional structure of halorhodopsin at 7 Å resolution. J Mol Biol 247:726–738

Heberle J, Fitter J, Sass HJ, Büldt G (2000) Bacteriorhodopsin: the functional details of a molecular machine are being resolved. Biophys Chem 85:229–248

Hein M, Wegener AA, Engelhard M, Siebert F (2003) Time-Resolved FTIR Studies of Sensory Rhodopsin II (NpSRII) from Natronobacterium pharaonis: Implications for Proton Transport and Receptor Activation. Biophys J 84:1208–1217

Helmreich EJM, Hofmann KP (1996) Structure and function of proteins in G-protein-coupled signal transfer. BBA-Rev Biomembranes 1286:285–322

Henderson R, Baldwin JM, Ceska TA, Zemlin F, Beckmann E, Downing KH (1990) Model for the structure of bacteriorhodopsin based on high- resolution electron cryo-microscopy. J Mol Biol 213:899–929

Henderson R, Unwin PNT (1975) Three-dimensional model of purple membrane obtained by electron microscopy. Nature 257:28–32

Hildebrand E, Dencher N (1975) Two photosystems controlling behavioural responses of *Halobacterium halobium*. Nature 257:46–48

Hippler-Mreyen S, Klare JP, Wegener AA, Seidel RP, Herrmann C, Schmies G, Nagel G, Bamberg E, Engelhard M (2003) Probing the Sensory Rhodopsin II Binding Domain of its Cognate Transducer by Calorimetry and Electrophysiology. J Mol Biol 330:1203–1213

Hoff WD, Jung K-H, Spudich JL (1997) Molecular Mechanism of Photosignaling by Archaeal Sensory Rhodopsins. Annu Rev Biophys Biomol Struct 26:223–258

Hohenfeld IP, Wegener AA, Engelhard M (1999) Purification of histidine tagged bacteriorhodopsin, *pharaonis* halorhodopsin and *pharaonis* sensory rhodopsin II functionally expressed in *Escherichia coli*. FEBS Lett 442:198–202

Hou SB, Brooun A, Yu HS, Freitas T, Alam M (1998) Sensory rhodopsin II transducer HtrII is also responsible for serine chemotaxis in the archaeon *halobacterium salinarum*. J Bacteriol 180:1600–1602

Hulko M, Berndt F, Gruber M, Linder JU, Truffault V, Schultz A, Martin J, Schultz JE, Lupas AN, Coles M (2006) The HAMP Domain Structure Implies Helix Rotation in Transmembrane Signaling. Cell 126:929–940

Ihara K, Amemiya T, Miyashita Y, Mukohata Y (1994) Met-145 is a key residue in the dark adaptation of bacteriorhodopsin homologs. Biophys J 67:1187–1191

Ihara K, Umemura T, Katagiri I, Kitajima-Ihara T, Sugiyama Y, Kimura Y, Mukohata Y (1999) Evolution of the Archaeal Rhodopsins: Evolution Rate Changes by Gene Duplication and Functional Differentiation. J Mol Biol 285:163–174

Inoue K, Sasaki J, Spudich JL, Terazima M (2007) Laser-Induced Transient Grating Analysis of Dynamics of Interaction between Sensory Rhodopsin II D75N and the HtrII Transducer. Biophys J 92:2028–2040

Jung KH, Trivedi VD, Spudich JL (2003) Demonstration of a sensory rhodopsin in eubacteria. Mol Microbiol 47:1513–1522

Jung KH, Spudich EN, Trivedi VD, Spudich JL (2001) An Archaeal Photosignal-Transducing Module Mediates Phototaxis in *Escherichia coli*. J Bacteriol 183:6365–6371

Kamikubo H, Kataoka M, Váró G, Oka T, Tokunaga F, Needleman R, Lanyi JK (1996) Structure of the N intermediate of bacteriorhodopsin revealed by X-ray diffraction. Proc Natl Acad Sci USA 93:1386–1390

Kataoka M, Kamikubo H, Tokunaga F, Brown LS, Yamazaki Y, Maeda A, Sheves M, Needleman R, Lanyi JK (1994) Energy coupling in an ion pump. The reprotonation switch of bacteriorhodopsin. J Mol Biol 243:621–638

Khorana HG (1993) Two light-transducing membrane proteins: Bacteriorhodopsin and the mammilian rhodopsin. Proc Natl Acad Sci USA 90:1166–1171

Kim KK, Yokota H, Kim S-H (1999) Four-helical-bundle structure of the cytoplasmic domain of a serine chemotaxis receptor. Nature 400:787–792

Kim S-H (1994) Frozen dynamic dimer model for transmembrane signaling in bacterial chemotaxis receptors. Protein Sci 3:159–165

Klare JP, Bordignon E, Engelhard M, Steinhoff HJ (2004a) Sensory rhodopsin II and bacteriorhodopsin: Light activated helix F movement. Photochem Photobiol Sci 3:543–547

Klare JP, Gordeliy VI, Labahn J, Büldt G, Steinhoff H-J, Engelhard M (2004b) The archaeal sensory rhodopsin II/transducer complex: a model for transmembrane signal transfer. FEBS Lett 564:219–224

Klare JP, Schmies G, Chizhov I, Shimono K, Kamo N, Engelhard M (2002) Probing the proton channel and the retinal binding site of *Natronobacterium pharaonis* sensory rhodopsin II. Biophys J 82:2156–2164

Klink BU, Winter R, Engelhard M, Chizhov I (2002) Pressure Dependence of the Photocycle Kinetics of Bacteriorhodopsin. Biophys J 83:3490–3498

Koch MHJ, Dencher NA, Oesterhelt D, Plöhn H-J, Rapp G, Büldt G (1991) Time-resolved X-ray diffraction study of structural changes associated with the photocycle of bacteriorhodopsin. EMBO J 10:521–526

Koch MK (2005) Investigations on halobacterial transducers with respect to membrane potential sensing and adaptive methylation. Fakultät für Chemie und Pharmazie der Ludwig-Maximilians-Universität München, München, pp 1–171

Kolbe M, Besir H, Essen LO, Oesterhelt D (2000) Structure of the light-driven chloride pump halorhodopsin at 1.8 Å resolution. Science 288:1390–1396

Korenstein R, Hess B, Kuschmitz D (1978) Branching reactions in the photocycle of bacteriorhodpsin. FEBS Lett 93:266–270

Kouyama T, Nishikawa T, Tokuhisa T, Okumura H (2004) Crystal Structure of the L Intermediate of Bacteriorhodopsin: Evidence for Vertical Translocation of a Water Molecule during the Proton Pumping Cycle. J Mol Biol 335:531–546

Kulcsár A, Groma GI, Lanyi JK, Váró G (2000) Characterization of the Proton-Transporting Photocycle of Pharaonis Halorhodopsin. Biophys J 79:2705–2713

Landau EM, Rosenbusch JP (1996) Lipidic cubic phases – a novel concept for the crystallization of membrane proteins. Proc Natl Acad Sci USA 93:14532–14535

Lanyi JK, Duschl A, Hatfield GW, May K, Oesterhelt D (1990) The primary structure of a halorhodopsin from *Natronobacterium pharaonis*. Structural, functional and evolutionary implications for bacterial rhodopsins and halorhodopsins. J Biol Chem 265:1253–1260

Lanyi JK, Váró G (1995) The photocycles of bacteriorhodopsin. Isr J Chem 35:365–385

Lanyi JK (2000) Crystallographic studies of the conformational changes that drive directional transmembrane ion movement in bacteriorhodopsin. BBA-Bioenergetics 1459:339–345

Lanyi JK, Schobert B (2003) Mechanism of Proton Transport in Bacteriorhodopsin from Crystallographic Structures of the K, L, M1, M2, and M2' Intermediates of the Photocycle. J Mol Biol 328:439–450

Lanyi JK (2004a) Bacteriorhodopsin. Annu Rev Physiol 66:665–688

Lanyi JK (2004b) X-ray diffraction of bacteriorhodopsin photocycle intermediates. Mol Membrane Biol 21:143–150

Lanyi JK, Schobert B (2004) Local-global conformational coupling in a heptahelical membrane protein: Transport mechanism from crystal structures of the nine states in the bacteriorhodopsin photocycle. Biochemistry 43:3–8

Lanyi JK, Schobert B (2006) Propagating Structural Perturbation Inside Bacteriorhodopsin: Crystal Structures of the M State and the D96A and T46V Mutants. Biochemistry 45:12003–12010

Lanyi JK, Schobert B (2007) Structural Changes in the L Photointermediate of Bacteriorhodopsin. J Mol Biol 365:1379–1392

Le Moual H, Koshland DE Jr (1996) Molecular Evolution of the C-terminal Cytoplasmic Domain of a Superfamiliy of Bacterial Receptors Involved in Taxis. J Mol Biol 261:568–585

Lozier RH, Bogomolni RA, Stoeckenius W (1975) Bacteriorhodopsin: a light-driven proton pump in *Halobacterium halobium*. Biophys J 15:955–962

Luecke H, Schobert B, Cartailler JP, Richter HT, Rosengarth A, Needleman R, Lanyi JK (2000) Coupling Photoisomerization of Retinal to Directional Transport in Bacteriorhodopsin. J Mol Biol 300:1237–1255

Luecke H, Schobert B, Lanyi JK, Spudich EN, Spudich JL (2001) Crystal structure of sensory rhodopsin II at 2.4 Ångstroms: Insights into color tuning and transducer interaction. Science 293:1499–1503

Luecke H, Schobert B, Richter HT, Cartailler JP, Lanyi JK (1999) Structural changes in bacteriorhodopsin during ion transport at 2 Ångstrom resolution. Science 286:255–261

Man D, Wang W, Sabehi G, Aravind L, Post AF, Massana R, Spudich EN, Spudich JL, Beja O (2003) Diversification and spectral tuning in marine proteorhodopsins. EMBO J 22:1725–1731

Man-Aharonovich D, Sabehi G, Sineshchekov OA, Spudich EN, Spudich JL, Beja O (2004) Characterization of RS29, a blue-green proteorhodopsin variant from the Red Sea. Photochem Photobiol Sci 3:459–462

Marwan W, Oesterhelt D (1990) Quantitation of photochromism of sensory rhodopsin-I by computerized tracking of Halobacterium halobium cells. J Mol Biol 215:277–285

Mathias G, Marx D (2007) Structures and spectral signatures of protonated water networks in bacteriorhodopsin. PNAS 104:6980–6985

Matsuno-Yagi A, Mukohata Y (1977) Two possible roles of bacteriorhodopsin; a comparative study of strains of Halobacterium halobium differing in pigmentation. Biochem Biophys Res Commun 78:237–243

Matsuno-Yagi A, Mukohata Y (1980) ATP synthesis linked to light-dependent proton uptake in a red mutant strain of Halobacterium lacking bacteriorhodopsin. Arch Biochem Biophys 199:297–303

Mogi T, Stern LJ, Marti T, Chao BH, Khorana HG (1988) Aspartic acid substitutions affect proton translocation by bacteriorhodopsin. Proc Natl Acad Sci USA 85:4148–4152

Mongodin EF, Nelson KE, Daugherty S, DeBoy RT, Wister J, Khouri H, Weidman J, Walsh DA, Papke RT, Sanchez Perez G, Sharma AK, Nesbo CL, MacLeod D, Bapteste E, Doolittle WF, Charlebois RL, Legault B, Rodriguez-Valera F (2005) The genome of Salinibacter ruber: Convergence and gene exchange among hyperhalophilic bacteria and archaea. Proc Natl Acad Sci USA 102:18147–18152

Moukhametzianov R, Klare JP, Efremov R, Baeken C, Göppner A, Labahn J, Engelhard M, Büldt G, Gordeliy VI (2006) Development of the signal in sensory rhodopsin and its transfer to the cognate transducer. Nature 440:115–119

Mukohata Y, Matsuno-Yagi A, Kaji Y (1980) Light-induced proton uptake and ATP synthesis by bacteriorhodopdin-depleted *Halobacterium*. In: Morishita H, Masui M (eds) Saline Environment. Business Center Academic Society, Tokyo, pp 31–38

Mukohata Y, Ihara K, Tamura T, Sugiyama Y (1999) Halobacterial rhodopsins. J Biochem 125:649–657

Mukohata Y, Kaji Y (1981) Light-induced membrane-potential increase, ATP synthesis, and proton uptake in Halobacterium halobium, R1mR catalyzed by halorhodopsin: Effects of N,N'-dicyclohexylcarbodiimide, triphenyltin chloride, and 3,5-di-tert-butyl-4-hydroxybenzylidenemalononitrile (SF6847). Arch Biochem Biophys 206:72–76

Müller K-H, Butt HJ, Bamberg E, Fendler K, Hess B, Siebert F, Engelhard M (1991) The reaction cycle of bacteriorhodopsin: An analysis using visible absorption, photocurrent and infrared techniques. Eur Biophys J 19:241–251

Müller K-H, Plesser T (1991) Variance reduction by simultaneous multi-exponential analysis of data sets from different experiments. Eur Biophys J 19:231–240

Nagel G, Szellas T, Kateriya S, Adeishvili N, Hegemann P, Bamberg E (2005a) Channelrhodopsins: directly light-gated cation channels. Biochem Soc Transact 33:863–866

Nagel G, Brauner M, Liewald JF, Adeishvili N, Bamberg E, Gottschalk A (2005b) Light Activation of Channelrhodopsin-2 in Excitable Cells of Caenorhabditis elegans Triggers Rapid Behavioral Responses. Curr Biol 15:2279–2284

Nagel G, Ollig D, Fuhrmann M, Kateriya S, Musti AM, Bamberg E, Hegemann P (2002) Channelrhodopsin-1: A Light-Gated Proton Channel in Green Algaee. Science 296:2395–2398

Nagel G, Szellas T, Huhn W, Kateriya S, Adeishvili N, Berthold P, Ollig D, Hegemann P, Bamberg E (2003) Channelrhodopsin-2, a directly light-gated cation-selective membrane channel. Proc Natl Acad Sci USA 100:13940–13945

Nagle JF, Zimányi L, Lanyi JK (1995) Testing BR photocycle kinetics. Biophys J 68:1490–1499

Neutze R, Pebay-Peyroula E, Edman K, Royant A, Navarro J, Landau EM (2002) Bacteriorhodopsin: a high-resolution structural view of vectorial proton transport. BBA-Biomembranes 1565:144–167

Ni BF, Chang M, Duschl A, Lanyi J, Needleman R (1990) An efficient system for the synthesis of bacteriorhodopsin in *Halobacterium halobium*. Gene 90:169–172

Nishikawa T, Murakami M, Kouyama T (2005) Crystal Structure of the 13-cis Isomer of Bacteriorhodopsin in the Dark-adapted State. J Mol Biol 352:319–328

Oesterhelt D (1995) Structure and function of halorhodopsin [review]. Isr J Chem 35:475–494

Oesterhelt D, Stoeckenius W (1971) Rhodopsin-like protein from the purple membrane of Halobacterium halobium. Nat New Biol 233:149–152

Oesterhelt D (1998) The structure and mechanism of the family of retinal proteins from halophilic archaea. Curr Opin Struct Biol 8:849–500

Olkhova E, Hutter MC, Lill MA, Helms V, Michel H (2004) Dynamic water networks in cytochrome C oxidase from Paracoccus denitrificans investigated by molecular dynamics simulations. Biophys J 86:1873–1889

Park SY, Borbat PP, Gonzalez-Bonet G, Bhatnagar J, Pollard AM, Freed JH, Bilwes AM, Crane BR (2006) Reconstruction of the chemotaxis receptor-kinase assembly. Nat Struct Mol Biol 13:400–407

Pebay-Peyroula E, Rummel G, Rosenbusch JP, Landau EM (1997) X-ray structure of bacteriorhodopsin at 2.5 Ångstroms from microcrystals grown in lipidic cubic phases. Science 277:1676–1681

Prado MM, Prado A, Fernandez R, Avalos J (2004) A gene of the opsin family in the carotenoid gene cluster of Fusarium fujikuroi. Curr Gen 46:47–58

Radzwill N, Gerwert K, Steinhoff H-J (2001) Time-resolved detection of transient movement of helices F and G in doubly spin-labeled bacteriorhodopsin. Biophys J 80:2856–2866

Rothschild KJ, Zagaeski M, Cantore WA (1981) Conformational Changes of Bacteriorhodopsin Detected by Fourier Transform Infrared Difference Spectroscopy. Biochem Biophys Res Commun 103:483–489

Royant A, Nollert P, Neutze R, Landau EM, Pebay-Peyroula E, Navarro J (2001a) X-ray structure of sensory rhodopsin II at 2.1 Å resolution. Proc Natl Acad Sci USA 98:10131–10136

Royant A, Edman K, Ursby T, Pebay-Peyroula E, Landau EM, Neutze R (2001b) Spectroscopic Characterization of Bacteriorhodopsins L-intermediate in 3D Crystals Cooled to 170 K. Photochem Photobiol 74:794–804

Rudolph J, Oesterhelt D (1995) Chemotaxis and phototaxis require a CheA histidine kinase in the archaeon Halobacterium salinarium. EMBO J 14:667–673

Rudolph J, Oesterhelt D (1996) Deletion Analysis of the *che* Operon in the Archaeon *Halobacterium salinarium*. J Mol Biol 258:548–554

Rudolph J, Nordmann B, Storch KF, Gruenberg H, Rodewald K, Oesterhelt D (1996) A family of halobacterial transducer proteins. FEMS Microbiol Lett 139:161–168

Ruiz-Gonzalez MX, Marin I (2004) New Insights into the Evolutionary History of Type 1 Rhodopsins. J Mol Evol 58:348–358

Sabehi G, Beja O, Suzuki MT, Preston CM, DeLong EF (2004) Different SAR86 subgroups harbour divergent proteorhodopsins. Environ Microbiol 6:903–910

Sabehi G, Massana R, Bielawski JP, Rosenberg M, DeLong EF, Beja O (2003) Novel Proteorhodopsin variants from the Mediterranean and Red Seas. Environ Microbiol 5:842–849

Sakmar TP (1998) Rhodopsin: A prototypical G protein-coupled receptor. Prog Nucl Acid Res Mol Biol 59:1–34

Saranak J, Foster KW (1997) Rhodopsin guides fungal phototaxis. Nature 387:465–466

Sasaki J, Brown LS, Chon Y-S, Kandori H, Maeda A, Needleman R, Lanyi JK (1995) Conversion of bacteriorhodopsin into a chloride ion pump. Science 269:73–75

Sass HJ, Schachowa IW, Rapp G, Koch MHJ, Oesterhelt D, Dencher NA, Büldt G (1997) The tertiary structural changes in bacteriorhodopsin occur between M states – X-ray diffraction and Fourier transform infrared spectroscopy. EMBO J 16:1484–1491

Sass HJ, Büldt G, Gessenich R, Hehn D, Neff D, Schlesinger R, Berendzen J, Ormos P (2000) Structural alterations for proton translocation in the M state of wild-type bacteriorhodopsin. Nature 406:649–653

Sato M, Kanamori T, Kamo N, Demura M, Nitta K (2002) Stopped-flow analysis on anion binding to blue-form halorhodopsin from Natronobacterium pharaonis: comparison with the anion-uptake process during the photocycle. Biochemistry 41:2452–2458

Sato Y, Hata M, Neya S, Hoshino T (2005) Computational analysis of the transient movement of helices in sensory rhodopsin II. Protein Sci 14:183–192

Scharf B, Engelhard M (1994) Blue halorhodopsin from *Natronobacterium pharaonis*: Wavelength regulation by anions. Biochemistry 33:6387–6393

Scherrer P, Mathew MK, Sperling W, Stoeckenius W (1989) Retinal Isomer Ratio in Dark-Adapted Purple Membrane and Bacteriorhodopsin Monomers. Biochemistry 28:829–834

Schilde C (1968) Rapid photoelectric effect in the algae Acetabularia. Z Naturforsch B 23:1369–1376

Schmies G, Lüttenberg B, Chizhov I, Engelhard M, Becker A, Bamberg E (2000) Sensory rhodopsin II from the haloalkaliphilic *Natronobacterium pharaonis*: Light-activated proton transfer reactions. Biophys J 78:967–976

Schobert B, Lanyi JK (1982) Halorhodopsin is a light-driven chloride pump. J Biol Chem 257:10306–10313

Schobert B, Brown LS, Lanyi JK (2003) Crystallographic Structures of the M and N Intermediates of Bacteriorhodopsin: Assembly of a Hydrogen-bonded Chain of Water Molecules Between Asp-96 and the Retinal Schiff Base. J Mol Biol 330:553–570

Schroll C, Riemensperger T, Bucher D, Ehmer J, Voller T, Erbguth K, Gerber B, Hendel T, Nagel G, Buchner E, Fiala A (2006) Light-Induced Activation of Distinct Modulatory Neurons Triggers Appetitive or Aversive Learning in *Drosophila* Larvae. Curr Biol 16:1741–1747

Schweiger U, Tittor J, Oesterhelt D (1994) Bacteriorhodopsin can function without a covalent linkage between retinal and protein. Biochemistry 33:535–541

Seidel R, Scharf B, Gautel M, Kleine K, Oesterhelt D, Engelhard M (1995) The primary structure of sensory rhodopsin II: A member of an additional retinal protein subgroup is coexpressed with its transducer, the halobacterial transducer of rhodopsin II. Proc Natl Acad Sci USA 92:3036–3040

Sharma AK, Spudich JL, Doolittle WF (2006) Microbial rhodopsins: functional versatility and genetic mobility. Trends Microbiol 14:463–469

Siebert F (1990) Resonance raman and infrared difference spectroscopy of retinal pigments. Meth Enzymol 189:123–137

Siebert F, Mäntele W, Kreutz W (1982) Evidence for the Protonation of 2 Internal Carboxylic Groups During the Photocycle of Bacteriorhodopsin – Investigation by Kinetic Infrared-Spectroscopy. FEBS Lett 141:82–87

Sineshchekov OA, Govorunova EG (2001) Rhodopsin receptors of phototaxis in green flagellate algaee. Biochemistry (Moscow) 66:1300–1310

Sineshchekov OA, Govorunova EG, Jung K-H, Zauner S, Maier UG, Spudich JL (2005a) Rhodopsin-Mediated Photoreception in Cryptophyte Flagellates. Biophys J 89:4310–4319

Sineshchekov OA, Spudich JL (2004) Light-induced intramolecular charge movements in microbial rhodopsins in intact *E. coli* cells. Photochem Photobiol Sci 3:548–554

Sineshchekov OA, Trivedi VD, Sasaki J, Spudich JL (2005b) Photochromicity of Anabaena Sensory Rhodopsin, an Atypical Microbial Receptor with a cis-Retinal Light-adapted Form. J Biol Chem 280:14663–14668

Sperling W, Schimz A (1980) Photosensory retinal pigments in Halobacterium halobium. Biophys Struct Mech 6:165–169

Spudich EN, Spudich JL (1982) Control of transmembrane ion fluxes to select halorhodopsin-deficient and other energy-transduction mutants of Halobacterium halobium. Proc Natl Acad Sci USA 79:4308–4312

Spudich EN, Sundberg SA, Manor D, Spudich JL (1986) Properties of a second sensory receptor protein in Halobacterium halobium phototaxis. Proteins 1:239–246

Spudich JL, Bogomolni RA (1984) Mechanism of colour discrimination by a bacterial sensory rhodopsin. Nature 312:509–513

Spudich JL, Luecke H (2002) Sensory rhodopsin II: functional insights from structure. Curr Opin Struc Biol 12:540–546

Spudich JL, Yang CS, Jung KH, Spudich EN (2000) Retinylidene proteins: Structures and functions from Archaea to Humans. Annu Rev Cell Dev Biol 16:365–392

Spudich JL (2006) The multitalented microbial sensory rhodopsins. Trends Microbiol 14:480–487

Spudich JL, Stoeckenius W (1979) Photosensory and Chemosensory Behaviour of *Halobacterium Halobium*. Photobiochem Photobiophys 1:43–53

Steinhoff HJ, Savitsky A, Wegener C, Pfeiffer M, Plato M, Möbius K (2000) High-field EPR studies of the structure and conformational changes of site-directed spin labeled bacteriorhodopsin. Biochim Biophys Acta 1457:253–262

Stock AM, Robinson VL, Goudreau PN (2000) Two-Component Signal Transduction. Annu Rev Biochem 69:183–215

Stoeckenius W, Lozier RH (1974) Light energy conversion in *Halobacterium halobium*. J Supramolec Struct 2:769–774

Subramaniam S, Gerstein M, Oesterhelt D, Henderson R (1993) Electron diffraction analysis of structural changes in the photocycle of bacteriorhodopsin. EMBO J 12:1–8

Subramaniam S, Henderson R (2000) Molecular mechanism of vectorial proton translocation by bacteriorhodopsin. Nature 406:653–657

Subramaniam S, Hirai T, Henderson R (2002) From structure to mechanism: electron crystallographic studies of bacteriorhodopsin. Philos Trans Royal Soc A 360:859–874

Subramaniam S, Lindahl I, Bullough P, Faruqi AR, Tittor J, Oesterhelt D, Brown L, Lanyi J, Henderson R (1999) Protein conformational changes in the bacteriorhodopsin photocycle. J Mol Biol 287:145–161

Sudo Y, Furutani Y, Spudich JL, Kandori H (2007) Early Photocycle Structural Changes in a Bacteriorhodopsin Mutant Engineered to Transmit Photosensory Signals. J Biol Chem 282:15550–15558

Sudo Y, Iwamoto M, Shimono K, Kamo N (2002) Tyr-199 and Charged Residues of pharaonis Phoborhodopsin Are Important for the Interaction with its Transducer. Biophys J 83:427–432

Sudo Y, Spudich JL (2006) Three strategically placed hydrogen-bonding residues convert a proton pump into a sensory receptor. PNAS 103:16129–16134

Takahashi T, Tomioka H, Kamo N, Kobatake Y (1985) A photosystem other than PS370 also mediates the negative phototaxis of *Halobacterium halobium*. FEMS Microbiol Lett 28:161–164

Takeda K, Matsui Y, Kamiya N, Adachi S, Okumura H, Kouyama T (2004) Crystal structure of the M intermediate of bacteriorhodopsin: Allosteric structural changes mediated by sliding movement of a transmembrane helix. J Mol Biol 341:1023–1037

Tittor J, Haupts U, Haupts C, Oesterhelt D, Becker A, Bamberg E (1997) Chloride and proton transport in bacteriorhodopsin mutant D85T – different modes of ion translocation in a retinal protein. J Mol Biol 271:405–416

Trivedi VD, Spudich JL (2003) Photostimulation of a sensory rhodopsin II/HtrII/Tsr fusion chimera activates CheA-autophosphorylation and CheY-phosphotransfer in vitro. Biochemistry 42:13887–13892

Tsuda M, Hazemoto N, Kondo M, Kamo N, Kobatake Y, Terayama Y (1982) Two photocycles in halobacterium halobium that lacks bacteriorhodopsin. Biochem Biophys Res Commun 108:970–976

Tsunoda SP, Ewers D, Gazzarrini S, Moroni A, Gradmann D, Hegemann P (2006) H^+-Pumping Rhodopsin from the Marine Algae Acetabularia. Biophys J 91:1471–1479

Váró G, Lanyi JK (1991a) Kinetic and spectroscopic evidence for an irreversible step between deprotonation and reprotonation of the Schiff base in the bacteriorhodopsin photocycle. Biochemistry 30:5008–5015

Váró G, Lanyi JK (1991b) Thermodynamics and energy coupling in the bacteriorhodopsin photocycle. Biochemistry 30:5016–5022

Váró G, Lanyi JK (1990) Pathways of the rise and decay of the M photointermediate(s) of bacteriorhodopsin. Biochemistry 29:2241–2250

Váró G, Zimányi L, Fan X, Sun L, Needleman R, Lanyi JK (1995) Photocycle of Halorhodopsin from *Halobacterium salinarium*. Biophys J 68:2062–2072

Venter JC, et al. (2004) Environmental genome shotgun sequencing of the Sargasso Sea. Science 304:66–74

Vogeley L, Luecke H (2006) Crystallization, X-ray diffraction analysis and SIRAS/molecular-replacenent phasing of three crystal forms of Anabaena sensory rhodopsin transducer. Acta Crystallogr F 62:388–391

Vogeley L, Sineshchekov OA, Trivedi VD, Sasaki J, Spudich JL, Luecke H (2004) Anabaena Sensory Rhodopsin: A Photochromic Color Sensor at 2.0 Å. Science 306:1390–1393

Vogeley L, Trivedi VD, Sineshchekov OA, Spudich EN, Spudich JL, Luecke H (2007) Crystal Structure of the Anabaena Sensory Rhodopsin Transducer. J Mol Biol 367:741–751

Vonck J (1996) A three-dimensional difference map of the N intermediate in the bacteriorhodopsin photocycle: Part of the F helix tilts in the M to N transition. Biochemistry 35:5870–5878

Wang H, Peca J, Matsuzaki M, Matsuzaki K, Noguchi J, Qiu L, Wang D, Zhang F, Boyden E, Deisseroth K, Kasai H, Hall WC, Feng G, Augustine GJ (2007) High-speed mapping of synaptic connectivity using photostimulation in Channelrhodopsin-2 transgenic mice. Proc Natl Acad Sci USA 104:8143–8148

Wang WW, Sineshchekov OA, Spudich EN, Spudich JL (2003) Spectroscopic and Photochemical Characterization of a Deep Ocean Proteorhodopsin. J Biol Chem 278:33985–33991

Waschuk SA, Bezerra AG Jr, Shi L, Brown LS (2005) Leptosphaeria rhodopsin: Bacteriorhodopsin-like proton pump from a eukaryote. Proc Natl Acad Sci USA 102:6879–6883

Wegener AA, Chizhov I, Engelhard M, Steinhoff HJ (2000) Time-resolved detection of transient movement of helix F in spin- labelled pharaonis sensory rhodopsin II. J Mol Biol 301:881–891

Wegener AA, Klare JP, Engelhard M, Steinhoff HJ (2001) Structural insights into the early steps of receptor-transducer signal transfer in archaeal phototaxis. EMBO J 20:5312–5319

Wolff EK, Bogomolni RA, Scherrer P, Hess B, Stoeckenius W (1986) Color discrimination in halobacteria: Spectroscopic characterization of a second sensory receptor covering the blue-green region of the spectrum. Proc Natl Acad Sci USA 83:7272–7276

Xie AH, Nagle JF, Lozier RH (1987) Flash spectroscopy of purple membrane. Biophys J 51:627–635

Yan B, Takahashi T, Johnson R, Spudich JL (1991) Identification of signaling states of a sensory receptor by modulation of lifetimes of stimulus-induced conformations: The case of sensory rhodopsin II. Biochemistry 30:10686–10692

Yang C-S, Spudich JL (2001) Light-Induced Structural Changes Occur in the Transmembrane Helices of the *Natronobacterium pharaonis* HtrII Transducer. Biochemistry 40:14207–14214

Yao VJ, Spudich JL (1992) Primary structure of an archaebacterial transducer, a methyl-accepting protein associated with sensory rhodopsin I. Proc Natl Acad Sci USA 89:11915–11919

Yoshida H, Sudo Y, Shimono K, Iwamoto M, Kamo N (2004) Transient movement of helix F revealed by photo-induced inactivation by reaction of a bulky SH-reagent to cysteine-introduced pharaonis phoborhodopsin (sensory rhodopsin II). Photochem Photobiol Sci 3:537–542

Zhai Y, Heijne WHM, Smith DW, Saier MH Jr (2001) Homologues of archaeal rhodopsins in plants, animals and fungi: structural and functional predications for a putative fungal chaperone protein. BBA-Biomembranes 1511:206–223

Zhang F, Wang LP, Brauner M, Liewald JF, Kay K, Watzke N, Wood PG, Bamberg E, Nagel G, Gottschalk A, Deisseroth K (2007) Multimodal fast optical interrogation of neural circuitry. Nature 446:633–639

Zhang W, Broourn A, Mueller MM, Alam M (1996) The primary structures of the Archaeon *Halobacterium Salinarium* blue light receptor sensory rhodopsin II and its transducer, a methyl-accepting protein. Proc Natl Acad Sci USA 93:8230–8235

Zimányi L, Lanyi JK (1997) Fourier transform Raman study of retinal isomeric composition and equilibration in halorhodopsin. J Phys Chem B 101:1930–1933

Life Close to the Thermodynamic Limit: How Methanogenic Archaea Conserve Energy

Uwe Deppenmeier[1] · Volker Müller[2] (✉)

[1]Department of Biological Sciences, University of Wisconsin-Milwaukee, 3209 N. Maryland Ave, Milwaukee, WI 53211, USA

[2]Molecular Microbiology & Bioenergetics, Institute of Molecular Biosciences, Goethe University Frankfurt am Main, 60438 Frankfurt/Main, Germany
vmueller@bio.uni-frankfurt.de

Abstract Methane-forming archaea are strictly anaerobic, ancient microbes that are widespread in nature. These organisms are commonly found in anaerobic environments such as rumen, anaerobic sediments of rivers and lakes, hyperthermal deep sea vents and even hypersaline environments. From an evolutionary standpoint they are close to the origin of life. Common to all methanogens is the biological production of methane by a unique pathway currently only found in archaea. Methanogens can grow on only a limited number of substrates such as $H_2 + CO_2$, formate, methanol and other methyl group-containing substrates and some on acetate. The free energy change associated with methanogenesis from these compounds allows for the synthesis of 1 (acetate) to a maximum of only 2 mol of ATP under standard conditions while under environmental conditions less than one ATP can be synthesized. Therefore, methanogens live close to the thermodynamic limit. To cope with this problem, they have evolved elaborate mechanisms of energy conservation using both protons and sodium ions as the coupling ion in one pathway. These energy conserving mechanisms are comprised of unique enzymes, cofactors and electron carriers present only in methanogens. This review will summarize the current knowledge of energy conservation of methanogens and focus on recent insights into structure and function of ion translocating enzymes found in these organisms.

1
Introduction

The methanogenic microbes belong to the domain *Archaea* and produce methane as the major end product of their metabolism. Methanogens are widespread in anoxic environments such as fresh water sediments, swamps, tundra areas, rice fields, intestinal tracts of ruminants and termites, and anaerobic digesters of sewage treatment plants (Garcia et al. 2000). Biological methanogenesis is an important process for the maintenance of the carbon cycle on Earth because methanogenic archaea catalyze the terminal step in the breakdown of organic material in many anaerobic environments. The substrates mentioned above are formed in the course of complex degradation processes of organic matter as performed by fermentative and syntrophic bacteria (Schink 1997). The products of methanogenesis, CH_4 and CO_2 are released from anaerobic habitats and can reenter the global carbon cycle. Hence, large amounts of these greenhouse gases reach the atmosphere and therefore the process of biological methane formation is also of great interest for the global ecology (Khalil and Rasmusen 1994; Reay 2003). Besides the ecological importance, the process of methanogenesis creates a combustible gas that can be used as an energy source for both domestic or industrial applications. Suitable technologies have yet to be developed for a controlled decomposition of renewable biomass to methane. This process is called biomethanation and has great potential as an important nonpolluting energy source in the face of dwindling supplies of fossil fuels (Kashyap et al. 2003). New aspects of methanogenesis concern microbial methane consumption in anoxic sediments which significantly impacts the global environment by reducing the flux of greenhouse gases (Hallam et al. 2004, Meyerdierks et al. 2005). Current models suggest that relatives of methane-producing archaea developed the capacity to reverse methanogenesis and thereby consume methane to produce cellular carbon and energy (Thauer and Shima 2006). Furthermore, it is thought that biological methane production is involved in the formation of methane hydrates (Brooks et al. 2000). These ice-like structures are found in ocean-floor sediments at water depths greater than about 500 meters. Gas hydrates are solids, composed of rigid cages of water molecules that trap CH_4 molecules (Kvenvolden 1999) and represent a potentially enormous natural gas resource (Wood et al. 2002).

Some methanogens grow under conditions that are, from a human perspective, extreme with respect to temperature or salinity. These conditions resemble those of the early days on Earth and therefore, methanogenic archaea are often considered to be very early life forms. The conversion of substrates like $H_2 + CO_2$ or acetate to methane is close to the thermodynamic limit and, as we will see later, methanogens have developed very elaborate mechanisms to conserve energy. Some of these mechanisms are only found in methanogenic archaea. Therefore, ever since methanogenic archaea have been isolated their unique life style and bioenergetics have attracted much attention.

Members of the orders *Methanobacteriales*, *Methanococcales*, *Methanomicrobiales* and *Methanopyrales* only use $H_2 + CO_2$ as substrates. Most of these species can also oxidize formate to form methane (Boone et al. 1993). The *Methanosarcinales* [e.g. *Methansarcina* (*Ms.*) species] are metabolically the most versatile methanogens, capable of growth on $H_2 + CO_2$, acetate, methanol and other methylated C_1 compounds such as methylamines (mono-, di-, or trimethylamine) and methylated thiols (dimethylsulfide, methanethiol or methylmercaptopropionate) (Deppenmeier 2002a). The genome of three species, *Ms. acetivorans*, *Ms. mazei* and *Ms. barkeri* have been published (Galagan et al. 2002; Deppenmeier et al. 2002). They are the largest yet sequenced genomes among species belonging to the domain *Archaea*, reflecting their wide range of metabolic capabilities which distinguish these methane-producing microbes from all other archaea. The genomic data reveal, for the first time, new insights into the metabolism and the cellular functions of this interesting group of microorganisms. The metabolic pathways of methane formation are unique and involve a number of unusual enzymes and coenzymes (Wolfe 1986; Thauer 1998; Deppenmeier et al. 1999). Here we describe the biochemistry of methane formation with special interest given to unique membrane-bound enzymes generating or using primary ion gradients.

2
The Process of Methanogenesis

Several unusual cofactors are involved in biological methane formation (Deppenmeier 2002a). Coenzyme B (HS-CoB, 7-mercaptoheptanoylthreonine phosphate) and coenzyme F_{420} (a 5-deazaflavin derivative with a mid point potential of -360 mV) function as electron carriers in the process of methanogenesis (Fig. 1). F_{420} is the central electron carrier in the cytoplasm of methanogens, which replaces nicotinamide adenine dinucleotides in many reactions (Walsh 1986). HS-CoB is the electron donor in the last reaction of methanogenesis. There is a third electron carrier involved in methane formation which has been discovered only recently (Abken et al. 1998,

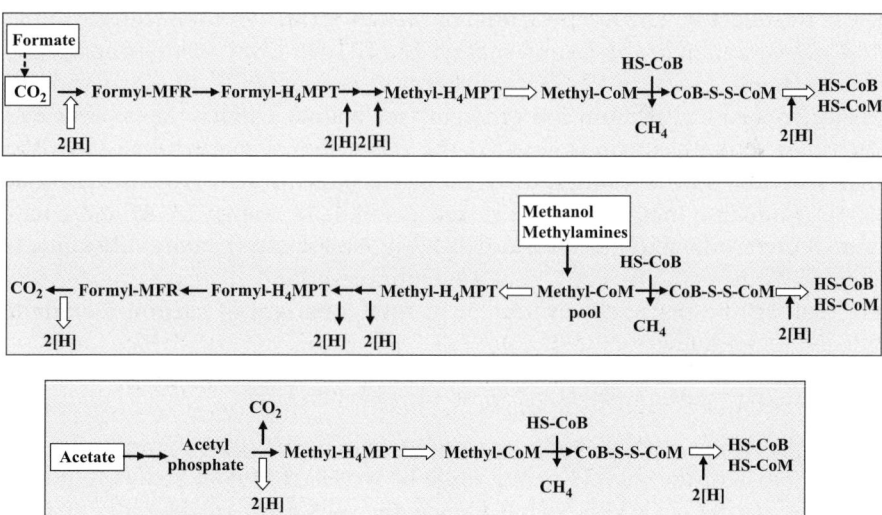

Fig. 1 The process of methanogenesis from formate or $H_2 + CO_2$ (top), methyl group-containing substrates (*middle*) or acetate (*bottom*). MFR, methanofuran; H_4MPT, tetrahydromethanopterin; CoM, coenzyme M, CoB, coenzyme B. *White arrows* denote reactions catalyzed by membrane-bound, ion-translocating enzymes

Beifuss and Tietze 2005). It is referred to as methanophenazine and represents a 2-hydroxy-phenazine derivative that is connected via an ether bridge to a pentaprenyl side chain (Beifuss et al. 2000). This hydrophobic cofactor has been isolated from the cytoplasmic membrane of *Methanosarcina* species (Deppenmeier 2004) and functions as a membrane integral electron carrier similar to quinones found in bacteria and eukarya. Coenzyme M (2-mercaptoethanesulfonate, HS-CoM), methanofuran (MFR), and tetrahydromethanopterin (H_4MPT) carry C_1-moieties of intermediates in the methanogenic pathways. In *Methanosarcina* species this cofactor is slightly modified and referred to as tetrahydrosarcinapterin (H_4SPT) (van Beelen et al. 1984). MFR is involved in CO_2 reduction and HS-CoM accepts methyl groups from methyl-H_4MPT, methanol and methylamines. The methylated form, methyl-S-CoM (2-methylthioethanesulfonate), is the central intermediate in methane formation (Fig. 1).

Methanogenesis from $H_2 + CO_2$, formate, methylated C_1-compounds and acetate proceeds by a central, and in most parts reversible, pathway (Deppenmeier 2002a). When cells grow on CO_2 in the presence of molecular hydrogen (Shima et al. 2002), carbon dioxide is bound to MFR and is then reduced to formyl-MFR (Fig. 1). This endergonic reaction is driven by the electrochemical ion gradient across the cytoplasmic membrane, which will be discussed in the following sections. In the next step the formyl group is transferred to H_4MPT and the resulting formyl-H_4MPT is stepwise reduced to methyl-H_4MPT. Reducing equivalents are derived from reduced F_{420} ($F_{420}H_2$), which is produced by the F_{420}-reducing hydrogenase using hydrogen as a reductant. The methyl group of methyl-H_4MPT is then transferred to HS-CoM by a membrane-bound methyl-H_4MPT:HS-CoM-methyltransferase. This exergonic reaction ($\Delta G_0' = -29$ kJ/mol) is coupled to the formation of an electrochemical sodium ion gradient (see below). Coenzyme B (HS-CoB) functions as the electron donor for the reduction of methyl-S-CoM in the final reaction that is catalyzed by methyl-coenzyme M reductase (Ermler 2005), producing methane and a mixed disulfide of coenzyme M and coenzyme B (heterodisulfide, CoM-S-S-CoB) (Fig. 1). Finally, the heterodisulfide is reduced by the catalytic activity of a membrane-bound electron transfer system that will be discussed in Sect. 3. In total, one mol of carbon dioxide is reduced to methane:

$$CO_2 + 4H_2 = CH_4 + 2H_2O \quad \Delta G_0' = -131 \text{ kJ/mol } CH_4 . \tag{1}$$

During growth on methylated C_1-compounds in the absence of molecular hydrogen, parts of the methyl-group must be oxidized to gain reducing power for the reduction of CoM-S-S-CoB. Hence, the substrates are degraded to CH_4 and CO_2. With methanol as a substrate one out of four methyl groups is oxidized to CO_2 and three methyl moieties are reduced to methane (Fig. 1):

$$4CH_3OH \rightarrow 3CH_4 + 1CO_2 + 2H_2O \quad \Delta G_0' = -106 \text{ kJ/mol } CH_4 . \tag{2}$$

Many *Methanosarcina* strains can grow on trimethylamine, dimethylamine and monomethylamine, the substrates being metabolized to CH_4, CO_2 and NH_3:

$$4CH_3-NH_3^+ + 2H_2O \rightarrow 3CH_4 + CO_2 + 4NH_4^+ \quad (3)$$
$$\Delta G_0' = -77 \text{ kJ/molCH}_4.$$

The methyl groups are channeled into the central pathway by substrate-specific soluble methyltransferases that catalyze the methyl group transfer to coenzyme M (Paul et al. 2000). The nonhomologous genes encoding the full-length methyltransferases for methylamine utilization each possess an in-frame UAG (amber) codon that does not terminate translation (Krzycki 2004). The amber codon is decoded by a dedicated tRNA, and corresponds to the novel amino acid pyrrolysine, indicating pyrrolysine as the 22nd genetically encoded amino acid (Krzycki 2005). The methyl group transfer to H_4MPT is catalyzed by the membrane-bound methyl-H_4MPT:HS-CoM-methyltransferase (see Sect. 4.5; Gottschalk and Thauer 2001). The next steps involve the stepwise oxidation of methyl-H_4MPT to formyl-H_4MPT. Reducing equivalents derived from the oxidation reactions are used for F_{420} reduction. After transfer of the formyl group to MFR the formyl-MFR dehydrogenase catalyzes the oxidation to CO_2 and MFR (Fig. 1). Hence, the oxidative branch in the methylotrophic pathway of methanogenesis is the reversal of CO_2 reduction to methyl-CoM as found in the CO_2-reducing pathway of methane formation. In the reductive branch of this pathway three out of four methyl groups are transferred to HS-CoM. Again, the HS-CoB-dependent reduction of methyl-S-CoM leads to the formation of CH_4 and CoM-S-S-CoB (Fig. 1).

Most of the methane in anaerobic food chains is derived from the methyl group of acetate. This process is referred to as the aceticlastic pathway of methanogenesis (Fig. 1) and is carried out only by the genera *Methanosarcina* and *Methanosaeta* (Ferry 1997). In *Methanosarcina* species, acetate is activated by conversion to acetyl-coenzyme A (CoA), which is cleaved by the nickel-containing CO dehydrogenase/acetyl-CoA synthase, yielding enzyme-bound methyl and carbonyl groups (Drennan et al. 2004). Following cleavage, the methyl group of acetate is transferred to H_4MPT. In the course of the reaction enzyme-bound CO is oxidized to CO_2 and the electrons are used for ferredoxin reduction. The resulting methyl-H_4MPT is converted to methane by the catalytic activities of the Na^+-translocating methyl-H_4MPT:HS-CoM methyltransferase and methyl-S-CoM reductase as described above (Fig. 1). As in the case of the other methanogenic pathways, an additional product of the reaction catalyzed by methyl-CoM reductase is the heterodisulfide CoM-S-S-CoB, which is reduced by a membrane-bound electron transport system to regenerate the reactive sulfhydryl forms of the coenzymes.

As mentioned above, all methanogens use methanogenesis for energy conservation. Recently however, evidence was presented that *Ms. acetivorans* is able to employ the acetyl-CoA pathway to conserve energy for growth (Rother and Metcalf 2004). This organism was isolated from a bed of decaying kelp, algae known to accumulate up to 10% CO in their float cells (Sowers et al. 1984; Abbott and Hollenberg 1976). In contrast to many other methanogens, this methanogen is devoid of significant hydrogen metabolism due to the lack of functional hydrogenases. With carbon monoxide the cells grew to high densities with a doubling time of 24 h. Surprisingly, acetate and formate, rather than methane, were the major metabolic end products of metabolism. Moreover, methane production decreased with increasing CO partial pressures, consistent with inhibition of methanogenesis by CO. In spite of these facts, methanogenesis was still required for growth on CO, because the potent methyl-CoM reductase inhibitor BES abolished growth on CO (Rother and Metcalf 2004). Not much is known about the biochemistry of CO as a growth substrate for *Ms. acetivorans*. However, mutations in the operon encoding phosphotransacetylase and acetate kinase failed to use CO as a growth substrate, indicating that these enzymes are required for acetate formation and ATP synthesis via substrate level phosphorylation.

3
Reactions and Compounds of the Methanogenic Electron Transport Chains

The synthesis of CH_4 and the formation of the heterodisulfide (CoM-S-S-CoB; Fig. 1) mark the end of all pathways leading to methane formation (Deppenmeier 2002b). In *Methanosarcina* species CoM-S-S-CoB is the electron acceptor of a branched respiratory chain and is reduced to the thiol-containing cofactors HS-CoM and HS-CoB by an enzyme referred to as heterodisulfide reductase (Fig. 2). The protein is membrane-bound and functions as a terminal respiratory reductase (Hedderich et al. 2005). The source of reducing equivalents necessary for the reduction of the heterodisulfide depends on the growth substrate. In the presence of hydrogen a membrane-bound hydrogenase (F_{420}-nonreducing hydrogenase) channels electrons to the heterodisulfide reductase. This electron transport system is referred to as H_2:heterodisulfide oxidoreductase system (Fig. 2a) (Ide et al. 1999). When *Methanosarcina* strains grow on methylated C_1 compounds part of the methyl groups are oxidized to CO_2 and electrons are transferred to coenzyme F_{420}. Under this growth condition the membrane-bound $F_{420}H_2$:heterodisulfide oxidoreductase system catalyzes the oxidation of $F_{420}H_2$ and the reduction of the heterodisulfide (Fig. 2b) (Deppenmeier et al. 1991). The key enzyme of this electron transport chain is the $F_{420}H_2$ dehydrogenase which is responsible for the oxidation of the reduced cofactor (Bäumer et al. 2000). In the aceticlastic pathway the heterodisulfide reductase reduces CoM-S-S-CoB which derives electrons from reduced ferredoxin by a third membrane-bound electron transport chain referred

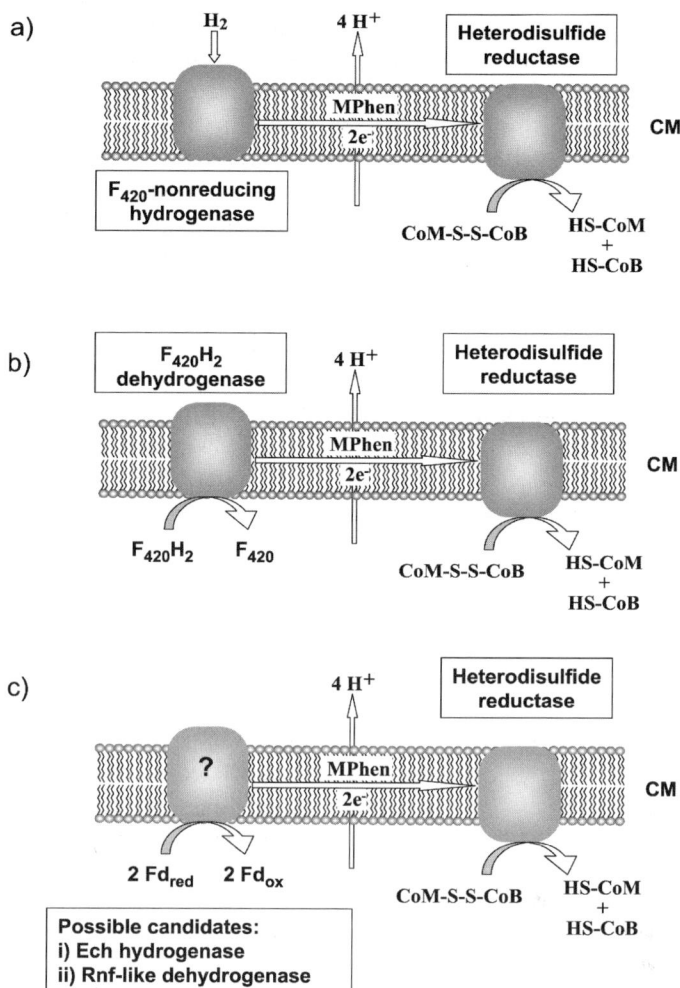

Fig. 2 Membrane-bound electron transport systems operating during methanogenesis from $H_2 + CO_2$ (**a**) methyl groups (**b**) or acetate (**c**). Mphen, methanophenazine; Fd, ferredoxin. For further explanations, see text

to as reduced ferredoxin:heterodisulfide oxidoreductase system (Fig. 2c) (Deppenmeier 2002a).

Using inverted vesicle preparations from *Ms. mazei*, it has been shown that electron transport is catalyzed by the H_2:heterodisulfide oxidoreductase system and the $F_{420}H_2$:heterodisulfide oxidoreductase system is coupled to proton translocation across the cytoplasmic membrane (Deppenmeier 2004). An A_1A_O ATP synthase (Sect. 5) catalyzes ATP synthesis from ADP + P_i thereby taking advantage of the electrochemical proton gradient (Müller et al. 1999). Additional experiments using uncouplers and ATP synthase inhibitors clearly demonstrated that ATP is formed by electron transport-driven phosphorylation of ADP (Deppenmeier et al. 1990, 1991).

After the elucidation of the enzymes involved in energy conservation the question arose how electrons are channeled from one protein complex to the other. In the cytoplasmic membrane of methanogens only tocopherolquinones in very low concentration were identified (Hughes and Tove 1982), which had obviously no function in the electron transport chain. A comprehensive search for other redox-active, lipid-soluble components indicated the presence of one major factor in the membranes of Ms. mazei. Detailed NMR analysis revealed the aromatic structure as a phenazine derivative connected at C-2 to an unsaturated side chain via an ether bridge (Abken et al. 1998; Beifuss et al. 2000). The lipophilic side chain is responsible for the anchorage in the membrane and consists of five isoprene units linked to each other in a head-to-tail manner. The redox active product was referred to as methanophenazine and was the first phenazine isolated from archaea. After completion of the total synthesis of methanophenazine (Beifuss et al. 2000), electron transport in Ms. mazei could be analyzed in more detail. It was shown that methanophenazine is reduced by molecular hydrogen or $F_{420}H_2$, catalyzed by the F_{420}-nonreducing hydrogenase and the $F_{420}H_2$ dehydrogenase, respectively (Fig. 2a,b). Furthermore, the membrane-bound heterodisulfide reductase was able to use dihydro-methanophenazine as an electron donor for the reduction of CoM-S-S-CoB (Bäumer et al. 2000; Ide et al. 1999). Therefore, the cofactor is able to mediate the electron transport between the membrane-bound enzymes. Hence, methanophenazine was characterized as the first phenazine derivative involved in the electron transport of biological systems. More features of the electron transport chains could be analyzed using the water soluble analogue 2-hydroxyphenazine (2-OH-phenazine). Washed inverted vesicles of this organism were found to couple electron transfer processes with the transfer of four protons across the cytoplasmic membrane. It was shown that 2-OH-phenazine is reduced by molecular hydrogen as catalyzed by the F_{420}-nonreducing hydrogenase. Furthermore, the membrane-bound heterodisulfide reductase was able to use dihydro-2-OH-phenazine as the electron donor for the reduction of CoM-S-S-CoB (Bäumer et al. 2000; Ide et al. 1999) according to:

$$H_2 + \text{2-OH-phenazine} + 2H^+_{in} \rightarrow \text{dihydro-2-OH-phenazine} + 2H^+_{out} \quad (4)$$

$$\text{dihydro-2-OH-phenazine} + \text{CoM-S-S-CoB} + 2H^+_{in} \rightarrow$$
$$\text{2-OH-phenazine} + \text{HS-CoM} + \text{HSCoB} + 2H^+_{out} . \quad (5)$$

There are two proton-translocating segments. The first one involves the F_{420}-nonreducing hydrogenase (Eq. 4) and the second one the heterodisulfide reductase (Eq. 5). Thus, the $H^+/2e^-$ stoichiometry of the electron transport chain adds up to four and supports the value of four $H^+/2e^-$ translocated in the overall electron transport from H_2 to the heterodisulfide (Deppenmeier et al. 1991). 2-OH-phenazine was also a mediator of electron transfer within the $F_{420}H_2$:heterodisulfide oxidoreductase system (Bäumer et al. 1998). It has been shown that reducing equivalents are transferred from $F_{420}H_2$ to 2-OH-phenazine by the membrane-bound $F_{420}H_2$ dehydrogenase. Also, this process is coupled to proton translocation across the cytoplasmic membrane exhibiting a stoichiometry of about two protons translocated per two electrons transferred (Bäumer et al. 2000). The second reaction of this electron transport system is again catalyzed by the heterodisulfide reductase that uses dihydro-2-OH-phenazine as the electron donor for CoM-S-S-CoB reduction. Just as in the H_2-dependent system, both partial reactions of the $F_{420}H_2$:heterodisulfide oxidoreductase system are coupled to the translocation of two protons (Deppenmeier 2004):

$$F_{420}H_2 + \text{2-OH-phenazine} + 2H^+_{in} \rightarrow$$
$$\text{dihydro-2-OH-phenazine} + F_{420} + 2H^+_{out} \quad (6)$$
$$\text{dihydro-2-OH-phenazine} + \text{CoM-S-S-CoB} + 2H^+_{in} \rightarrow$$
$$\text{2-OH-phenazine} + \text{HS-CoM} + \text{HSCoB} + 2H^+_{out}. \quad (7)$$

As mentioned above, the oxidation of H_2 and $F_{420}H_2$ as well as the reduction of CoM-S-S-CoB are catalyzed by membrane-bound electron transport chains that couple the redox reaction with the translocation of protons. The mid-point potentials of the electron carriers were determined to -420 mV for $H_2/2H^+$, -360 mV for $F_{420}H_2/F_{420}$ (Walsh 1984), -165 mV for methanophenazine/dihydromethanophenazine and -143 mV for CoM-S-S-CoB/HS-CoM + HS-CoB (Tietze et al. 2003). Thus, the change of free energy ($\Delta G^{o\prime}$) coupled to the H_2- and $F_{420}H_2$-dependent methanophenazine reduction (Eqs. 4 + 6) is -49.2 kJ/mol and -42.1 kJ/mol, respectively. Taking into account that the membrane potential Δp is about -180 mV (Peinemann 1989), the translocation of two protons per reaction cycle is feasible. In contrast, the change of free energy for the dihydromethanophenazine-dependent reduction of the heterodisulfide (Eq. 7) under standard conditions is only -4.2 kJ/mol. However, the reductive demethylation of methyl-CoM as catalyzed by the methyl-CoM reductase, is highly exergonic (-32 kJ/mol). Hence, it is tempting to speculate that the formation of CH_4 is the driving force for proton translocation at the second coupling site of the membrane-bound electron transport systems.

There is another enzyme system that has to be discussed in respect to methanogenesis from $H_2 + CO_2$ or methanol. A novel hydrogenase (Ech) was discovered in acetate-grown cells of *Ms. barkeri* (Meuer et al. 1999) that is homologous to hydrogenase 3 and 4 from *E. coli* and to the CO-induced hydrogenase from *Rhodospirillum rubrum* (Sauter et al. 1992; Fox et al. 1996). The enzyme catalyzed the H_2-dependent reduction of a two [4Fe – 4S] ferredoxin and is also able to perform the reverse reaction, namely hydrogen formation from reduced ferredoxin (Fig. 2c). The Ech hydrogenase could be involved in the formation and degradation of formyl-MFR (Stojanowic and Hedderich 2004). In the CO_2-reducing pathway H_2 is oxidized by the membrane-bound Ech hydrogenase and electrons are transferred to a ferredoxin, which in turn is used by the formyl-MFR dehydrogenase to catalyze the reduction of CO_2. Taking into account the low hydrogen pressures found in the natural environments of methanogens, the H_2-dependent reduction of the ferredoxin is an endergonic process (Hedderich and Forzi 2005). It has been proposed that the electrochemical proton gradient is the energy source for the Ech hydrogenase to drive this reduction (Hedderich and Forzi 2005; Hedderich 2004). Thus, H_2-dependent ferredoxin reduction is enabled by the influx of protons or sodium ions through the enzyme. During growth on methanol the formyl-MFR dehydrogenase and the Ech hydrogenase may function in the reverse direction. In this case, the oxidation of formyl-MFR is coupled to the formation of reduced ferredoxin. The Ech hydrogenase could then oxidize reduced ferredoxin and release H_2. The overall process might be coupled to the translocation of protons or sodium ions and could contribute to the formation of the electrochemical ion gradient (Meuer et al. 2002). Molecular hydrogen is not an end product but is used as an electron donor for the H_2:heterodisulfide oxidoreductase system (Deppenmeier 2004).

The smallest change of free energy $\Delta G_0' = -36$ kJ/molCH_4 of all substrates is coupled to methane formation from acetate. Therefore, the organisms must possess efficient energy-conserving systems to cope with this thermodynamic limitation. In *Methanosarcina* strains one ATP is used in acetate activation to form acetyl-CoA. As mentioned above the reactions of the aceticlastic pathway of methanogenesis results in the formation of the intermediates methyl-H_4MPT and reduced ferredoxin (Fd_{red}). The methyl moiety is transferred to H_4MPT and then to CoM by the Na^+-motive methyltetrahydromethanopterin:coenzyme M methyltransferase (Sect. 4.5). Reduced ferredoxin is used as the electron donor for the reduction of CoM-S-S-CoB catalyzed by the Fd_{red}:heterodisulfide oxidoreductase (Fig. 2c). It is very possible that this membrane-bound electron transport system is also able to generate an electrochemical proton gradient (Deppenmeier 2004):

$$2Fd_{red} + \text{CoM-S-S-CoB} \rightarrow$$
$$2Fd_{ox} + \text{HS-CoM} + \text{HS-CoB} \quad (\Delta G_0' \sim -50 \text{ kJ/mol}).$$
(8)

The composition of this third electron transport system is still a matter of debate and two scenarios are discussed:
a) It has been proposed that the Ech hydrogenase and the heterodisulfide oxidoreductase are involved in this process (Hedderich et al. 1999). Molecular hydrogen would be an intrinsic intermediate formed by the Ech hydrogenase at the expense of reduced ferredoxin. H_2 would be reoxi-

dized by the F_{420}-nonreducing hydrogenase and the electrons channeled via methanophenazine to the heterodisulfide reductase. Studies using *Ms. barkeri* mutants lacking the Ech hydrogenase confirmed these results (Meuer et al. 2002; Stojanowic and Hedderich 2004).

b) The genome of *Ms. acetivorans* does not contain genes encoding a functional Ech hydrogenase (Galagan et al. 2002), suggesting alternative electron transport components involved in the transfer of electrons to CoM-S-S-CoB. Compared to methanol-grown cells, acetate-grown *Ms. acetivorans* synthesized greater amounts of subunits of the potential ion-translocating Rnf electron transport complex previously characterized from bacteria (Li et al. 2006). Combined with sequence and physiological analyses, these results suggest that *Ms. acetivorans* replaces the H_2-evolving Ech hydrogenase complex with the Rnf complex. The subunits of the Rnf complex from *R. capsulatus* (Rc-Rnf) have been previously characterized (Jouanneau et al. 1998; Kumagai et al. 1997). In *Ms. acetivorans* the Rnf complex is presumably composed of a membrane-bound subcomplex containing a cytochrome *c* subunit that functions as an ion channel for the translocation of either protons or sodium and catalyzes the electron transfer to methanophenazine (Li et al. 2006). A membrane associated subcomplex containing FeS clusters is thought to interact with reduced ferredoxin and an electron-transfer module probably mediates electron transfer between the subcomplexes mentioned before. Subunits of the Rnf complex are similar to subunits of the Na^+-translocating NADH-quinone-reductase (Nqr) from *Vibrio* species (Steuber et al. 2002) indicating that Rnf-type enzymes catalyze ion transport coupled to electron transport. However, it is important to note that ion transport by Rnf-type enzymes has not yet been demonstrated. It is also interesting to note that the genomes of the freshwater organisms *Ms. mazei* and *Ms. barkeri* do not contain genes homologous to the Rnf complex. This may indicate that there may be differences in the pathway for acetate conversion to methane especially in the ferredoxin-dependent electron-transport chains in *Methanosarcina* species.

4
Structure and Function of Ion-Translocating Enzymes

Comprehensive reviews have been published recently that describe the features of the proton translocating proteins in great detail ($F_{420}H_2$ dehydrogenase, Deppenmeier 2004; NiFe hydrogenases, Vignais and Colbeau 2004; Ech hydrogenase, Hedderich and Forzi 2006; heterodisulfide reductase, Hedderich et al. 2005). Therefore, we discuss these enzymes only briefly and will focus on the methyltetrahydromethanopterin:coenzyme M methyltransferase and the A_1A_O ATP synthase from methanogens.

4.1
$F_{420}H_2$ Dehydrogenase

Genes encoding the $F_{420}H_2$ dehydrogenases are found in all genomes of methylotrophic methanogens sequenced so far, and the protein was purified from several archaea (Abken and Deppenmeier 1997; Haase et al. 1992; Kunow et al. 1994). In methanogens the enzyme catalyzes the $F_{420}H_2$-dependent reduction of phenazine derivatives, thereby transferring two protons across the cytoplasmic membrane (Fig. 3a) (Bäumer et al. 2000). Thus, the enzyme represents a novel kind of a proton-translocating complex in methanogenic archaea (Deppenmeier et al. 2002). The protein from Ms. mazei is encoded by the *fpo* cluster that comprises 12 genes which were designated *fpo* A, B, C, D, H, I, J, K, L, M, N, O (Bäumer et al. 2000). The gene *fpoF* is not part of the *fpo* operon and is located elsewhere on the chromosome. The enzyme subunits are highly homologous to proton translocating NADH dehydrogenases of respiratory chains from bacteria (NDH-1) and eukarya (complex I). In this context it is important to note that the $F_{420}H_2$ dehydrogenase and bacterial NADH dehydrogenases have some in-

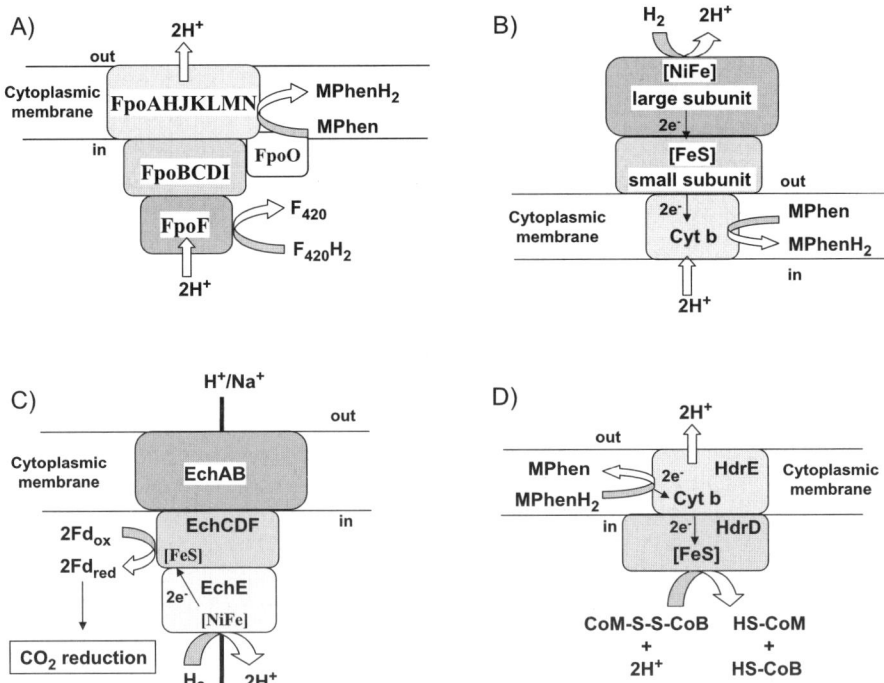

Fig. 3 Ion-translocating redox enzymes in methanogens. **A** F_{420} dehydrogenase. **B** F_{420} non-reducing hydrogenase. **C** Ech hydrogenase. **D** Heterodisulfide reductase

teresting features in common (Friedrich and Scheide 2000). Both enzymes have a complex subunit composition and contain flavin and iron-sulfur centers. They are the initial enzymes of membrane-bound electron transport systems. The electron donors $F_{420}H_2$ and NADH are similar in that both cofactors are reversible hydride donors with comparable mid-point potentials. The enzymes use small hydrophobic electron acceptors, namely quinones in the case of NADH dehydrogenases, and methanophenazine in the case of $F_{420}H_2$ dehydrogenases. Moreover, both enzymes show redox-driven proton-translocating activity (Friedrich et al. 2005). The overall similarity of the $F_{420}H_2$ dehydrogenase to complex I allows for speculation on the reaction mechanism of the archaeal enzyme. The proteins are essentially composed of three subcomplexes (Fig. 3a). The electron donors NADH and $F_{420}H_2$ are oxidized by the input module, which is formed by NuoEFG (complex I from *E. coli*) and FpoF ($F_{420}H_2$ dehydrogenase from *Ms. mazei*), respectively. Electrons are channeled to the membrane integral subcomplex (NuoAHIJKLMN, *E. coli*; FpoAHJKLMN, *Ms. mazei*) that is involved in the reduction of the species-specific electron acceptor (ubiquinone, *E. coli*; methanophenazine, *Ms. mazei*). The membrane-associated module (NuoBCDI, *E. coli*; FpoBCDI, *Ms. mazei*) connects the above-mentioned subunits and catalyzes electron transfer between the modules.

4.2
F_{420}-nonreducing Hydrogenase

The membrane-bound F_{420}-nonreducing hydrogenases from *Ms. mazei* and *Ms. barkeri* were purified and found to be composed of a small and a large subunit (Deppenmeier et al. 1992; Kemner and Zeikus 1994). The genes, arranged in the order *vhoG* and *vhoA* were identified as those encoding the small and the large subunit of the NiFe hydrogenases (Fig. 3b). The third gene in the *vho* operon (*vhoC*) encodes a membrane-spanning cytochrome *b* (Deppenmeier et al. 1995). The crystal structure of the highly homologous nickel-iron hydrogenase from *Desulfovibrio gigas* revealed that the large subunit harbors the binuclear [NiFe]-active site, which is coordinated by two conserved CxxC motifs (Volbeda et al. 1995). This polypeptide is likely co-translocated with the small subunit to the periplasmic side of the cytoplasmic membrane. The small subunit contains three iron-sulfur clusters, forming an electron transfer line from the [NiFe]-center of the large subunit to the prosthetic heme groups of the *b*-type cytochrome. Hence, the third subunit acts as the primary electron acceptor of the core enzyme (Bernhard et al. 1997).

A proposed model for electron and proton transfer within the F_{420}-nonreducing hydrogenase comprises hydrogen oxidation by the bimetallic Ni/Fe center of the large subunit, thereby separating electrons and protons (Fig. 3b) (Deppenmeier 2004). Protons are released on the outside of the cytoplasmic membrane, and the iron-sulfur clusters in the small subunit accept the electrons. In the next step, electrons are transferred to the heme groups of the cytochrome *b* subunit, which then completes the reaction by accepting two protons from the cytoplasm for the reduction of methanophenazine. Thus, the overall reaction would lead to the production of two scalar protons.

4.3
Ech Hydrogenase

A membrane-bound [NiFe]-hydrogenase was isolated from *Ms. barkeri* (Meuer et al. 1999) that belongs to the class of energy-converting [NiFe] hydrogenases (Fig. 3c). These enzymes are found in several anaerobic or facultatively anaerobic microorganisms (Hedderich and Forzi 2006). From growth characteristics of *Rhodospirillum rubrum* and from cell-suspension experiments with *Ms. barkeri*, it can be concluded that the [NiFe]-hydrogenases in these organisms probably act as a proton or sodium ion pump (Hedderich and Forzi 2006). The purified Ech hydrogenase consists of six subunits encoded by genes organized in the *echABCDEF* operon (Kuenkel et al. 1998; Meuer et al. 1999). The six subunits show a striking amino acid sequence similarity to the catalytic core of complex I (Nuo BCDIHL) and the $F_{420}H_2$ dehydrogenase (FpoBCDIHL) (Hedderich 2004). The evolution of complex I and energy-converting hydrogenases has been discussed in recent reviews (Friedrich and Scheide 2000; Brandt et al. 2003; Hedderich 2004; Vignais and Colbeau 2004).

The obligate hydrogenotrophic methanogens *Methanococcus (Mc.) maripaludis* and *Methanothermobacter (Mt.) marburgensis* contain genes for two separate multisubunit energy-conserving hydrogenases, Eha and Ehb, which are composed of 20 and 17 subunits, respectively (Tersteegen and Hedderich 1999; Hendrickson et al. 2004). Some of the subunits are also homologous to the NADH-ubiquinone oxidoreductase. Like the Ech hydrogenase from *Methanosarcina* species, these enzymes may be necessary to reduce low-potential ferredoxins for the reduction of CO_2 as catalyzed by the formyl-MFR reductase. A deletion in the *ehb* operon of *Mc. maripaludis* showed the mutant strain had severely impaired growth in minimal medium. Both acetate and yeast extract were necessary to restore growth to nearly wild-type levels, suggesting that Ehb is involved in anabolic CO_2 assimilation (Porat et al. 2006).

4.4
Heterodisulfide Reductase

As mentioned above the heterodisulfide reductase has a key function in the energy metabolism of methanogens (Fig. 3d). The enzyme catalyzes the reduction of CoM-S-S-CoB generated in the final step of methanogenesis. Two types of heterodisulfide reductases were characterized from distantly related methanogens (Hedderich et al. 1990; Heiden et al. 1994; Simianu et al. 1998). In *Mt. marburgensis*, and other obligate hydrogenotrophic methanogens the enzyme is a soluble iron-sulfur flavoprotein composed of three subunits HdrA, HdrB and HdrC (Hedderich et al. 1994). HdrA contains a FAD-binding motif and four binding motifs for [4Fe – 4S] clusters. HdrC contains two binding motifs for [4Fe – 4S] clusters. Evidence has been presented that the catalytic center is formed by subunits HdrC and HdrB. The second type of heterodisulfide reductase is found in *Methanosarcina* species (Fig. 3d) (Heiden et al. 1994; Simianu et al. 1998). It is composed of two subunits, a membrane-anchoring *b*-type cytochrome (HdrE) and a hydrophilic iron-sulfur protein (HdrD). The latter subunit is regarded as a hypothetical fusion protein of subunits HdrC and HdrB of the *Mt. marburgensis* enzyme (Kuenkel et al. 1997) and comprises the catalytic center in *Methanosarcina* species. The active site harbors a [4Fe-4S] cluster, which is directly involved in the disulfide cleavage reaction (Hedderich et al. 2005). The two types of heterodisulfide reductases use different physiological electron donors. HdrDE receives reducing equivalents from the reduced methanophenazine pool via its *b*-type cytochrome subunit whereas HdrABC forms a complex with the F_{420}-nonreducing hydrogenase (Mvh) which is located in the cytoplasm after cell lysis (Setzke et al. 1994). The mechanism of energy conservation in obligate hydrogenotrophic methanogens is still unknown.

4.5
Structure and Function of the Methyltetrahydromethanopterin: Coenzyme M Methyltransferase, a Primary Sodium Ion Pump

Growth as well as methanogenesis is strictly sodium ion dependent (Perski et al. 1981, 1982) and the methyltetrahydromethanopterin:coenzyme M methyltransferase was identified to be responsible for the Na^+ dependence of the pathway (Müller et al. 1988b; Gärtner et al. 1993; Weiss et al. 1994; Lienard et al. 1996). The enzyme couples the methyltransfer from methyltetrahydromethanopterin to CoM with vectorial transport of Na^+ across the cytoplasmic membrane, and is the first example of a methyltransferase catalyzing ion transport across a membrane. Because the central pathway is reversible, this enzyme functions as a generator of a sodium ion potential during methanogenesis from CO_2 or acetate (Fig. 1), but the reaction is endergonic and driven by the sodium ion potential in the course of methyl-group oxidation (Müller et al. 1988a). Unlike the cytochromes and the resulting differences in the electron-transport chains, the methyltransferase is found in every methanogen and there is no reason to assume different reaction mechanisms.

The energetics of the methyltransferase was first investigated using cell suspensions of *Ms. barkeri* and the substrate combination H_2 + HCHO. Upon addition of the substrate, sodium ions were actively extruded from the cytoplasm, resulting in the generation of a transmembrane Na^+ gradient of – 60 mV (Müller et al. 1988b). A Na^+/formaldehyde stoichiometry of 3 to 4 was determined using cell suspensions (Kaesler and Schönheit 1989), but proteoliposomes containing the purified methyltransferase from *Ms. mazei* catalyzed an electrogenic Na^+ transport with a stoichiometry of 1.7 mol Na^+ per mol methyl-H_4MPT demethylated (Lienard et al. 1996).

The methyltransferase was purified from *Mt. thermoautotrophicus* and *Ms. mazei* Göl (Gärtner et al. 1993; Lienard et al. 1996). Six subunits were found in the *Ms. mazei* enzyme, with apparent molecular masses of 34, 28, 20, 13, 12 and 9 kDa. It contains a [4Fe–4S] cluster with a E_0' of – 215 mV and a "base on" cobamide with a standard reduction potential of – 426 mV for the $Co^{2+}/^{1+}$ couple (Lu et al. 1995; Lienard et al. 1996). In *Mt. thermoautotrophicus*, eight subunits were found, with apparent molecular masses of 34 (MtrH), 28 (MtrE), 24 (MtrC), 23 (MtrA), 21 (MtrD), 13 (MtrG), 12.5 (MtrB), and 12 kDa (MtrF) (Fig. 4). The purified enzyme contains two mol corrinoid, eight mol nonheme iron, and eight mol acid-labile sulfur (Gärtner et al. 1993; Harms et al. 1995). The encoding genes have been sequenced from a number of methanogens, and they are organized in an operon in the order *mtrED-CBAFGH*. Hydrophobicity plots indicate that all of the subunits except MtrA and MtrH are hydrophobic and potentially membrane-bound. The membrane localization of MtrD was confirmed experimentally for *Ms. mazei* Göl, *Mt. thermoautotrophicus* and *Methanocaldococcus (Mc.) jannaschii* (Ruppert et al. 2001). This subunit may be directly involved in Na^+ transport (Lienard and Gottschalk 1998).

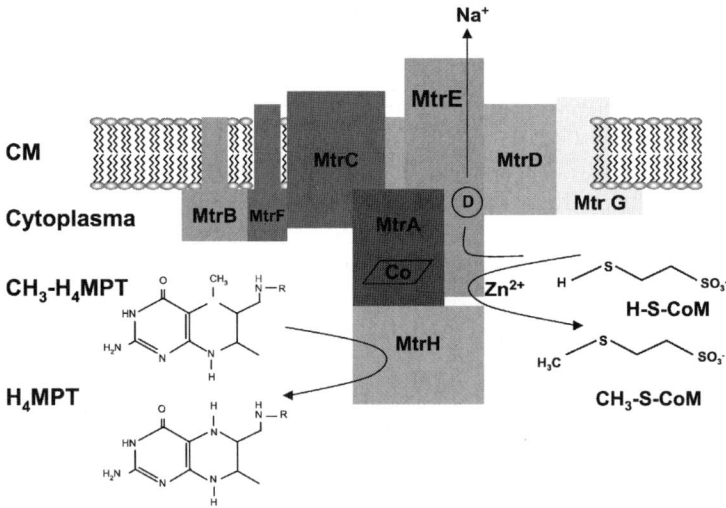

Fig. 4 The Na^+-motive methyltetrahydromethanopterin:coenzyme M methyltransferase of methanogenic archaea. For explanations see text. Adapted from Gottschalk and Thauer 2001

The corrinoid that catalyzes the methyltransfer is Coα-[α-(5-hydroxybenzimidazolyl)]-cobamide (factor III) (Poirot et al. 1987; Gärtner et al. 1993, 1994). The cofactor in its super-reduced Co(I) form accepts the methyl group from methyltetrahydromethanopterin, giving rise to a methyl-Co(III) intermediate. In the second partial reaction, this methyl-Co(III) is subjected to a nucleophilic attack, probably by the thiolate anion of CoM, to give rise to methyl-CoM and regenerated Co(I) (Fischer et al. 1992; Gärtner et al. 1994).

$$CH_3 - H_4MPT + E : Co(I) \rightarrow H_4MPT + E : CH_3 - Co(III) \tag{9}$$

$$E : CH_3 - Co(III) + HS - CoM \rightarrow CH_3\text{-S-CoM} + E : Co(I) . \tag{10}$$

Reaction 9 has a free-energy change of -15 kJ/mol and is not stimulated by sodium ions. On the other hand, demethylation of the enzyme-bound corrinoid (Eq. 10) is also accompanied by a free-energy change of -15 kJ/mol, and this reaction was sodium-ion-dependent, with half-maximal activity obtained at approximately 50 μM Na^+. This finding indicates that the demethylation of the enzyme-bound corrinoid is coupled to sodium-ion translocation (Weiss et al. 1994) (Fig. 4).

MtrA was overexpressed, purified from *E. coli* and successfully reconstituted with cobalamin. EPR spectroscopic studies indicate that the cobalamin is in the "base off" form and that the axial ligand is a histidine residue of MtrA (Harms and Thauer 1996). From this observation, a hypothetical mechanism was formulated for coupling the methyltransfer reaction to ion transport via a long-range conformational change in the protein (Harms and Thauer 1996). It is known that cob(II)alamin and cob(III)alamin, but not cob(I)alamin, carry an axial ligand. Methylation of cob(I)alamin gives rise to a methylcob(III)alamin, which is then able to ligate the histidine residue; demethylation leads to a reversal of this reaction. It is easily conceivable that binding and dissociation of the histidine residue with the corrinoid lead to a conformational change in the hydrophilic part of the enzyme. This change is then transmitted to the membrane-bound subunits, giving rise to Na^+ transport.

5
ATP Synthesis in Methanogens

Methanogens are the only microorganisms known to produce two primary ion gradients, $\Delta\mu_{Na^+}$ and $\Delta\mu_{H^+}$, at the same time (Schäfer et al. 1999; Müller et al. 2005a). They are, therefore, confronted with the problem of coupling both ion gradients to the synthesis of ATP. How this is achieved is still a matter for debate. Inhibitor studies using whole cells suggested the presence of two distinct ATP synthases in *Mt. thermoautotrophicus* and *Ms. mazei*. It was suggested that they have a F_1F_O ATP synthase that couples with Na^+

and a A_1A_O ATP synthase that couples with H^+ (Becher and Müller 1994; Smigan et al. 1994, 1995). However, later it turned out that the genomes of these two species only encode a A_1A_O ATP synthase. The same is true for all other methanogens sequenced so far, except two that contain potential F_1F_O ATP synthase genes in addition to the A_1A_O ATPase. *Ms. barkeri* strain MS contains a gene cluster that potentially encodes proteins with similarity to subunits of a F_1F_O ATP synthase; however, the order of the genes is different from any other F_1F_O ATP synthase gene cluster, the deduced γ subunit is very unusual and presumably nonfunctional, and no gene encoding subunit δ was found. Since a *m*RNA transcript could not be detected in cells grown on methanol it is doubtful that the F_1F_O-like genes are expressed in *Ms. barkeri* (Lemker and Müller, unpublished). A gene cluster potentially encoding proteins with similarity to subunits of a F_1F_O ATP synthase is also present in *Ms. acetivorans* but whether it indeed encodes an active ATPase has not been addressed. Most likely, the F_1F_O ATPase genes present in these two methanogens have arisen from horizontal gene transfer.

The presence of only the A_1A_O ATP synthase poses the question whether the enzyme translocates Na^+, H^+ or both. A definite answer can not be given at the moment due to the lack of a functional proteoliposome system. Although the first entire A_1A_O ATPase has been recently purified (Lingl et al. 2003), so far it could not be reconstituted in a coupled state into liposomes. Also, Na^+ dependence of ATP hydrolysis could not be addressed due to the lack of a suitable buffer system. However, in every methanogen sequenced so far, the membrane embedded rotor subunit, subunit *c*, that catalyzes the ion transport has a motif similar to the Na^+ binding motif identified experimentally in F_1F_O ATP synthase and V_1V_O ATPases (Müller 2004; Meier et al. 2005; Murata et al. 2005). This makes it highly likely that the A_1A_O ATP synthase can use both, Na^+ and H^+. If not, the electrochemical sodium gradient across the cytoplasmic membrane of methanogens such as *Ms. mazei* or *Ms. barkeri* could be converted to a proton potential by action of a Na^+/H^+ antiporter. Interestingly, the genome of *Ms. mazei* encodes three such secondary transporters but their function in cellular bioenergetics has not been studied.

5.1
Structure and Function of the A_1A_O ATP Synthase from Methanogens

The ATP synthases/ATPases arose from a common ancestor and, therefore, A_1A_O ATP synthases share properties with both the eukaryal V_1V_O ATPase and the F_1F_O ATP synthase as present in bacteria, chloroplasts and mitochondria. The overall subunit composition and the primary sequence of the major subunits A and B is more closely related to V_1V_O ATPases than to F_1F_O ATP synthases but their function clearly is to synthesize ATP. Therefore, archaeal ATP synthases are very unique, ancient energy converters and phylogenetic analyses clearly revealed that they form a separate class of ATPases, the A_1A_O ATP synthases/ATPases (Müller et al. 1999).

The A_1A_O ATP synthases from methanogens are hitherto the best investigated specimen of A_1A_O ATP synthases. The A_1A_O ATP synthase has at least nine subunits ($A_3:B_3:C:D:E:F:H:a:c_x$) (Müller and Grüber 2003; Müller et al. 2005a,b). The A_1 complex of archaeal ATP synthases has a pseudohexagonal arrangement of six peripheral globular masses, reflecting the major subunits A and B, as proposed from two dimensional images of the thermoacidophilic archaea *Sulfolobus acidocaldarius* and *Ms. mazei* Göl (Lübben et al. 1988; Wilms et al. 1996). Despite several attempts over the years, so far only one A_1A_O ATP synthase, from *Mc. jannaschii*, could be purified without loss of subunits (Lingl et al. 2003). *Mc. jannaschii* is a hyperthermophile that grows optimally at 85 °C. The ATPase was solubilized by Triton-X-100 and purified by gel filtration and ion exchange chromatography to apparent homogeneity. The first projected structure of an intact A_1A_O ATP synthase was determined by electron microscopy of single particles at a resolution of 1.8 nm (Coskun et al. 2004a). The enzyme has an overall length of 25.9 nm and is organized in an A_1 headpiece (9.4 × 11.5 nm), and a membrane domain A_O (6.4 × 10.6 nm), that are linked by a central stalk about 8 nm in length (Fig. 5). A part of the central stalk is surrounded by a col-

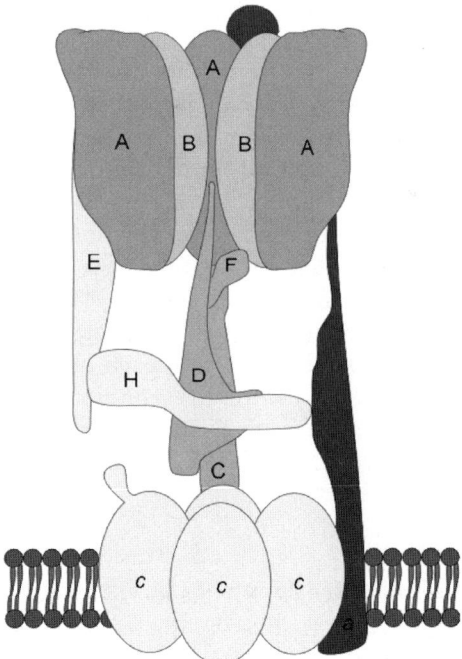

Fig. 5 Structure and subunit topology of A_1A_O ATP synthases. The cartoon shows the enzyme from *Mc. jannaschii* with only five monomers of the rotor subunit

lar. The collar is connected to the top of the A_1 portion via a peripheral stalk, and in addition, there is a second peripheral stalk that connects the A_O with the A_1 domain.

The overall structure of an A_1 subcomplex was obtained by small angle X-ray scattering of an ABCDF subcomplex heterologously produced in *E. coli*. It is asymmetric, with a headpiece that is approximately 94 Å long and 92 Å wide and a stalk with a length of approximately 84 Å and 60 Å in diameter (Grüber et al. 2001; Lemker et al. 2001, 2002) (Fig. 5). Global structural alterations occur in the A_1 ATPase due to nucleotide binding. Subunits C and F are exposed stalk subunits, whereas subunit D is the functional homolog of the γ subunit of F_1F_O ATPases (Grüber et al. 2001; Coskun et al. 2002).

Recently, high-resolution structures of the noncatalytic A and the catalytic B subunits of A_1A_O ATP synthases were obtained. The structure of subunit A was determined at a resolution of 2.55 Å and shown to contain four domains (Maegawa et al. 2006). One represents an insertion of about 90 amino acids that is absent in the homologous β subunit of F_1F_O ATP synthases and corresponds to the "knob-like structure" seen in electron micrographs suggested to be involved in connecting the peripheral stalk to the AB-assembly (Coskun et al. 2004a,b). The noncatalytic subunit B binds ADP and ATP, but with a weaker affinity than subunit A. The overall structure, as determined at 1.5 Å resolution, is similar to that of the related α subunit of F_1F_O ATP synthases; however, like in the V_1V_O ATPases, the P-loop is missing in subunit B of the methanoarchaeal ATP synthase (Schäfer et al. 2006b). The first low-resolution shape of subunit F of the A_1A_O ATP synthase from the archaeon *Ms. mazei* Gö1 in solution was determined by small angle X-ray scattering (Schäfer et al. 2006a). The protein is monomeric and has an elongated shape, divided in a main globular part with a length of about 4.5 nm, and a hook-like domain of about 3.0 nm in length. Subunit D can be cross-linked to the catalytic A subunit depending on nucleotide binding. This interaction between A and D involves the N- and C-termini of subunit D (Coskun et al. 2002), whose secondary structures are predicted to be α helical (Wilms et al. 1996), as described for both termini of subunit γ of F_1. Cross-linking studies provide evidence that subunit B and F interact with each other and the contact surface of B–F could be mapped in the high-resolution structure of subunit B of the A1AO ATP synthase. Furthermore, D–E, A–H, and A–B–D crosslinks were obtained in the intact A_1A_O ATP synthase. Taken together, these data suggest the topology of subunits depicted in Fig. 5.

5.2
The Unique Membrane-Embedded Rotor of A_1A_0 ATP Synthases

The A_O domain contains only two membrane-intrinsic subunits, *a* and *c* (cf. Fig. 5). Subunit *a* is the stator and subunit *c* builds the rotor of this membrane-embedded motor. Rotor subunits have been purified and characterized from some archaea and in almost every case they were shown to be of $M_r \approx 8000$ with two transmembrane helices (Müller 2004). This size corresponds to the size of the c subunit from F_1F_O ATP synthases and was until now assumed to be the reason for the F_1F_O-like properties of the A_1A_O ATP synthases, i.e. their function as ATP synthases. However, *Mt. thermautotrophicus* has a duplicated and *Mc. maripaludis* and *M. jannaschii* have triplicated c subunits. Apparently, these c subunits arose by gene duplication and triplication, respectively, with subsequent fusion of the genes (Ruppert et al. 1999, 2001; Müller et al. 2005b, Lewalter and Müller 2006). In the case of *Mt. thermautotrophicus*, the ion binding site is conserved in helix two and four, but in *Mc. jannaschii* and *Mc. maripaludis* it is only conserved in helix four and six, in helix two it is substituted by a glutamine residue. The genome sequence of *Methanopyrus kandleri* revealed another extraordinary feature: the A_1A_O ATP synthase genes are located in one cluster, and the gene encoding the c subunit is 13-times the size of the gene encoding an 8-kDa c subunit. The sequence predicts a c subunit of 97.5 kDa comprising 13 covalently linked hairpin domains (Slesarev et al. 2002)! These domains have a highly conserved sequence (55.9 to 86.3%), and the ion-binding site is conserved in helix two of every hairpin domain. However, post-transcriptional and post-translational modifications can not be excluded and, therefore, the extraordinary size of the c subunit has to be verified by other means.

It should be mentioned that the extraordinary variation in c subunits in archaea is not restricted to methanogens. The pyrococci *Pyrococcus furiosus*, *Pyrococcus horikoshii* and *Pyrococcus abyssi* are anaerobic archaea that have a fermentative metabolism. Interestingly, their c subunit genes arose by duplication and subsequent fusion of a precursor gene coding for one hairpin (Müller 2004). The duplicated c subunit with two covalently linked hairpins has an ion binding site in hairpin two but not in one. Therefore, the c subunit of the A_1A_O ATP synthases from pyrococci is identical to the 16-kDa c subunit of eukaryal V_1V_O ATPases. This finding gives further evidence that 16-kDa c subunit with only one ion binding site in two hair pins are not an exclusive feature of eukarya.

The rotor stoichiometry has not been solved for any A_1A_O ATP synthase but two important conclusions regarding the structure of the rotor and the function of the enzyme can be drawn. For structural considerations, it is assumed for the sake of simplicity that the rotor contains 12 hairpins. This would accommodate 12 copies of the 8-kDa c subunits from most archaea, six of the one from *Mt. thermautotrophicus* and pyrococci, and four of *M. jannaschii*. These rotors are multimeric, but the number of subunits decreases in this order. A comparison to the optimal and maximal growth temperatures reveals a striking correlation of the number of rotor subunits to the optimal and maximal growth temperatures. The higher the growth temperature of the organisms the fewer the number of subunits per rotor. The extreme is encountered in the presumably monomeric rotor of *M. kandleri* that thrives at 110 °C. It should be kept in mind that the rotor subunits are embedded into the membrane and are shielded from heat protective mechanisms present in the cytoplasm. Therefore, they are directly exposed to the heat and it is easily conceivable that the increase of covalently-linked rotor subunits increases the stability and supports the function of the rotor in the cytoplasmic membrane at high temperatures.

For the function of the enzymes, the number of ion-translocating residues per rotor unit is important. The capability to synthesize ATP is directly dependent on the number of ions translocated per ATP synthesized. According to $\Delta G_P = - n \cdot F \cdot \Delta p$, a phosphorylation potential (ΔG_P) of $\sim$ 50 to 70 kJ/mol is sustained by the use of n = 3–4 ions/ATP at a physiological electrochemical ion potential of – 180 mV (Δp). Assuming a rotor with 12 ion-translocating groups and a catalytic domain with three $\alpha\beta$/AB pairs and thus three ATP binding sites, this gives exactly the number of four ions required thermodynamically for ATP synthesis. This is apparently realized in most archaeal ATPases found to date. However, a special case is the rotor of *Mc. jannaschii* and *Mc. maripaludis* that have only eight ion-binding sites (assuming 12 hairpin domains per rotor). Apparently, 2.6 carboxyl groups per catalytic center are already sufficient for ATP synthesis.

6
Concluding Remarks

Archaea are truly fascinating microbes that live under conditions that are, from a human perspective, extreme. Some of them are "ancient" and developed in the early history of life. During evolution, they kept their ecological niches and their physiological properties. Therefore, we have a window through which we can glance at very early life processes, including mechanisms of energy conservation. Methanogens are hitherto the best investigated archaea with respect to energy metabolism, they have very unique enzymes involved in energy conservation only found in methanogens, they employ Na^+- and H^+-based energetics and their ATP synthases have an outstanding variety of rotor subunits. Unfortunately, there is no high-resolution structure of any of the energy conserving enzymes available, this is still a challenging task for future studies. The structure-function analyses are still in their infancy, but genetic techniques have improved and heterologous expression systems may lead to the quantities required for structural analyses. The emerging pictures of the $F_{420}H_2$ dehydrogenase and the ATP synthase shows that the structure determination of energy-conserving enzymes in methanogens is well on its way and promises interesting new structures in the future.

Acknowledgements We are indebted to our coworkers for their excellent work, to our academic teacher Gerhard Gottschalk for introducing us to the art of anaerobes, and the Deutsche Forschungsgemeinschaft for continuous financial support.

References

Abbott IA, Hollenberg GJ (1976) Marine algae of california. Stanford Univ Press, Stanford, CA, USA

Abken HJ, Deppenmeier U (1997) Purification and properties of an $F_{420}H_2$ dehydrogenase from *Methanosarcina mazei* Göl. FEMS Microbiol Lett 154:231–237

Abken HJ, Tietze M, Brodersen J, Bäumer S, Beifuss U, Deppenmeier U (1998) Isolation and characterization of methanophenazine and the function of phenazines in membrane-bound electron transport of *Methanosarcina mazei* Göl. J Bacteriol 180:2027–2032

Bäumer S, Murakami E, Brodersen J, Gottschalk G, Ragsdale SW, Deppenmeier U (1998) The $F_{420}H_2$:heterodisulfide oxidoreductase system from *Methanosarcina* species. FEBS Lett 428:295–298

Bäumer S, Ide T, Jacobi C, Johann A, Gottschalk G, Deppenmeier U (2000) The $F_{420}H_2$ dehydrogenase from *Methanosarcina mazei* Göl is a redox-driven proton pump closely related to NADH dehydrogenases. J Biol Chem 275:17968–17973

Becher B, Müller V (1994) $\Delta\mu Na^+$ drives the synthesis of ATP via an $\Delta\mu Na^+$-translocating F_1F_O-ATP synthase in membrane vesicles of the archaeon *Methanosarcina mazei* Göl. J Bacteriol 176:2543–2550

Beifuss U, Tietze M, Bäumer S, Deppenmeier U (2000) Methanophenazine: Structure, total synthesis, and function of a new cofactor from methanogenic archaea. Angew Chem Int Ed 39:2470–2472

Beifuss U, Tietze M (2005) Methanophenazine and other natural biologically active phenazines. Top Curr Chem 244:77–113

Bernhard M, Benelli B, Hochkoeppler A, Zannoni D, Friedrich B (1997) Functional and structural role of the cytochrome *b* subunit of the membrane-bound hydrogenase complex of *Alcaligenes eutrophus* H16. Eur J Biochem 248:179–186

Boone DR, Whitman WB, Rouviere PE (1993) Diversity and taxonomy of methanogens. In: Ferry JG (ed) Methanogenesis. Chapman & Hall, New York, pp 35–80

Brandt U, Kerscher S, Drose S, Zwicker K, Zickermann V (2003) Proton pumping by NADH:ubiquinone oxidoreductase. A redox driven conformational change mechanism? FEBS Lett 545:9–17

Breas O, Guillou C, Reniero F, Wada E (2002) The global methane cycle: Isotopes and mixing ratios, sources and sinks. Isot Environ Healt S 37:257–379

Brooks JM, Bryant WR, Bernard BB, Cameron NR (2000) The nature of gas hydrates on the Nigerian continental slope Gas hydrates. In: Challenges for the future. Ann NY Acad Sci 912:76–93

Coskun Ü, Grüber G, Koch MH, Godovac-Zimmermann J, Lemker T, Müller V (2002) Crosstalk in the A_1-ATPase from *Methanosarcina mazei* Göl due to nucleotide-binding. J Biol Chem 277:17327–17333

Coskun Ü, Chaban YL, Lingl A, Müller V, Keegstra W, Boekema EJ, Grüber G (2004a) Structure and subunit arrangement of the A-type ATP synthase complex from the archaeon *Methanococcus jannaschii* visualized by electron microscopy. J Biol Chem 279:38644–38648

Coskun Ü, Radermacher M, Müller V, Ruiz T, Grüber G (2004b) Three-dimensional organization of the archaeal A_1-ATPase from *Methanosarcina mazei* Göl. J Biol Chem 279:22759–22764

Deppenmeier U, Blaut M, Mahlmann A, Gottschalk G (1990) Reduced coenzyme $F_{420}H_2$-dependent heterodisulfide oxidoreductase: a proton-translocating redox system in methanogenic bacteria. Proc Natl Acad Sci USA 87:9449–9453

Deppenmeier U, Blaut M, Gottschalk G (1991) H_2 : heterodisulfide oxidoreductase, a second energy-conserving system in the methanogenic strain Göl. Arch Microbiol 155:272–277

Deppenmeier U, Blaut M, Schmidt B, Gottschalk G (1992) Purification and properties of a F_{420}-nonreactive membrane-bound hydrogenase from *Methanosarcina* strain Göl. Arch Microbiol 157:505–511

Deppenmeier U, Blaut M, Lentes S, Herzberg C, Gottschalk G (1995) Analysis of the *vhoGAC* and *vhtGAC* operons from *Methanosarcina mazei* strain Göl, both encoding a membrane-bound hydrogenase and a cytochrome *b*. Eur J Biochem 227:261–269

Deppenmeier U, Lienard T, Gottschalk G (1999) Novel reactions involved in energy conservation by methanogenic archaea. FEBS Lett 457:291–297

Deppenmeier U, Johann A, Hartsch T, Merkl R, Schmitz RA, Martinez-Arias R, Henne A, Wiezer A, Bäumer S, Jacobi C, Bruggemann H, Lienard T, Christmann A, Bomeke M, Steckel S, Bhattacharyya A, Lykidis A, Overbeek R, Klenk HP, Gunsalus RP, Fritz HJ, Gottschalk G (2002) The genome of *Methanosarcina mazei*: Evidence for lateral gene transfer between bacteria and archaea. J Mol Microbiol Biotechnol 4:453–461

Deppenmeier U (2002a) The unique biochemistry of methanogenesis. Prog Nucl Acid Res Mol Biol 71:223–283

Deppenmeier U (2002b) Redox-driven proton translocation in methanogenic Archaea. Cell Mol Life Sci 59:1–21

Deppenmeier U (2004) The membrane-bound electron transport system of *Methanosarcina* species. J Bioenerg Biomembr 36:55–64

Drennan CL, Doukov TI, Ragsdale SW (2004) The metalloclusters of carbon monoxide dehydrogenase/acetyl-CoA synthase: a story in pictures. J Biol Inorg Chem 9:511–515

Ermler U (2005) On the mechanism of methyl-coenzyme M reductase. Dalton Trans 21:3451–3458

Ferry JG (1997) Enzymology of the fermentation of acetate to methane by *Methanosarcina thermophila*. Biofactors 6:25–35

Fischer R, Gärtner P, Yeliseev A, Thauer RK (1992) N^5-methyltetrahydromethanopterin:coenzyme M methyltransferase in methanogenic archaebacteria is a membrane protein. Arch Microbiol 158:208–217

Fox JD, Kerby RL, Roberts GP, Ludden PW (1996) Characterization of the CO-induced, CO-tolerant hydrogenase from *Rhodospirillum rubrum* and the gene encoding the large subunit of the enzyme. J Bacteriol 178:1515–1524

Friedrich T, Scheide D (2000) The respiratory complex I of bacteria, archaea and eukarya and its module common with membrane-bound multisubunit hydrogenases. FEBS Lett 479:1–5

Friedrich T, Stolpe S, Schneider D, Barquera B, Hellwig P (2005) Ion translocation by the *Escherichia coli* NADH:ubiquinone oxidoreductase (complex I). Biochem Soc Trans 33:836–839

Galagan JE, Nusbaum C, Roy A, Endrizzi MG, Macdonald P, FitzHugh W, Calvo S, Engels R, Smirnov S, Atnoor D, Brown A, Allen N, Naylor J, Stange-Thomann N, DeArellano K, Johnson R, Linton L, McEwan P, McKernan K, Talamas J, Tirrell A, Ye WJ, Zimmer A, Barber RD, Cann I, Graham DE, Grahame DA, Guss AM, Hedderich R, Ingram-

Smith C, Kuettner HC, Krzycki JA, Leigh JA, Li WX, Liu JF, Mukhopadhyay B, Reeve JN, Smith K, Springer TA, Umayam LA, White O, White RH, de Macario EC, Ferry JG, Jarrell KF, Jing H, Macario AJL, Paulsen I, Pritchett M, Sowers KR, Swanson RV, Zinder SH, Lander E, Metcalf WW, Birren B (2002) The genome of *Methanosarcina acetivorans* reveals extensive metabolic and physiological diversity. Genome Res 12:532–542

Garcia JL, Patel BKC, Ollivier B (2000) Taxonomy, phylogenetic, and ecological diversity of methanogenic Archaea. Anaerobe 6:205–226

Gärtner P, Ecker A, Fischer R, Linder D, Fuchs G, Thauer RK (1993) Purification and properties of N^5-methyltetrahydromethanopterin:coenzyme M methyltransferase from *Methanobacterium thermoautotrophicum*. Eur J Biochem 213:537–545

Gärtner P, Weiss DS, Harms U, Thauer RK (1994) N^5-methyltetrahydromethanopterin:coenzyme M methyltransferase from *Methanobacterium thermoautotrophicum*—catalytic mechanism and sodium ion dependence. Eur J Biochem 226:465–472

Gottschalk G, Thauer RK (2001) The Na^+-translocating methyltransferase complex from methanogenic archaea. Biochim Biophys Acta 1505:28–36

Grüber G, Svergun DI, Coskun Ü, Lemker T, Koch MH, Schägger H, Müller V (2001) Structural insights into the A_1 ATPase from the archaeon, *Methanosarcina mazei* Göl. Biochemistry 40:1890–1896

Haase P, Deppenmeier U, Blaut M, Gottschalk G (1992) Purification and characterization of $F_{420}H_2$-dehydrogenase from *Methanolobus tindarius*. Eur J Biochem 203:527–531

Hallam SJ, Putnam N, Preston CM, Detter JC, Rokhsar D, Richardson PM, DeLong EF (2004) Reverse methanogenesis: testing the hypothesis with environmental genomics. Science 305:1457–1462

Harms U, Weiss DS, Gärtner P, Linder D, Thauer RK (1995) The energy conserving N^5-methyltetrahydromethanopterin:coenzyme M methyltransferase complex from *Methanobacterium thermoautotrophicum* is composed of eight different subunits. Eur J Biochem 228:640–648

Harms U, Thauer RK (1996) Methylcobalamin:coenzyme M methyltransferase isoenzymes MtaA and MtbA from *Methanosarcina barkeri*—Cloning, sequencing and differential transcription of the encoding genes, and functional overexpression of the *mtaA* gene in *Escherichia coli*. Eur J Biochem 235:653–659

Hedderich R, Berkessel A, Thauer RK (1990) Purification and properties of heterodisulfide reductase from *Methanobacterium thermoautotrophicum* (strain Marburg). Eur J Biochem 193:255–261

Hedderich R, Koch J, Linder D, Thauer RK (1994) The heterodisulfide reductase from *Methanobacterium thermoautotrophicum* contains sequence motifs characteristic of pyridine nucleotide-dependent thioredoxin reductases. Eur J Biochem 225:253–261

Hedderich R, Klimmek O, Kröger A, Dirmeier R, Keller M, Stetter KO (1998) Anaerobic respiration with elemental sulfur and with disulfides. FEMS Microbiol Rev 22:353–381

Hedderich R (2004) Energy-converting [NiFe] hydrogenases from archaea and extremophiles: ancestors of complex I. J Bioenerg Biomembr 36:65–75

Hedderich R, Hamann N, Bennati M (2005) Heterodisulfide reductase from methanogenic archaea: a new catalytic role for an iron-sulfur cluster. Biol Chem 386:961–970

Hedderich R, Forzi L (2005) Energy-converting [NiFe] hydrogenases: more than just H_2 activation. J Mol Microbiol Biotechnol 10:92–104

Heiden S, Hedderich R, Setzke E, Thauer RK (1994) Purification of a two-subunit cytochrome *b*-containing heterodisulfide reductase from methanol-grown *Methanosarcina barkeri*. Eur J Biochem 221:855–861

Hendrickson EL, Kaul R, Zhou Y, Bovee D, Chapman P, Chung J, de Macario EC, Dodsworth JA, Gillett W, Graham DE, Hackett M, Haydock AK, Kang A, Land ML, Levy R, Lie TJ, Major TA, Moore BC, Porat I, Palmeiri A, Rouse G, Saenphimmachak C, Soll D, Van Dien S, Wang T, Whitman WB, Xia Q, Zhang Y, Larimer FW, Olson MV, Leigh JA (2004) Complete genome sequence of the genetically tractable hydrogenotrophic methanogen *Methanococcus maripaludis*. J Bacteriol 186:6956–6969

Hughes PE, Tove SB (1982) Occurrence of alpha-tocopherolquinone and alpha-tocopherolquinol in microorganisms. J Bacteriol 151:1397–1402

Ide T, Bäumer S, Deppenmeier U (1999) Energy conservation by the H_2:heterodisulfide oxidoreductase from *Methanosarcina mazei* Göl: identification of two proton-translocating segments. J Bacteriol 181:4076–4080

Jouanneau Y, Jeong HS, Hugo N, Meyer C, Willison JC (1998) Overexpression in *Escherichia coli* of the *rnf* genes from *Rhodobacter capsulatus* – characterization of two membrane-bound iron-sulfur proteins. Eur J Biochem 251:54–64

Kaesler B, Schönheit P (1989) The role of sodium ions in methanogenesis. Formaldehyde oxidation to CO_2 and 2 H_2 in methanogenic bacteria is coupled with primary electrogenic Na^+ translocation at a stoichiometry of 2–3 Na^+/CO_2. Eur J Biochem 184:223–232

Kashyap DR, Dadhich KS, Sharma SK (2003) Biomethanation under psychrophilic conditions: a review. Bioresour Technol 87:147–153

Kemner JM, Zeikus JG (1994) Purification and characterization of membrane-bound hydrogenase from *Methanosarcina barkeri* MS. Arch Microbiol 161:47–54

Khalil MAK, Rasmussen RA (1994) Global emission of methane during the last several centuries. Chemosphere 29:833–842

Krzycki JA (2004) Function of genetically encoded pyrrolysine in corrinoid-dependent methylamine methyltransferases. Curr Opin Chem Biol 8:484–491

Krzycki JA (2005) The direct genetic encoding of pyrrolysine. Curr Opin Microbiol 8:706–712

Kumagai H, Fujiwara T, Matsubara H, Saeki K (1997) Membrane localization, topology, and mutual stabilization of the *rnf*ABC gene products in *Rhodobacter capsulatus* and implications for a new family of energy coupling NADH oxidoreductases. Biochemistry 36:5509–5521

Künkel A, Vaupel M, Heim S, Thauer RK, Hedderich R (1997) Heterodisulfide reductase from methanol-grown cells of *Methanosarcina barkeri* is not a flavoenzyme. Eur J Biochem 244:226–234

Kunow K, Linder D, Stetter KO, Thauer RK (1994) $F_{420}H_2$:quinone oxidoreductase from *Archaeoglobus fulgidus*. Eur J Biochem 223:503–511

Kvenvolden KA (1999) Potential effects of gas hydrate on human welfare. Proc Natl Acad Sci USA 96:3420–3426

Lemker T, Ruppert C, Stöger H, Wimmers S, Müller V (2001) Overproduction of a functional A_1 ATPase from the archaeon *Methanosarcina mazei* Göl in *Escherichia coli*. Eur J Biochem 268:3744–3750

Lemker T, Schmid R, Grüber G, Müller V (2003) Subcomplexes of the heterologously produced archaeal A_1 ATPase from *Methanosarcina mazei* Göl: subunit composition and redox modulation. FEBS Lett 544:206–209

Lewalter K, Müller V (2006) Bioenergetics of archaea: ancient energy conserving mechanisms developed in the early history of life. Biochim Biophys Acta 1757:437–445

Li O, Li L, Rejtar T, Lessner DJ, Karger BL, Ferry JG (2006) Electron transport in the pathway of acetate conversion to methane in the marine archaeon *Methanosarcina acetivorans*. J Bacteriol 188:702–710

Lienard T, Gottschalk G (1998) Cloning and expression of the genes encoding the sodium translocating N^5-methyltetrahydromethanopterin:coenzyme M methyltransferase of the methylotrophic archaeon *Methanosarcina mazei* Göl. FEBS Lett 425:204–208

Lienard T, Becher B, Marschall M, Bowien S, Gottschalk G (1996) Sodium ion translocation by N^5-methyltetrahydromethanopterin:coenzyme M methyltransferase from *Methanosarcina mazei* Göl reconstituted in ether lipid liposomes. Eur J Biochem 239:857–864

Lingl A, Huber H, Stetter KO, Mayer F, Kellermann J, Müller V (2003) Isolation of a complete A_1A_O ATP synthase comprising nine subunits from the hyperthermophile *Methanococcus jannaschii*. Extremophiles 7:249–257

Lu WP, Becher B, Gottschalk G, Ragsdale SW (1995) Electron paramagnetic resonance spectroscopic and electrochemical characterization of the partially purified N^5-methyltetrahydromethanopterin:coenzyme M methyltransferase from *Methanosarcina mazei* Göl. J Bacteriol 177:2245–2250

Lübben M, Lünsdorf H, Schäfer G (1988) Archaebacterial ATPase: studies on subunit composition and quaternary structure of the F_1-analogous ATPase from *Sulfolobus acidocaldarius*. Biol Chem Hoppe Seyler 369:1259–1266

Maegawa Y, Morita H, Iyaguchi D, Yao M, Watanabe N, Tanaka I (2006) Structure of the catalytic nucleotide-binding subunit A of A-type ATP synthase from *Pyrococcus horikoshii* reveals a novel domain related to the peripheral stalk. Acta Crystallogr D Biol Crystallogr 62:483–488

Meier T, Polzer P, Diederichs K, Welte W, Dimroth P (2005) Structure of the rotor ring of F-Type Na^+-ATPase from *Ilyobacter tartaricus*. Science 308:659–662

Meuer J, Bartoschek S, Koch J, Künkel A, Hedderich R (1999) Purification and catalytic properties of Ech hydrogenase from *Methanosarcina barkeri*. Eur J Biochem 265:325–335

Meuer J, Kuettner HC, Zhang JK, Hedderich R, Metcalf WW (2002) Genetic analysis of the archaeon *Methanosarcina barkeri* Fusaro reveals a central role for Ech hydrogenase and ferredoxin in methanogenesis and carbon fixation. Proc Natl Acad Sci USA 99:5632–5637

Meyerdierks A, Kube M, Lombardot T, Knittel K, Bauer M, Glockner FO, Reinhardt R, Amann R (2005) Insights into the genomes of archaea mediating the anaerobic oxidation of methane. Environ Microbiol 7:1937–1951

Müller V, Blaut M, Gottschalk G (1988a) The transmembrane electrochemical gradient of Na^+ as driving force for methanol oxidation in *Methanosarcina barkeri*. Eur J Biochem 172:601–606

Müller V, Winner C, Gottschalk G (1988b) Electron transport-driven sodium extrusion during methanogenesis from formaldehyde + H_2 by *Methanosarcina barkeri*. Eur J Biochem 178:519–525

Müller V, Grüber G (2003) ATP synthases: structure, function and evolution of unique energy converters. Cell Mol Life Sci 60:474–494

Müller V (2004) An exceptional variability in the motor of archaeal A_1A_O ATPases: from multimeric to monomeric rotors comprising 6–13 ion binding sites. J Bioenerg Biomembr 36:115–125

Müller V, Lemker T, Lingl A, Weidner C, Coskun U, Grüber G (2005a) Bioenergetics of archaea: ATP synthesis under harsh environmental conditions. J Mol Microbiol Biotechnol 10:167–180

Müller V, Lingl A, Lewalter K, Fritz M (2005b) ATP synthases with novel rotor subunits: new insights into structure, function and evolution of ATPases. J Bioenerg Biomembr 37:455–460

Murata T, Yamato I, Kakinuma Y, Leslie AG, Walker JE (2005) Structure of the rotor of the V-Type Na$^+$-ATPase from *Enterococcus hirae*. Science 308:654–659

Paul L, Ferguson DJ, Krzycki JA (2000) The trimethylamine methyltransferase gene and multiple dimethylamine methyltransferase genes of Methanosarcina barkeri contain in-frame and read-through amber codons. J Bacteriol 182:2520–2529

Peinemann S (1989) Kopplung von ATP-Synthese und Methanogenese in Vesikelpräparationen des methanogenen Bakteriums Stamm Göl. PhD Thesis, University of Göttingen, Germany

Perski HJ, Moll J, Thauer RK (1981) Sodium dependence of growth and methane formation in *Methanobacterium thermoautotrophicum*. Arch Microbiol 130:319–321

Perski HJ, Schönheit P, Thauer RK (1982) Sodium dependence of methane formation in methanogenic bacteria. FEBS Lett 143:323–326

Poirot CM, Kengen SW, Valk E, Keltjens JT, van der Drift C, Vogels GD (1987) Formation of methylcoenzyme M from formaldehyde by cell free extracts of *Methanobacterium thermoautotrophicum*. Evidence for the involvement of a corrinoid-containing methyltransferase. FEMS Microbiol Lett 40:7–13

Porat I, Kim W, Hendrickson EL, Xia QW, Zhang Y, Wang TS, Taub F, Moore BC, Anderson IJ, Hackett M, Leigh JA, Whitman WB (2006) Disruption of the operon encoding Ehb hydrogenase limits anabolic CO_2 assimilation in the archaeon *Methanococcus maripaludis*. J Bacteriol 188:1373–1380

Rother M, Metcalf WW (2004) Anaerobic growth of *Methanosarcina acetivorans* C2A on carbon monoxide: An unusual way of life for a methanogenic archaeon. Proc Natl Acad Sci USA 101:16929–16934

Ruppert C, Kavermann H, Wimmers S, Schmid R, Kellermann J, Lottspeich F, Huber H, Stetter KO, Müller V (1999) The proteolipid of the A_1A_O ATP synthase from *Methanococcus jannaschii* has six predicted transmembrane helices but only two proton-translocating carboxyl groups. J Biol Chem 274:25281–25284

Ruppert C, Schmid R, Hedderich R, Müller V (2001) Selective extraction of subunit D of the Na$^+$-translocating methyltransferase and subunit c of the A_1A_O ATPase from the cytoplasmic membrane of methanogenic archaea by chloroform/methanol and characterization of subunit c of *Methanothermobacter thermoautotrophicus* as a 16-kDa proteolipid. FEMS Microbiol Lett 195:47–51

Sauter M, Böhm R, Böck A (1992) Mutational analysis of the operon (*hyc*) determining hydrogenase-3 formation in *Escherichia coli*. Mol Microbiol 6:1523–1532

Schäfer G, Engelhard M, Müller V (1999) Bioenergetics of the Archaea. Microbiol Mol Biol Rev 63:570–620

Schäfer I, Rössle M, Biukovic G, Müller V, Grüber G (2006a) Structural and functional analysis of the coupling subunit F in solution and topological arrangement of the stalk domains of the methanogenic A_1A_O ATP synthase. J Bioenerg Biomembr 38:83–92

Schäfer IB, Bailer SM, Düser MG, Börsch M, Bernal RA, Stock D, Grüber G (2006b) Crystal structure of the archaeal A_1A_O ATP synthase subunit B from *Methanosarcina mazei* Göl: Implications of nucleotide-binding differences in the major A_1A_O subunits A and B. J Mol Biol 358:725–740

Schink B (1997) Energetics of syntrophic cooperation in methanogenic degradation. Microbiol Mol Biol Rev 61:262–280

Setzke E, Hedderich R, Heiden S, Thauer RK (1994) H_2:heterodisulfide oxidoreductase complex from *Methanobacterium thermoautotrophicum*. Composition and properties. Eur J Biochem 220:139–148

Shima S, Warkentin E, Thauer RK, Ermler U (2002) Structure and function of enzymes involved in the methanogenic pathway utilizing carbon dioxide and molecular hydrogen. J Biosci Bioeng 93:519–530

Simianu M, Murakami E, Brewer JM, Ragsdale SW (1998) Purification and properties of the heme- and iron-sulfur containing heterodisulfide reductase from *Methanosarcina thermophila*. Biochemistry 37:10027–10039

Slesarev AI, Mezhevaya KV, Makarova KS, Polushin NN, Shcherbinina OV, Shakhova VV, Belova GI, Aravind L, Natale DA, Rogozin IB, Tatusov RL, Wolf YI, Stetter KO, Malykh AG, Koonin EV, Kozyavkin SA (2002) The complete genome of hyperthermophile *Methanopyrus kandleri* AV19 and monophyly of archaeal methanogens. Proc Natl Acad Sci USA 99:4644–4649

Smigan P, Majernik A, Greksak M (1994) Na^+-driven ATP synthesis in *Methanobacterium thermoautotrophicum* and its differentiation from H^+-driven ATP synthesis by rhodamine 6G. FEBS Lett 347:190–194

Smigan P, Majernik A, Polak P, Hapala I, Greksak M (1995) The presence of H^+ and Na^+-translocating ATPases in *Methanobacterium thermoautotrophicum* and their possible function under alkaline conditions. FEBS Lett 371:119–122

Sowers KR, Baron SF, Ferry JG (1984) *Methanosarcina acetivorans* sp. nov., an acetotrophic methane-producing bacterium isolated from marine sediments. Appl Environ Microbiol 47:971–978

Steuber J, Rufibach M, Fritz G, Neese F, Dimroth P (2002) Inactivation of the Na^+-translocating NADH:ubiquinone oxidoreductase from *Vibrio alginolyticus* by reactive oxygen species. Eur J Biochem 269:1287–1292

Stojanowic A, Hedderich R (2004) CO_2 reduction to the level of formylmethanofuran in *Methanosarcina barkeri* is non-energy driven when CO is the electron donor. FEMS Microbiol Lett 235:163–167

Tersteegen A, Hedderich R (1999) *Methanobacterium thermoautotrophicum* encodes two multisubunit membrane-bound [NiFe] hydrogenases—transcription of the operons and sequence analysis of the deduced proteins. Eur J Biochem 264:930–943

Thauer RK (1998) Biochemistry of methanogenesis: a tribute to Marjory Stephenson. Microbiology 144:2377–2406

Thauer RK, Shima S (2006) Biogeochemistry: Methane and microbes. Nature 440:878–879

Tietze M, Beuchle A, Lamla I, Orth N, Dehler M, Greiner G, Beifuss U (2003) Redox potentials of methanophenazine and CoB-S-S-CoM, factors involved in electron transport in methanogenic archaea. Chembiochem 4:333–335

Van Beelen P, Labro JF, Keltjens JT, Geerts WJ, Vogels GD, Laarhoven WH, Guijt W, Haasnoot CA (1984) Derivatives of methanopterin, a coenzyme involved in methanogenesis. Eur J Biochem 139:359–365

Vignais PM, Colbeau A (2004) Molecular biology of microbial hydrogenases. Curr Issue Mol Biol 6:159–188

Volbeda A, Charon MH, Piras C, Hatchikian EC, Frey M, Fontecilla-Camps JC (1995) Crystal structure of the nickel-ion hydrogenase from *Desulfovivrio gigas*. Nature 373:580–587

Walsh C (1986) Naturally occuring 5-deazaflavin coenzymes: Biological redox roles. Acc Chem Res 19:216–221

Weiss DS, Gärtner P, Thauer RK (1994) The energetics and sodium-ion dependence of N^5-methyltetrahydromethanopterin:coenzyme M methyltransferase studied with cob(I)alamin as methyl acceptor and methylcob(III)alamin as methyl donor. Eur J Biochem 226:799–809

Wilms R, Freiberg C, Wegerle E, Meier I, Mayer F, Müller V (1996) Subunit structure and organization of the genes of the A_1A_O ATPase from the archaeon *Methanosarcina mazei* Gö1. J Biol Chem 271:18843–18852

Wolfe RS (1985) Unusual coenzymes of methanogenesis. TIBS 10:396–399

Wood WT, Gettrust JF, Chapman NR, Spence GD, Hyndman RD (2002) Decreased stability of methane hydrates in marine sediments owing to phase-boundary roughness. Nature 420:656–660

ATP Synthesis by Decarboxylation Phosphorylation

Peter Dimroth (✉) · Christoph von Ballmoos

Institute of Microbiology, ETH Zürich,
Wolfgang-Pauli-Strasse 10, 8093 Zürich, Switzerland
dimroth@micro.biol.ethz.ch

Abstract Adenosine triphosphate (ATP) is used as a general energy source by all living cells. The free energy released by hydrolyzing its terminal phosphoric acid anhydride bond to yield ADP and phosphate is utilized to drive various energy-consuming reactions. The ubiquitous F_1F_0 ATP synthase produces the majority of ATP by converting the energy stored in a transmembrane electrochemical gradient of H^+ or Na^+ into mechanical rotation. While the mechanism of ATP synthesis by the ATP synthase itself is universal, diverse biological reactions are used by different cells to energize the membrane. Oxidative phosphorylation in mitochondria or aerobic bacteria and photophosphorylation in plants are well-known processes. Less familiar are fermentation reactions performed by anaerobic bacteria, wherein the free energy of the decarboxylation of certain metabolites is converted into an electrochemical gradient of Na^+ ions across the membrane (decarboxylation phosphorylation). This chapter will focus on the latter mechanism, presenting an updated survey on the Na^+-translocating decarboxylases from various organisms. In the second part, we provide a detailed description of the F_1F_0 ATP synthases with special emphasis on the Na^+-translocating variant of these enzymes.

1
Introduction

Adenosine triphosphate (ATP) is used as a general source of chemical energy by all living cells. The free energy released in hydrolyzing its terminal phosphoric anhydride bond to yield ADP and phosphate is utilized to drive various energy-consuming reactions, e.g., biosyntheses, membrane transport, regulatory networks, mechanical movements, nerve conduction, etc. Accordingly, the demand for ATP is impressive, amounting to a daily turnover of 50 kg in a human on average. In order to maintain a constant supply of ATP and to complete the cell energy cycle, its terminal phosphoric anhydride bond has to be continuously regenerated from ADP and phosphate. Individual cells may dispose of several distinct catabolic reaction sequences leading to ATP synthesis by substrate-level phosphorylation in the cytoplasmic compartment of the cell. The underlying mechanism involves storage of metabolic energy as an energy-rich phosphate bond and transfer of the phosphate moiety to ADP to yield ATP.

By far the greatest amount of ATP, however, is synthesized at the membrane by the ubiquitous F_1F_0 ATP synthase. This multicomponent protein complex is able to convert the free energy stored in an electrochemical gradient of protons or Na^+ ions across the membrane into the high-energy phosphoric anhydride bond of ATP. While the mechanism of ATP synthesis by the ATP synthase itself is universal, distinct reactions are used by different cells or organelles to energize the membrane. In photosynthesis by plants, the thylakoid membrane of chloroplasts is energized by light to generate an electrochemical gradient of protons, and consequently ATP synthesis in these organelles is termed photophosphorylation. In mitochondria or aerobic bacteria, oxidation reactions drive proton flux through the respiratory chain complexes to charge the membrane, and ATP synthesis in these organelles or cells is therefore termed oxidative phosphorylation. Several anaerobic bacteria perform fermentation reactions, in which the free energy of the decarboxylation of a certain metabolite is converted into an electrochemical gradient of Na^+ ions across the membrane. The thus stored energy is used to drive ATP synthesis in a process termed decarboxylation phosphorylation (for reviews see Dimroth 1997, 2004; Buckel 2001). In this chapter we will focus on the latter mechanism. We will first give a state-of-the-art overview of the Na^+-translocating decarboxylases and succeed with a detailed description of the F_1F_0 ATP synthases. Special emphasis will be given to Na^+ translocation and torque generation by the Na^+-translocating variant of these enzymes.

2
Fermentation Pathways with Na^+-Transport Decarboxylases (NaT-DC)

2.1
Fermentation of Citrate

The fermentation of tri- or dicarboxylic acids by various bacteria includes a decarboxylation reaction which is coupled to Na^+ translocation across the membrane (Dimroth 2004). This mode of energy conversion (Bush and Saier 2002) was first recognized for oxaloacetate decarboxylase of *Klebsiella pneumoniae* (Dimroth 1980, 1982a,b). The membrane-bound enzyme complex is induced during anaerobic growth on citrate, where it catalyzes a specific step of citrate metabolism. After uptake into the cell by the Na^+-dependent citrate carrier CitS (Pos and Dimroth 1996), citrate is cleaved by citrate lyase into acetate and oxaloacetate. The latter is subsequently converted to CO_2 and pyruvate by the membrane-bound oxaloacetate decarboxylase sodium ion pump. Pyruvate is converted to acetyl-CoA and formate by pyruvate formate lyase, and part of the formate is cleaved to CO_2 and H_2 by formate hydrogen lyase. Acetyl-CoA is further metabolized to acetyl phosphate by phosphotransacetylase, and acetate kinase converts acetyl phosphate and ADP to acetate and ATP. A peculiarity of this pathway is the mode of synthesis of NAD(P)H for biosynthetic reactions. While most bacteria growing fermentatively are faced with the problem of getting rid of reducing equivalents (mainly NADH) formed by the oxidation of the growth substrates, citrate fermentation by *K. pneumoniae* includes no oxidative steps and NADH is not generated. However, as the mean oxidation status of citrate is above average for cellular components, the cells require reducing equivalents for citrate assimilation. It has been shown that hydrogen generated in the formate hydrogen lyase reaction provides the reducing equivalents for NAD(P)H formation from NAD(P)$^+$ by a soluble or a membrane-bound hydrogenase (Steuber et al. 1999).

Overall, the citrate fermentation pathway produces 1 mol ATP per mol citrate by substrate-level phosphorylation in the acetate kinase reaction, and in addition, an electrochemical gradient of Na$^+$ ions is generated by the oxaloacetate decarboxylase. The chemical concentration gradient of Na$^+$ is used to drive citrate uptake by the electroneutral Hcitrate^{2-}/Na$^+$/H$^+$ symporter (CitS) (Pos and Dimroth 1996), and the electrical component of the $\Delta\mu$Na$^+$ is assumed to contribute driving force for ATP synthesis by the H$^+$-translocating F$_1$F$_0$ ATP synthase, which is constitutively expressed in this organism.

In *K. pneumoniae*, the structural genes for oxaloacetate decarboxylase are part of the gene cluster for citrate fermentation, which comprises two operons (*citS* and *citC* operons) with divergent orientation (Bott and Dimroth 1994; Bott et al. 1995; Bott 1997). The *citS* operon includes the genes for the citrate carrier (*citS*), the three structural genes for oxaloacetate decarboxylase (*oadGAB*), and a two-component regulatory system (*citAB*). The genes of the *citC* operon encode citrate lyase ligase (*citC*), the three subunits of citrate lyase (*citDEF*), and an enzyme involved in the biosynthesis of its 2'-(5"-phosphoribosyl)-3'-dephospho-CoA prosthetic group (*citG*) (Schneider et al. 2000a,b). An additional gene required for the prosthetic group biosynthesis (*citX*) has been found at a distant location, clustered with the two-component regulatory system (*citYZ*) and another citrate carrier (*citW*) (Schneider et al. 2002).

Data bank searches revealed genes for oxaloacetate decarboxylase in numerous anaerobic bacteria and in some archaea. Some of these organisms even possess several oxaloacetate decarboxylase genes. In *Salmonella typhimurium*, two copies of the *oad* genes have been identified, one being inserted into the citrate fermentation operon and expressed during anaerobic growth on citrate, and the other being associated with genes required for tartrate fermentation and expressed under anaerobic growth on tartrate (Wifling and Dimroth 1989; Woehlke et al. 1992; Woehlke and Dimroth 1994). *Vibrio cholerae* also has two copies of the *oad* genes; the *oad-2* genes are part of the citrate fermentation operon and are expressed during anaerobic growth on citrate (Dahinden et al. 2005a). The *oad-1* genes are not associated with genes for a specific fermentation pathway and conditions for their expression are unknown. The archaeon *Archaeoglobus fulgidus* contains a cluster of three *oad* genes. The gene for the oxaloacetate specific carboxyltransferase subunit is not part of this cluster but found at a separate location of the genome (Dahinden and Dimroth 2004). Furthermore, the genome contains a gene for a methylmalonyl-CoA specific carboxyltransferase. It is conceivable that these different carboxyltransferases form sodium ion translocating decarboxylase complexes with different substrate specificity, and share the three proteins that are encoded by the *oad* gene cluster.

2.2
Fermentation of Succinate or Lactate

Methylmalonyl-CoA decarboxylase is another member of the NaT-DC enzyme family. It has been characterized in lactate fermenting *Veillonella parvula* (Hilpert and Dimroth 1982, 1983; Huder and Dimroth 1993) and in succinate fermenting *Propionigenium modestum* (Bott et al. 1997). The decarboxylase converts (S)-methylmalonyl-CoA to propionyl-CoA and CO_2 and conserves energy by coupling the exergonic decarboxylation reaction to the transport of Na^+ ions across the membrane. In both bacteria, methylmalonyl-CoA decarboxylation is an essential step in the metabolism of succinate to propionate and CO_2 (Dimroth and Schink 1998; Schink and Pfennig 1982). In spite of these similarities, only *P. modestum* but not *V. parvula* is able to grow from the conversion of succinate to propionate and CO_2. In *V. parvula*, the conversion of succinate to propionate is part of the lactate fermentation pathway, which produces sufficient ATP for growth by substrate-level phosphorylation. ATP production by decarboxylation phosphorylation is therefore not mandatory in this organism, and its inability to grow on succinate may in fact be due to the lack of a suitable F_1F_0 ATP synthase (Denger and Schink 1982).

Propionigenium modestum was the first bacterium shown to gain energy for growth exclusively from a decarboxylation reaction and is thus the paradigm for the mechanism of decarboxylation phosphorylation (Fig. 1) (Hilpert and Dimroth 1982; Hilpert et al. 1984). Succinate metabolism starts with the transfer of the CoA moiety from propionyl-CoA to succinate to yield succinyl-CoA and propionate. Rearrangement of the carbon skeleton of succinyl-CoA with a B12-dependent enzyme leads to (R)-methylmalonyl-CoA, which is subsequently converted to (S)-methylmalonyl-CoA. The membrane-bound (S)-methylmalonyl-CoA decarboxylase conserves energy by coupling the decarboxylation reaction to the transport of Na^+ ions across the membrane. The product propionyl-CoA is used to activate a new succinate molecule to succinyl-CoA. The energy stored in the electrochemical gradient of Na^+ ions is subsequently used to drive ATP synthesis by a Na^+-translocating F_1F_0 ATP synthase. Given the bioenergetic demands for ATP synthesis in growing bacteria (~60 kJ/mol) (Thauer et al. 1977) and the free energy of the decarboxylation reaction (~ − 20 kJ/mol) (Schink and Pfennig 1982), it is clear that approximately three rounds of succinate degradation are required to support the synthesis of one molecule of ATP. The bioener-

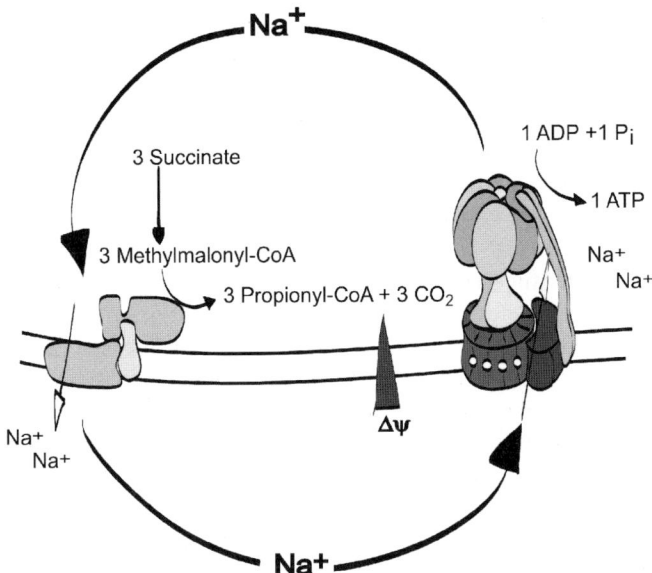

Fig. 1 ATP synthesis by decarboxylation phosphorylation in *P. modestum*

getic relationships require a proper adjustment of the ratios between chemical reaction and Na^+ translocation of the two membrane-bound complexes operating in tandem. Methylmalonyl-CoA decarboxylase has been shown to translocate one Na^+ ion electrogenically and one Na^+ ion electroneutrally (in exchange for H^+) across the membrane per decarboxylation event (Hilpert and Dimroth 1991; DiBerardino and Dimroth 1996). For the synthesis of one molecule of ATP, 3.3 Na^+ ions need to traverse the ATP synthase electrogenically (Stahlberg et al. 2001; Dimroth and Cook 2004). These ratios thus perfectly match the requirements for ATP synthesis by decarboxylation phosphorylation in *P. modestum*.

2.3
Fermentation of Malonate

Anaerobic malonate degrading bacteria are yet another example for the in vivo ATP synthesis by decarboxylation phosphorylation. A representative of this class of organisms is *Malonomonas rubra* which grows fermentatively on malonate by converting it to acetate and CO_2 (Dehning and Schink 1989). Malonate uptake into the cell has been shown to be catalyzed by a transporter consisting of two different subunits (MadL and MadM). The transported species is Hmalonate⁻ and this is taken up in an electroneutral symport with one Na^+ ion into the cell (Schaffitzel et al. 1998). Free malonate is chemically inert, and therefore the malonate degrading enzyme machinery (Hilbi et al.

1992; Dimroth and Hilbi 1997) harbors an activation module with a built-in acetyl thioester residue on a 2'-(5''-phosphoribosyl)-3'-dephospho-CoA prosthetic group on an acyl carrier protein (acetyl-ACP) (Berg et al. 1996, 1997). This device serves to transfer the ACP moiety from acetate to malonate in the first partial reaction, thus generating the activated malonyl-ACP species and acetate. The degradation continues with the transfer of the free carboxyl group of malonate to the prosthetic biotin group of a small biotin carrier protein (Berg and Dimroth 1998). The protein with the carboxybiotin residue diffuses to a membrane-bound decarboxylase, where the carboxybiotin is decarboxylated. This reaction is coupled to the transport of two Na^+ ions out of the cell and a simultaneous uptake of one proton into the cell. As a result, one Na^+ ion is exported from the cell electrogenically (together with a positive charge), but the other is exported electroneutrally (in exchange for H^+). It can be envisaged that the latter recycles into the cell during malonate uptake by the electroneutral $Hmalonate^-/Na^+$ symporter (Schaffitzel et al. 1998). The electrogenically exported Na^+ ions are thought to energize ATP synthesis. Details of the ATP synthesizing enzyme of this bacterium have not yet been elucidated. The bioenergetic premises of ATP synthesis in *M. rubra*, however, are similar to those in *P. modestum*, requiring about three decarboxylation events to synthesize one molecule of ATP. *M. rubra* and *P. modestum* are two examples where ATP is exclusively synthesized by decarboxylation phosphorylation. These organisms are distinct with respect to the dicarboxylic acid degraded and the appropriate enzyme equipment needed to decarboxylate its specific substrate, but share the conservation of chemical energy as an electrochemical Na^+ ion gradient.

2.4
Fermentation of Glutarate

Decarboxylation phosphorylation is also the ATP-generating mechanism in *Pelospora glutarica*, a strictly anaerobic bacterium growing on glutarate as sole carbon and energy source (Matthies and Schink 1992a; Matthies et al. 2000). The metabolism of glutarate starts with its activation to glutaryl-CoA by CoA transfer from acetyl-CoA. Subsequently, the glutaryl-CoA is oxidized by a NAD^+-dependent dehydrogenase to glutaconyl-CoA. The key energy conserving reaction is the decarboxylation of glutaconyl-CoA to crotonyl-CoA by a membrane-bound biotin-containing Na^+ pump that shares many properties with other members of the NaT-DC enzyme family. Reduction of crotonyl-CoA to butyryl-CoA with NADH regenerates the NAD^+ consumed in the glutaryl-CoA dehydrogenase reaction, and CoA transfer to acetate produces butyrate and generates the acetyl-CoA consumed in the activation of glutarate. Part of the butyryl-CoA isomerizes to isobutyryl-CoA, before the CoA moiety is transferred to acetate, yielding acetyl-CoA and isobutyrate (Matthies and Schink 1992b). Glutaconyl-CoA decarboxylase is also a key energy converting reaction in the fermentation of glutamate by *Acidaminococcus fermentans*, and the enzyme from this source has been well characterized biochemically (Buckel and Semmler 1983; Buckel 2001).

3
Structure and Mechanism of the NaT-DC Enzymes

The NaT-DC enzyme family includes oxaloacetate decarboxylase, methylmalonyl-CoA decarboxylase, glutaconyl-CoA decarboxylase, and malonate decarboxylase. These are multisubunit protein complexes, composed of water-soluble and membrane-intrinsic components. The biotin prosthetic group of these enzymes accepts the CO_2 moiety from the substrate at the water-exposed carboxyltransferase site and delivers it to the membrane-bound decarboxylase site, where the decarboxylation is coupled to Na^+ ion transport across the membrane. Many fundamental investigations have been performed with the oxaloacetate decarboxylase, and we shall give a detailed account of this enzyme below.

3.1
Oxaloacetate Decarboxylase Na^+ Pump

As described above, oxaloacetate decarboxylase of *K. pneumoniae*, *S. typhimurium*, or *V. cholerae* is encoded by the three structural genes *oadG*, *oadA*, and *oadB*, which are part of the *citS* operon within the citrate fermentation gene cluster (Woehlke et al. 1992). *oadG* codes for the γ subunit (8.9 kDa) which consists of an N-terminal hydrophobic α helix serving as a membrane anchor and a water-soluble C-terminal domain, which is linked to the membrane domain by a flexible linker peptide consisting mainly of proline and alanine residues. The γ subunit plays an essential role in the complex formation of the decarboxylase, but appears to have no catalytic function by itself (Schmid et al. 2002a). Near the C terminus, the γ subunit contains a Zn^{2+} metal ion (Dimroth and Thomer 1983; DiBerardino and Dimroth 1995). *oadA* codes for the water-soluble α subunit (63.5 kDa) which has a three domain structure. The N-terminal domain, which harbors the carboxyltransferase catalytic site, is bound via a flexible proline/alanine-rich linker peptide to the association domain, which promotes complex formation with the γ subunit (Dahinden et al. 2005b). At its C terminus, another proline/alanine linker connects the association domain to the biotin domain with the prosthetic group bound to a lysine residue, 35 amino acid residues before the C terminus. The crystal structure of the carboxyltransferase shows a dimer of $α_8β_8$ barrels with an active site Zn^{2+} ion at the bottom of a deep cleft that is liganded by an aspartate and two histidine residues (Studer et al. 2007). *oadB* codes for the β subunit (44.9 kDa), a very hydrophobic integral membrane protein that folds into a block of three

N-terminal membrane-spanning α helices, a hydrophobic linker peptide, inserting into the membrane but not traversing it, and a block of six C-terminal membrane-spanning α helices (Jockel et al. 1999).

A model of structure and function of the oxaloacetate decarboxylase Na^+ pump is depicted in Fig. 2. The catalytic cycle starts with the transfer of the carboxylic group from oxaloacetate to the biotin prosthetic group on the enzyme. The carboxyl transfer reaction is proposed to follow a two-step sequence. First, the Zn^{2+} ion is thought to coordinate the carboxylate in position 4 of oxaloacetate to facilitate its transfer to the ε-amino group of a conserved, essential, active site lysine residue. The biotin subsequently accepts the carboxyl group from the carbamoyl-lysine intermediate (Studer et al. 2007). Evidence for a carbamoyl-lysine was found in the crystal structure of the 5 S subunit of transcarboxylase which catalyzes exactly the same reaction as OadA (Hall et al. 2004). The carboxybiotin thus formed switches from the carboxyltransferase catalytic site to the decarboxylase site on OadB (Dimroth and Thomer 1983, 1993). It has been proposed that the Zn^{2+} on the C-terminal tail of the γ subunit coordinates the carboxyl group of carboxybiotin during the transfer to protect the chemically labile compound from spontaneous decarboxylation (Studer et al. 2007). At OadB the decarboxylation takes place, and the free biotin group is regenerated. During this Na^+-dependent reaction, a periplasmically derived proton is consumed and two sodium ions are translocated from the cytoplasm into the periplasm (DiBerardino and Dimroth 1996). Essential residues for this reaction have been identified by site-specific mutagenesis of OadB (Jockel et al. 2000a,b; Schmid et al. 2002b). Based on these studies and other biochemical investigations, a model for the reaction mechanism was proposed which is shown

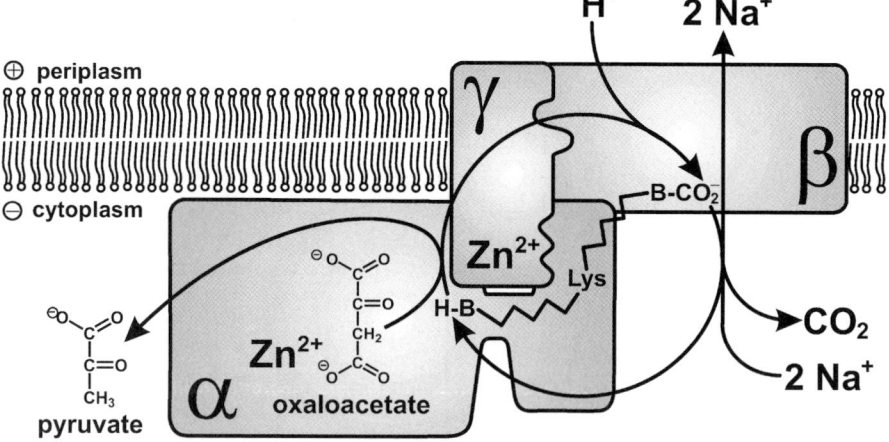

Fig. 2 Model of structure and function of the oxaloacetate decarboxylase Na^+ pump. (B-H, biotin; $B-CO_2^-$, carboxybiotin; Lys, biotin-binding lysine residue)

in Fig. 3. The model predicts that a number of highly conserved and functionally indispensable residues on helices IV and VIII and on region IIIa of OadB are involved in the ion translocation mechanism. At the decarboxylase site on OadB, the carboxybiotin is thought to form a stable complex, possibly with the side chain of R389 at the cytoplasmic surface of helix VIII. Site-

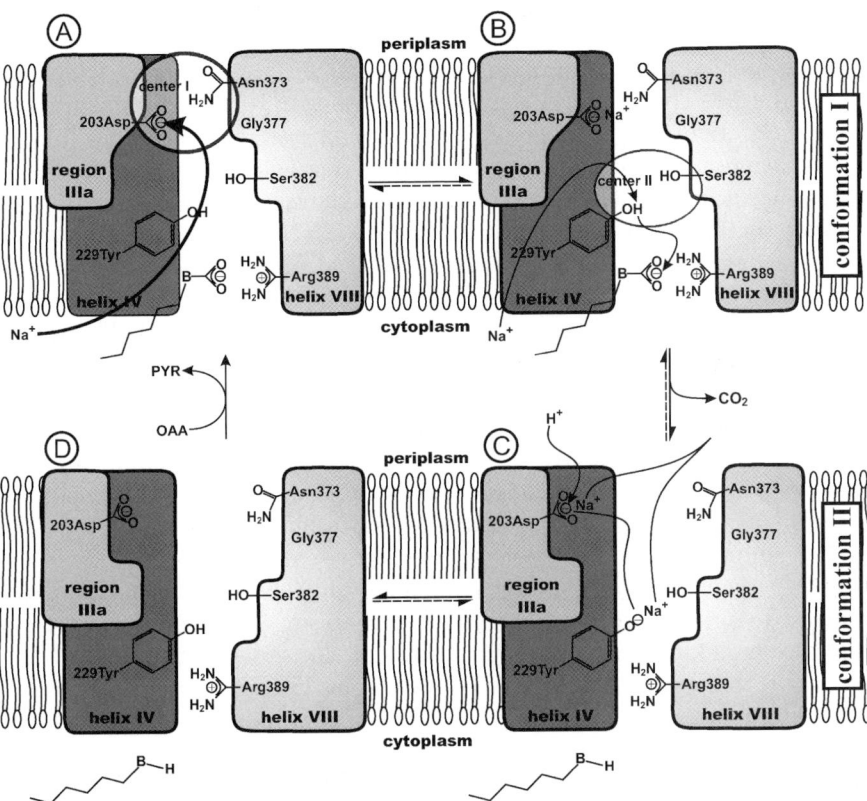

Fig. 3 Model for decarboxylation of carboxybiotin and vectorial Na^+ transport. The appropriate locations of important residues in helix IV, VIII, and in region IIIa of the β subunit and their participation in the Na^+ transport mechanism are depicted. *Panel A* shows the empty binding site region with enzyme-bound carboxybiotin ($B-CO_2^-$), exposing the Na^+ binding sites toward the cytoplasm. In *panel B*, the first Na^+ binding site at the D203-N373 pair has been occupied and the second Na^+ enters the Y229-S382 site, displacing the phenolic proton from Y229 with a rearrangement of the hydrogen-bonding network. The proton delivered to the carboxybiotin causes the immediate decarboxylation of this acid labile compound. This elicits the conformational change (B→C) exposing the Na^+ binding sites toward the periplasm. The ions are released into this reservoir and a proton enters the periplasmic channel and restores the hydroxyl group of Y229. In *panel D*, the Na^+ binding sites are empty and exposed toward the periplasm and the biotin is not modified (B→H). Upon carboxylation of the biotin, the protein switches back into the conformation where the Na^+ binding sites are exposed toward the cytoplasm (D→A)

directed sulfhydryl labeling with methanethiosulfonate reagents has identified helix VIII to align the channel for Na^+ and H^+ conductance across the membrane (Wild et al. 2003). Evidently, the proton which moves from the periplasmic reservoir through this channel must reach the carboxybiotin near the cytoplasmic surface to account for the consumption of a proton in the decarboxylation reaction. According to the model (Schmid et al. 2002b), the Na^+ channel is initially open to the cytoplasm. In this conformation, the two different Na^+ binding sites are of high affinity ($K_s \sim 1$ mM). The first Na^+ is thought to bind at a site near the periplasmic surface (center I), which includes the side chains of D203 and probably also of N373. Subsequently, the second Na^+ binds to the Y229 and S382 including site (center II). At this integral membrane location (center II) the Na^+ ion will only be tolerated after charge balancing, i.e., after dissociation and removal of a proton from the site. The proton from the phenolic hydroxyl group of Y229 is therefore assumed to dissociate as the Na^+ ion is approaching and to move to the carboxybiotin, where it is consumed during decarboxylation of this acid-labile compound. Concomitantly, the biotin prosthetic group leaves the site and OadB changes its conformation. This exposes the Na^+ binding sites toward the periplasm and simultaneously decreases their Na^+ binding affinities. The Na^+ ions dissociate into the periplasmic reservoir, while a proton enters the periplasmic channel and restores the hydroxyl group of Y229. Hence, each decarboxylation event is coupled to the transport of two Na^+ ions from the cytoplasm to the periplasm and the consumption of a periplasmically derived proton.

The amino acid sequences of the membrane-embedded β subunits of the NaT-DC family are very similar, and therefore these partial enzymes are thought to act by a common reaction mechanism. Distinct primary structures are found, however, for the carboxyltransferase subunits/domains, suggesting that these partial enzymes operate by different mechanisms. The crystal structures of the carboxyltransferase of glutaconyl-CoA decarboxylase (Wendt et al. 2003) and oxaloacetate decarboxylase (Studer et al. 2007) are indeed not related to each other and indicate a different reaction mechanism for each of them.

4
ATP Synthesis Energized by an Electrochemical Na^+ Ion Gradient

4.1
H^+- and Na^+-Translocating ATP Synthases

The mechanism of decarboxylation phosphorylation is completed by the synthesis of ATP, utilizing the electrochemical gradient of Na^+ ions as energy source. The enzyme responsible for ATP synthesis is a Na^+-translocating F_1F_0 ATP synthase. F_1F_0 ATP synthases are ubiquitous from bacteria to plants and animals, but usually use H^+ rather than Na^+ as the coupling ion (Boyer 1993; Capaldi and Aggeler 2002; Dimroth et al. 2006). This is reasonable, since the oxidation of nutrients in animals or bacteria or photosynthesis in plants leads to the formation of an electrochemical proton gradient across the membrane. Some bacteria with a fermentative metabolism, e.g., *P. modestum*, *M. rubra*, or *P. glutarica*, synthesize ATP exclusively by decarboxylation phosphorylation, while others can synthesize ATP also by substrate-level phosphorylation. Detailed studies have been performed with the closely related ATP synthases from *P. modestum* and *Ilyobacter tartaricus*. The *P. modestum* enzyme was the first F_1F_0 ATP synthase found to act with Na^+ and not with H^+ as the physiological coupling ion (Laubinger and Dimroth 1987, 1988, 1989). The advantage of this coupling ion specificity is to provide a direct link to the Na^+-translocating decarboxylase, completing the Na^+ cycle without an intermediate interconversion of a Na^+ into a H^+ gradient with a Na^+/H^+ antiporter. The ATP synthases of *P. modestum*, *I. tartaricus*, or *Escherichia coli* are encoded by eight structural genes that are organized in an operon (Krumholz et al. 1992; Kaim et al. 1992). The first gene of the operon, which is termed i gene, encodes an integral membrane protein that is not part of the ATP synthase complex. In an i gene deletion clone of *E. coli*, the cells were still able to synthesize ATP by the F_1F_0 ATP synthase, albeit at a reduced level (Gay 1984). This suggests that the i gene product is not essential for the production of a functional ATP synthase complex, but that it might help to make the assembly more efficient. Downstream of the i gene are the structural genes for the ATP synthase, starting with the gene for subunit a. This is followed by the genes for the other membrane-bound subunits c and b and by the genes for the water-soluble subunits δ, α, γ, β, and ε. The similarity between the *atp* operons from *P. modestum* and *E. coli* allowed the construction of hybrid ATP synthases harboring parts from either bacterium (Kaim and Dimroth 1993, 1994). The hybrids were constructed in the *E. coli* strain CM1470 that carries a deletion in the genes for the ATP synthase subunits a, c, b, δ, and part of α. As this strain does not form a functional ATP synthase, it is unable to grow by oxidative phosphorylation with succinate as carbon source. This defect could be complemented by transformation with plasmid pAP42 which harbors those ATP synthase genes from *P. modestum* that were lacking on the chromosome of *E. coli* CM1470. As a result of homologous recombination, the *P. modestum* genes were integrated into the chromosome of *E. coli* CM1470, yielding the new strain *E. coli* PEF42, which synthesizes a functional ATP synthase hybrid and is thus able to grow on succinate minimal medium. These results impressively buttress the common origin of H^+- and Na^+-translocating F_1F_0 ATP synthases. The Na^+-translocating ATP synthases occur exclusively in anaerobic bacteria. Other representatives of this class were found in *Acetobacterium woodii* (Müller et al. 2001) and in *Clostridium paradoxum* (Ferguson et al. 2006).

The F_1F_0 ATP synthases are nanosize rotary engines, organized as an assembly of two entities, F_1 and F_0, which are connected by a central and a peripheral stalk (Fig. 4). Each of these protein complexes functions as a reversible rotary motor and exchanges energy with the opposite motor through mechanical rotation of the central stalk. During ATP synthesis, the electrochemical ion gradient fuels the membrane-embedded F_0 motor to rotate the central stalk in its intrinsic direction. Conversely, ATP hydrolysis by the F_1 motor causes reverse rotation of the shaft, which converts the F_0 motor into an ion pump. Under normal circumstances, the F_0 motor generates the larger torque and drives the F_1 motor in the ATP synthesis direction. However, in fermenting bacteria, when the respiratory enzymes are not active, the F_1 motor hydrolyzes ATP to use the F_0 motor as the generator of the indispensable membrane potential. An interesting example is the anaerobic thermoalkaliphilic bacterium *C. paradoxum*

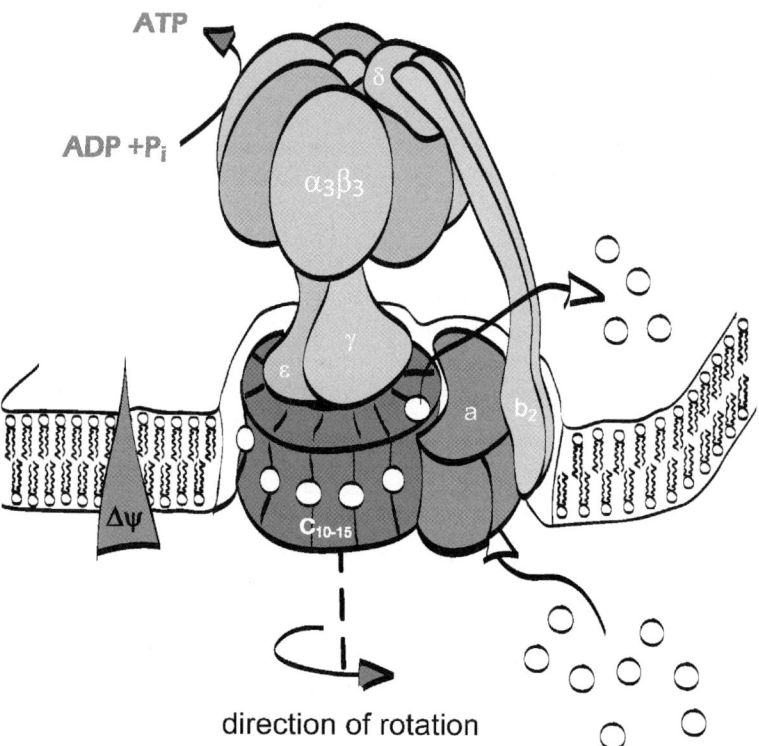

Fig. 4 Cartoon of structure and function of a bacterial ATP synthase. F_1 (subunits $\alpha_3\beta_3\gamma\delta\epsilon$) and F_0 (subunits ab_2c_{10-15}) are two motors that exchange energy by rotational coupling. The rotary subunits are $\gamma\epsilon c_{10-15}$ and the membrane anchored and cytoplasmic stator subunits are $ab_2\alpha_3\beta_3\delta$. During ATP synthesis, coupling ions (*shown as white circles*) pass through the F_0 motor from the periplasm to the cytoplasm, inducing rotation and enabling the F_1 motor to synthesize ATP

(Ferguson et al. 2006), which generates ATP in the fermentation of glucose to acetate and CO_2, and uses these acids to keep the cytoplasmic pH near neutral in spite of the alkaline environment. The physiological function of the F_1F_0 ATPase of this organism is to generate a membrane potential by hydrolyzing some of the ATP and in order to not compromise pH homeostasis, the ATPase uses Na^+ and not H^+ as the coupling ion.

4.2
The F_1 Motor

For the F_1 motor, no differences between H^+- and Na^+-translocating ATP synthases have been reported so far. The motor consists of a hexameric assembly of alternating α and β subunits around a central coiled-coil γ subunit. The F_1 complex is intrinsically asymmetric, owing to different interactions of the central γ subunit with each of the catalytic β subunits, and provides them with different conformations and nucleotide affinities at their catalytic sites. On rotation of the γ subunit, the conformations of the β subunits change sequentially such that each β subunit adopts the same conformations of varying affinity during one rotational cycle. As a result, three molecules of ATP are synthesized. This model, known as the binding change mechanism (Boyer 1993), explains a wealth of biochemical and kinetic data. Important details of this mechanism, however, are still under active investigation. The rotational model is consistent with the crystal structure of F_1, which shows a marked asymmetry in the conformations and nucleotide occupancy of the catalytic β subunits (Abrahams et al. 1994). Based on the structure, the rotational catalysis was experimentally proven by a variety of biochemical and spectroscopic techniques. Most convincingly, the rotation of a micrometer-sized fluorescent actin filament attached to the central shaft γ subunit has been directly visualized by video microscopy of single F_1 molecules (Noji et al. 1997). Rotation was also observed when the actin filament was attached to subunit ε or to the c_{10-15} oligomer (Kato-Yamada et al. 1998; Pänke et al. 2000; Sambongi et al. 1999; Tsonuda et al. 2001). These studies confirmed several cross-linking studies, which have been important to define subunits γ, ε, and c_{10-15} as the rotor subunits and subunits α_3, $\beta_3\delta$, a, and b_2 as the stator subunits (Capaldi et al. 2000). It was found that cross-links between subunits γ, ε, and c did not block ATP-driven proton translocation, whereas cross-links between subunits γ or ε and α or β led to an inhibition of the enzyme. Consistent with these data is a medium-resolution electron density map of an F_1 complex from yeast with an attached c_{10} ring, which shows a physical connection between the γ and ε subunits and the c ring at its cytoplasmic loops (Stock et al. 1999).

Single-molecule experiments with small magnetic or gold beads or with fluorescent dyes instead of the large actin filament provided insights into mechanistic details of the enzyme. Rotation in the fully coupled Na^+-ATP synthase was observed both in synthesis and hydrolysis mode (Kaim et al. 2002).

Rotation of the rotor was counterclockwise in the ATP hydrolysis direction and clockwise in the ATP synthesis direction, when viewed from the F_0 domain (Diez et al. 2004; Itoh et al. 2004), and performed up to 700 revolutions per second (Nakanishi-Matsui et al. 2006). In F_1, the γ subunit rotates in steps of 120° for each ATP molecule hydrolyzed. Each 120° step can be further divided into four stages. In the ATP binding dwell, an ATP molecule binds to the empty β_1 site and elicits a rapid 80° substep rotation of the γ subunit. In the following catalytic dwell, ATP is thought to be cleaved at the β_2 site and ADP and/or phosphate is released from the β_3 site. This initiates a 40° substep rotation of the γ subunit completing its 120° rotation (Yasuda et al. 2001; Nishizaka et al. 2004; Shimabukuru et al. 2003). The rotational behavior of F_1F_0 resembled that of F_1, indicating that friction in the F_0 motor is negligible during ATP-driven rotation. Tributyltin chloride, a specific inhibitor of the ion access route in subunit a (von Ballmoos et al. 2004), inhibited rotation by 96% (Ueno et al. 2005), in accordance with strict coupling between mechanical and ion translocation events. When the performance of the F_1 motor was probed by sophisticated single-molecule experiments in femtoliter-sized chambers, the hydrolysis of three ATP molecules per revolution was directly observed and showed very high mechanochemical coupling (Rondelez et al. 2005). If the γ subunit of F_1 was forced to rotate in the ATP synthesis direction, ATP synthesis from ADP and phosphate was observed. Interestingly, the mechanochemical coupling efficiency was low for an F_1 subcomplex lacking the ε subunit, but reached more than 70% after the reconstitution with this protein (Rondelez et al. 2005). The ε subunit thus has an important role in the synthesis of ATP, but the mechanism for this function has not yet been elucidated.

4.3
The F_0 Motor

The F_0 motor is a membrane-bound protein complex consisting of an oligomeric ring of c subunits, a single a subunit, and a dimer of b subunits, which flank the c ring laterally (Mellwig and Böttcher 2003; Rubinstein et al. 2003). Early on, the c subunit has attracted much interest, because of its very hydrophobic character and its small size, which facilitated its purification by extraction into organic solvent mixtures. Based on amino acid sequencing and specific labeling studies with a hydrophobic diazirine derivative, subunit c was predicted to fold as a helical hairpin, consisting of two membrane spanning α helices and a cytoplasmic connecting loop (Hoppe et al. 1984). Importantly, the C-terminal α helix contains a conserved acidic residue, approximately in the middle of the membrane, which plays a profound role in proton translocation. Accordingly, mutagenesis of the respective aspartic acidic residue in *E. coli* (D61) to asparagine, or its chemical substitution with the specific F-ATPase inhibitor dicyclohexylcarbodiimide (DCCD), results in the impairment of proton translocation (Hoppe et al. 1982; Sebald et al. 1980). Hence, the D61 residues are thought to extract protons from one and deliver them to the other side of the membrane, as the c ring rotates. Accordingly, the F_0 motor of *P. modestum*, which conducts Na^+, harbors specific binding sites for Na^+ on its c ring (Kluge and Dimroth 1992; Kaim et al. 1997). Biochemical evidence for these sites was obtained by DCCD labeling experiments (Kluge and Dimroth 1993a,b; 1994). The ATP synthase of *P. modestum* was specifically labeled at cE65, which is equivalent to cD61 of *E. coli*. The chemical modification with DCCD consumes a proton, and consequently the reaction rate increased from alkaline to acidic pH values, following a titration curve with an apparent pK of 6.5, which reflects the pK of cE65 (Kluge and Dimroth 1993a). Importantly, Na^+ ions protected from this modification and shifted the pK of cE65 into the acidic range, indicating a competition of Na^+ and H^+ binding to the same site.

The Na^+ binding sites on the c ring were confirmed by mutational studies. First, random mutagenesis of the c ring and selection for Na^+-independent growth on succinate minimal medium led to the identification of a double mutant in subunit c, which abolished Na^+ binding and made the holoenzyme to a more efficient H^+-translocating ATP synthase (Kaim and Dimroth 1995). As these c-subunit mutations are near the C terminus, in considerable distance to the ion binding glutamate 65, they must exert their effect via a long-distance conformational change at the binding site. Second, using site-specific mutagenesis, the Na^+ binding site was identified to be contributed by the triad cQ32, cE65, and cS66 (Kaim et al. 1997). These three residues are conserved in all Na^+-translocating F_1F_0 ATP synthases. In the high-resolution structure of the c ring from *I. tartaricus*, the side chains of these three amino acids were indeed seen to be Na^+ binding ligands (Meier et al. 2005). The topography of the binding site within the middle of the membrane was identified by photocrosslinking experiments using a photoactivatable DCCD derivative attached to cE65 (von Ballmoos et al. 2002a). This became specifically bound to the fatty acid side chains of the phospholipids. Quenching experiments with a fluorescent DCCD derivative attached to cE65 and phospholipids carrying a spin label at different positions of the fatty acid side chains were consistent with this membrane location (von Ballmoos et al. 2002b).

4.4
Subunit C Structures

Two different structures of the *E. coli* c subunit were solved by NMR at pH 5 and 8, respectively, in an organic solvent mixture (Girvin et al. 1998; Rastogi and Girvin 1999). In the pH 5 structure, which is thought to present the conformation during exposure to the lipids, D61 has an inward facing orientation. In the pH 8 structure, however, which is thought to represent the conformation in the interface with subunit a, D61 is exposed to the surface. The latter conformation matches cross-linking data between subunits c and a (Jiang and Fillingame 1998). To account for these results, a c ring model was proposed, in which the C-terminal helix performs a large 140° swiveling versus the N-terminal helix (Fillingame et al. 2003) in the interface with subunit a. This relocates the proton binding D61 to the surface to permit loading or unloading of the site from and to subunit a. NMR investigations of the *P. modestum* c subunit revealed different secondary structures in SDS and in the organic solvent mixture, and neither was similar to one of the *E. coli* c subunit structures (Matthey et al. 1999, 2002). Therefore, an artifactual folding of the protein due to the unphysiological environment of a c monomer in an organic solvent mixture or a harsh detergent could not be excluded.

In the native ATP synthase, the c subunits assemble into an oligomeric ring, the stoichiometry of which varies depending on the species. Rings with ten monomers exist in yeast mitochondria (Stock et al. 1999), the thermophilic bacterium PS3 (Mitome et al. 2004), and possibly in *E. coli* (Jiang et al. 2001), whereas rings found in the bacterium *I. tartaricus* (Stahlberg et al. 2001) and *P. modestum* (Meier et al. 2003) have 11 monomers. The archaeon *Methanopyrus kandleri* harbors a gene in which 13 c subunits are fused (Lolkema et al. 2003), suggesting that this organism produces a corresponding ring with 13 fused c subunits. Furthermore, ring stoichiometries of 14 and 15 have been found in chloroplasts (Seelert et al. 2000) and the cyanobacterium *Spirulina platensis* (Pogoryelov et al. 2005), respectively. An even larger ring with 20 hairpins resulting from ten double-size subunits and harboring ten ion binding sites exists in the V-type ATPase from *Enterococcus hirae* (Murata et al. 2005). The number of subunits in each ring also indicates the number of ions transported across the membrane during each cycle of the ATP synthase. Consequently, as the F_1 motor contains three catalytic sites and synthesizes three molecules of ATP per cycle, a variation in the number of c subunits and ion binding sites automatically leads to different H^+ (Na^+) to ATP ratios. ATP synthases with large c rings have a high H^+ (Na^+) to ATP ratio, which would be advantageous for ATP synthesis at low ion motive force. Conversely, ATP synthases with small c rings might prevail in organisms with constantly high ion motive force: the low H^+ (Na^+) to ATP ratio of these enzymes results in a more efficient use of energy. A mismatch to the threefold symmetry of the F_1 motor created by F_0 motors with 10, 11, 13, or 14 c subunits has been regarded as functionally important (Stock et al. 1999; Murata et al. 2004) in order to prevent the enzyme from becoming trapped in deep energy minima. However, recent data from the c_{15} ring of the *S. platensis* ATP synthase show that symmetry mismatch is not mandatory for function (Pogoryelov et al. 2005).

4.5
Structure of the C Ring From *I. tartaricus*

The Na^+-translocating ATP synthases of *P. modestum* or *I. tartaricus* have the distinct advantage of containing an oligomeric c ring of excessive stability (Laubinger and Dimroth 1988; Neumann et al. 1998; Meier and Dimroth 2002). This facilitated the purification of the c ring from the *I. tartaricus* ATP synthase in high yield and purity (Meier et al. 2003). The isolated c ring could be reconstituted with subunits a and b to form a functional F_0 moiety, indicating that it retained its native conformation (Wehrle et al. 2002a). The initial investigations on the c ring structure were performed by atomic force microscopy and electron microscopy after reconstitution into lipid vesicles and crystallization in two dimensions (Stahlberg et al. 2001; Meier et al. 2003). The structure was solved at medium resolution (4 Å) by electron crystallography (Vonck et al. 2002) and at high resolution (2.4 Å) by X-ray crystallography (Meier et al. 2004) (Fig. 5). In the structure of the oligomer, each c ring is composed of 11 subunits. Of these, each monomer is folded as a helical hairpin with the loop at the cytoplasmic side and the termini at the periplasmic side, as was predicted earlier by independent methods (Girvin et al. 1998; Hoppe et al. 1984). The structure (Fig. 5) shows a cylindrical, hourglass-shaped protein complex which has a height of ∼70 Å and protrudes from the membrane on either side. Its outer diameter is ∼40 Å in the middle and ∼50 Å at the top

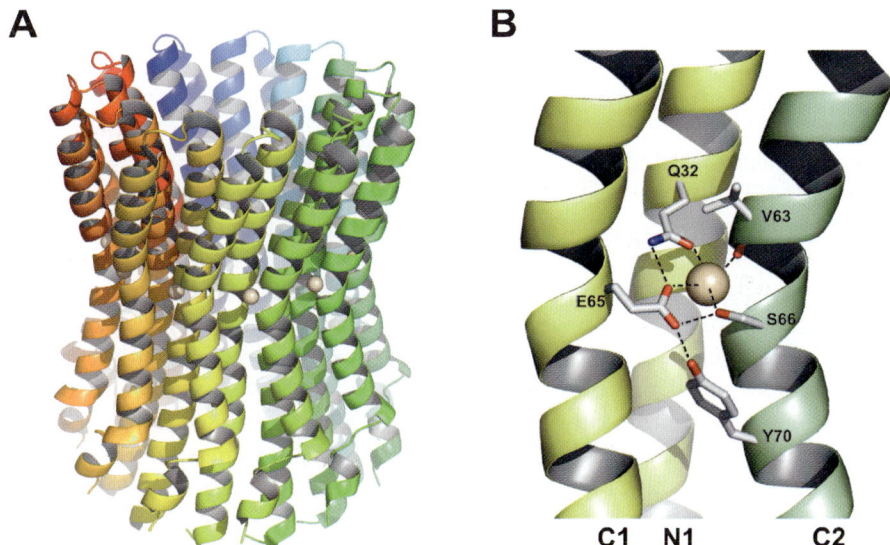

Fig. 5 Structure of the *I. tartaricus* c_{11} ring in ribbon form. **A** Individual subunits are shown in different colors. The *spheres* indicate the bound Na^+ ions. **B** Close-up of the Na^+ binding site formed by the inner (N1) and outer helix (C1) of one c subunit and the outer helix (C2) of the neighboring c subunit. Na^+ coordination and selected hydrogen bonds are indicated with *dashed lines*. The structure shows the locked conformation. During opening, the side chain of Y70 might relocate into a cavity underneath the binding site, thus destabilizing the hydrogen-bonding network and allowing unloading and loading of the binding site to and from subunit a

and bottom. The 11 N-terminal helices are closely spaced within an inner ring surrounding a cavity of $\sim$17 Å. The tight helix packing leaves no space for side chains and is accounted for by a highly conserved motif of four glycine residues in the inner, N-terminal helix (Vonck et al. 2002). Each N-terminal helix is connected to a C-terminal helix by a loop formed by the highly conserved peptide R45, N46, P47 which is exposed to the cytoplasmic surface (Watts et al. 1995; Hermolin et al. 1999). The C-terminal helices pack into the grooves formed between N-terminal helices, producing the outer ring. All helices show a bend of about 20° in the middle of the membrane (at P28 and E65 in the N-terminal and C-terminal helices, respectively) causing the narrow part of the hourglass shape. Moreover, the bend tilts the helices in the cytoplasmic half out of the plane by $\sim$10°, yielding a right-handed twisted packing. When the c ring is viewed from the cytoplasm, it rotates counter-clockwise during ATP synthesis against the drag imposed by the F_1 motor components. Thus, the resulting torque might decrease the bend and increase the interhelical distance in the cytoplasmic part of the c ring, depending on the energies involved. Such a conformational change under load might serve to store elastic energy in the c ring, adding to that described for the central and peripheral stalk subunits (Junge et al. 2001). The internal surface of the c ring is very hydrophobic and was shown by photocross-linking experiments to be filled with phospholipids in the natural environment of the membrane (Oberfeld et al. 2006).

Perhaps the most instructive feature of the c ring structure is the arrangement of the Na^+ binding sites. Eleven bound Na^+ ions are seen in the c ring structure near the middle of the membrane facing toward the outer surface of the c ring (Fig. 5) (Meier et al. 2004), which confirms previous cross-linking data (von Ballmoos et al. 2002a). Each of the 11 Na^+ ions is bound at the interface of an N-terminal and two C-terminal helices. The coordination sphere is formed by side chain oxygen atoms of Q32 and E65 of one subunit and the side chain oxygen atom of S66 and the backbone carbonyl oxygen atom of V63 of the neighboring subunit (Fig. 5). An intriguing observation is that E65 acts not only as one of the Na^+ binding ligands but also as the recipient of hydrogen bonds from the side chains of Q32, S66, and Y70. This arrangement generates a stable, locked conformation of the binding site, from which the horizontal transfer of Na^+ to subunit a is prevented. This implies that the present locked conformation of the binding site converts to an open one at the subunit a/c interface in order to allow the horizontal ion transfer to and from subunit a.

The a subunit abuts the c ring laterally and is implicated in providing access to the ion binding site at the c subunit in the subunit a/c interface from either one or both sides of the membrane, depending on the model. Subunit a is an extremely hydrophobic protein containing five to six transmembrane helices (Jäger et al. 1998; Long et al. 1998; Valiyaveetil and Fillingame 1998). Of particular interest is the interface of the a subunit with the c ring, which has to guarantee the stability of the a/c complex while allowing an almost frictionless rotation between these protein components, as revealed by single-molecule experiments (Ueno et al. 2005). These peculiarities have impeded structural determinations of subunit a. So far, biochemical and mutational studies suggest that the universally conserved aR226 residue (*P. modestum* numbering), which is localized approximately at the same level in the membrane as the cE65, is important in ion translocation.

5
Mechanism of the F_0 Motor

5.1
The Proton Motor

The F_0 motor converts the energy stored in an electrochemical ion gradient into mechanical rotation. Hence, the torque generating mechanism is an integral part of the transport of the coupling ions across the membrane. Theoretical considerations and a large body of experimental data have led to the proposal of several models for the F_0 motor function, which are considerably distinct in detail but share common features as well (Aksimentiev et al. 2004; Feniouk et al. 2004; Xing et al. 2004). The initial model of the H^+-driven motor of the *E. coli* ATP synthase (Junge et al. 1997; Vik and Antonio 1994) predicts that only a protonated, uncharged cD61 site can move from the subunit a/c interface into the lipid phase. To account for sufficient torque generation and to prevent proton leakage in the subunit a/c interface, the initial model was extended by positioning the essential positive stator arginine between the access and the exit pathway of the ion (Elston et al. 1998). The model predicts that inlet and outlet channels are located in a noncoaxial manner in subunit a (Fig. 6). The ion is proposed to enter the site through the inlet channel from the low-pH reservoir and to exit through the outlet channel into the high-pH reservoir, after performing an almost complete rotation. The unprotonated site is located between the two channels, and without an external driving force the motor is in an idling mode, and the site shuttles between the channels with equal probability to either side. In the presence of a ΔpH, the proton concentration in each of the separate aqueous access channels is unequal, and the site will be protonated more frequently at the position of the channel with higher proton concentration. The protonated site is now able to move out of the a/c subunit interface into the lipid phase. Simultaneously, a new site enters the interface, where its ion is displaced by the stator arginine. The negatively charged empty site acts as a ratchet, due to the large energy penalty of its backward movement into the lipid phase. In this model, the pH gradient determines the direction of rotation and simultaneously acts as the main driving force for the generation of torque.

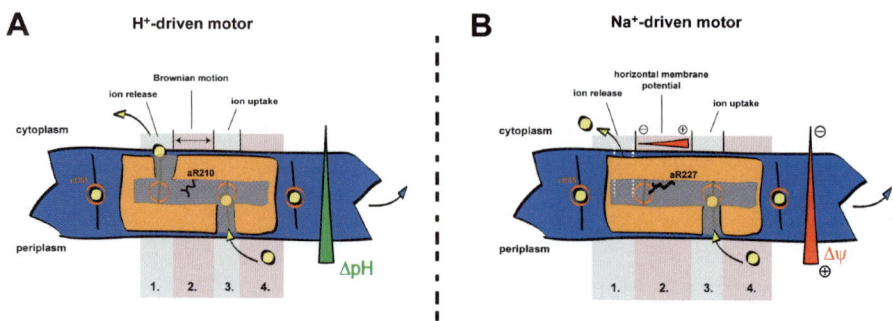

Fig. 6 Model for torque generation in the H^+- and Na^+-translocating F_0 motor. **A** Two-channel model with a ratchet-type mechanism for H^+-dependent enzymes. The crucial events during ion translocation in ATP synthesis direction in the a/c interface can be divided into four zones. (1) The occupied rotor site enters the interface and releases its coupling ion through the outlet channel into the cytoplasm with high pH. Deprotonation prevents the backward rotation into the lipid phase acting as a molecular ratchet. (2) The negative charge of the binding site is compensated by the stator arginine. In this functionally symmetric state, Brownian back and forth motions toward either channel are possible. (3) As the inlet channel, which is in contact with the periplasm where the pH is low, contains more protons than the outlet channel, which is in contact with the cytoplasm, the binding site is more frequently protonated from the periplasm. Therefore, the ΔpH determines the direction of rotation. (4) The loaded binding site can now move out of the interface into the lipid bilayer, whereby the next binding site enters the interface and experiences the events described in (1). **B** Push-and-pull model for Na^+-dependent enzymes. (1) In ATP synthesis direction, an occupied rotor site enters the interface from the left and releases its bound Na^+ ion toward the cytoplasm. This process is aided by the stator arginine. (2) The arginine compensates for the now negatively charged empty site. The horizontal component of the membrane potential, however, pulls the arginine to the left and pushes the glutamate to the right. Therefore, the electrical component of the ion motive force determines the direction of rotation from left to right. (3) The hydration of the binding site within the inlet channel stabilizes this conformation and allows loading of the binding site from the periplasm. Movement of the binding site from zone 2 to 3 pulls the next rotor site into the a/c interface as described in (1). (4) The binding site that has been occupied from the periplasm is allowed to rotate out of the interface into the lipid phase. The event is aided by a push mechanism during the events described in (2)

Experiments with engineered cysteines within helices II, IV, and V show good accessibility of these residues from the periplasmic side to approximately the middle of the membrane, which would be in accord with a periplasmic inlet channel. From the cytoplasmic surface, cysteines were only accessible within helix IV, and those cysteines were very close to the membrane surface (Angevine and Fillingame 2003; Angevine et al. 2003). Conclusive evidence for a cytoplasmic access channel within subunit a is therefore not available so far.

5.2
The Sodium Motor

In contrast to the proton-translocating ATP synthases, where the path of the ion across the membrane is intrinsically difficult to establish, the Na^+-translocating counterparts provide unique experimental options for these investigations. For example, (1) Na^+ concentrations on either side of the membrane can be manipulated at will without compromising the stability of the enzyme, (2) Na^+ translocation can be conveniently measured with the radioactive isotope $^{22}Na^+$ (Dimroth 1982b) or with a Na^+-binding fluorophore (von Ballmoos and Dimroth 2004), and (3) screening methods can be used to select for mutants with impaired or modified Na^+ binding characteristics. Many details of the ion translocation through the F_0 motor and of the mechanical rotation connected to it have therefore been explored with the Na^+-translocating ATP synthase of *P. modestum*. These experiments support a model where the ion binding sites on the c ring are accessible from the periplasmic surface through a corridor in subunit a, and from the cytoplasmic surface through a route that is either c ring-intrinsic or within the interface between the c ring and subunit a (Dimroth et al. 2006).

5.2.1
The Ion Path

In the following we will describe some of the experiments that have led to the model of the sodium F_0 motor and we will then give a detailed description of this model. *E. coli* deletion mutants in the *atp* operon are unable to grow on a nonfermentable carbon source such as succinate. This defect could be complemented by expressing the *E. coli* F_1 ($-\delta$) together with the *P. modestum* F_0 ($+\delta$) genes on a plasmid and thus allowed the construction and investigation of mutants in the F_0 genes of the Na^+-dependent enzyme (Kaim and Dimroth 1993, 1994). The parent *E. coli/P. modestum* hybrid ATP synthase showed the same Na^+-dependent growth characteristics as the F_1F_0 ATP synthase of *P. modestum*. In a random mutagenesis approach, mutants with a Na^+-independent growth phenotype could be isolated, which could be grouped into two categories. Mutants of the first category contained a double mutation in the C-terminal tail of subunit c (Kaim and Dimroth 1995). The mutation affected the geometry of the binding site in such a way that Na^+ ions were no longer able to bind, whereas the ability for Li^+ or H^+ binding was retained. The second group of mutants contained a triple mutation in subunit a, which affected the ion specificity of the periplasmic access route to the binding site, so that Na^+ access was abolished whereas Li^+ or H^+ access was retained (Kaim and Dimroth 1998a). The mutant cells could grow on succinate only in the absence, but not in the presence, of Na^+ ions, suggesting that sodium ions were unable to get access to the binding site through the periplasmic entrance channel, and simultaneously prevented protons from reaching the binding sites via this route. Furthermore, Na^+ ions inhibited the ATP hydrolysis activity of the mutant ATP synthase, and one Na^+ per F_1F_0 became firmly occluded in the subunit a/c interface after ATP addition (Kaim et al. 1998; Kaim and Dimroth 1998a). Hence, after ATP-driven rotation of the binding site into the interface, any further rotation is prevented if the Na^+ ion cannot be released from the binding site into the periplasm due to blockage of this route by the subunit a triple mutation. Besides by mutation, the periplasmic access route for Na^+ was also blocked by the inhibitor tributyltin chloride (von Ballmoos et al. 2004). A photoactivatable derivative of this compound was found to specifically label subunit a. Sodium ions were protected from this labeling, indicating a competition between tributyltin chloride and Na^+ for the binding at the same site. The most obvious candidate for this site is the Na^+ access pathway. This supposition is in accord with biochemical data, showing that tributyltin chloride not only blocks the hydrolysis of ATP, but also transport of sodium ions across the membrane.

So far, neither inhibitors nor mutants have been found for the cytoplasmic access route to the binding site, and it is not resolved as to whether this route is localized entirely in the c ring or in the interface between the c ring and subunit a. Although several lines of evidence compellingly demonstrate that cytoplasmic access to the binding site can occur outside the subunit a/c interface in a time scale of seconds (von Ballmoos et al. 2002; Meier et al. 2003), it is not certain whether this ion path can account for physiological rates in the millisecond timescale. Among these evidences are (1) the demonstration of Na^+ or H^+ binding to the c subunit sites within the isolated c ring; (2) ATP-driven occlusion of $^{22}Na^+$ by the a subunit triple mutant, even after partial modification of the c ring with DCCD (Kaim and Dimroth 1998b)—this observation compulsorily requires that the radioactive $^{22}Na^+$ enters a site between the DCCD-modified c subunit and the a subunit; and (3) the exchange of Na^+ between both sides of the membrane with partially DCCD-modified c subunits (Kaim and Dimroth 1998b). This result requires shuttling of the rotor between the periplasmic access channel in the subunit a/c interface and a position outside the a/c interface, where the ions must exchange between the cytoplasm and the binding site.

5.2.2
The Driving Force

Experiments have also been performed to determine the driving forces required for torque generation. According to Mitchell's chemiosmotic hypothesis, the chemical concentration gradient of the coupling ion and the membrane potential contribute equally to the driving force in a system at thermodynamic equilibrium. However, under physiological conditions, far from thermodynamic equilibrium, each individual component can contribute kinetically unequally to the driving force. It was found that in the absence of external energy, the F_0 motor, either isolated or connected to the F_1 motor, performs an idling motion of the rotor versus the stator, which causes Na^+ ions to exchange between the exterior and interior compartments separated by the membrane (Kluge and Dimroth 1992; Kaim and Dimroth 1998b). Exchange by the isolated F_0 motor is not converted into unidirectional rotation coupled to unidirectional ion flux by applying large Na^+ concentration gradients, but only with a threshold membrane potential of about 40 mV (Kluge and Dimroth 1992; Dimroth et al. 2003). Hence, the membrane potential is the kinetically indispensable driving force for torque generation and cannot be compensated by a Na^+ concentration gradient. These results were corroborated by ATP synthesis experiments with reconstituted ATP synthases from *P. modestum, E. coli,* or chloroplasts. With each of these enzymes, ATP synthesis was only observed after applying a certain membrane potential and not with large Na^+ or H^+ concentration gradients (Kaim and Dimroth 1999).

It is clear that the principal torque generating and ion translocation events occur in the subunit a/c interface, and it would therefore be of great interest to learn more about the critical parts of this protein assembly. As mentioned above, probably the most important residue for function in subunit a is the universally conserved arginine 226 (*P. modestum* numbering) near the middle of the penultimate transmembrane helix. Mutants of this residue in the *P. modestum* a subunit have provided important insights into its role for the operation of the F_0 motor (Wehrle et al. 2002b). All the evidence is consistent with an attribution of the positively charged residue to repel the Na^+ ions from approaching binding sites into the appropriately positioned release route to either the periplasmic or the cytoplasmic surface, depending on the direction of rotation. The most conservative R226K mutant was completely inactive in ATP-driven ion transport at pH 7.0, similar to the corresponding *E. coli* mutant (Valiyaveetil and Fillingame 1997), but interestingly attained catalytic power between pH 8 and 9. In the R226H mutation, the activity profile was shifted into the acidic range, with maximal activities between pH 6.5 and 7.5. ATP synthase with an R226A substitution catalyzed Na^+-dependent ATP hydrolysis, which was completely inhibited by DCCD, but not coupled to Na^+ transport. Taken together, these results suggest that a positive charge at position 226 is necessary to accomplish the dissociation of the Na^+ ions from approaching rotor sites. Furthermore, an arginine with its delocalized positive charge appears to be better suited for an unrestricted diffusion of the negatively charged empty rotor site than a lysine with a positive point charge. Hence, at neutral pH, when the lysine is completely protonated, it electrostatically attracts a negatively charged rotor site so powerfully that the charge has to be attenuated by its partial deprotonation at higher pH values, in order for the rotor site to escape.

5.2.3
The Model for Torque Generation

On the basis of these experimental data, a model for torque generation by the Na^+-translocating F_0 motor was proposed which is in accordance with mathematical calculations (Dimroth et al. 1999, 2006; Xing et al. 2004). In this model, the binding sites in the rotor/stator interface have to interact in a coordinated manner with the coupling ion, the positive stator charge (R226), and the membrane potential, as shown schematically in Fig. 6 for the ATP synthesis direction. During its journey through the interface with the stator, a rotor site moves along a sequence of states as follows (Xing et al. 2004). As the occupied site (site 1) enters the a/c interface, it is in the vicinity of the positive stator charge, which forces the bound sodium ion to dissociate. The negative charge of site 1 is now compensated by the positive stator charge. In the idling mode, when no external energy source is applied, the rotor can shuttle back and forth against the stator up to a certain angle. However, in the presence of a membrane potential (cytoplasmic negative), the movement becomes directional (see below), bringing the empty site 1 in juxtaposition with the periplasmic access route of subunit a. Upon hydration of the site, the free energy drops to trap it at its present position, until a Na^+ ion has been bound. So far the movement of site 1 has simultaneously pulled the next occupied rotor site (site 2) into the a/c interface. Site 2 now follows the same events as site 1 before, to release its Na^+ ion and to be pulled into juxtaposition with the stator charge, and this movement pushes the occupied site 1 out of the interface into the lipid phase. Accordingly, two rotor sites operate together in a push and pull fashion to achieve rotation of the rotor versus the stator.

To account for the essence of the membrane potential for torque generation (see above), it can be assumed that only part of it drops vertically between the periplasmic channel entrance and its terminus in the middle of the membrane and that between the latter and the terminus of the cytoplasmic access route, the membrane potential drops in a horizontal direction. This horizontal potential causes the stator charge to orient toward the incoming rotor sites and the charged rotor site to become attracted to the periplasmic entrance channel. Hence, the membrane potential acts as the main kinetic driving force for unidirectional rotation and ATP synthesis.

The Na^+ and the H^+ motors are distinct in their use of the ion release pathway, which is thought to be located in subunit a in the proton motor, whereas evidence now suggests the presence of an ion release pathway in the c ring of the sodium motor. The other main difference is in the energy source that drives each F_0 motor out of the relaxed, idling mode into unidirectional rotation to produce torque. Apart from these differences, both models are similar. Each motor operating in the direction of ATP synthesis performs a molecular cycle, in which a specific binding site on the c ring captures a coupling ion from the side with high electrochemical potential and releases it to the side with low electrochemical potential. As loading and unloading of the binding site can only be accomplished at defined positions at the subunit a/c interface, ion translocation is closely connected to the rotation of the c ring. Hence, the mechanical rotation is linked to the ion translocation events in the F_0 motor components, and is transmitted through the camshaft-like rotating central stalk into the ATP synthesizing F_1 motor.

Acknowledgements We wish to thank Pius Dahinden for preparing Figs. 2 and 3. Work in the authors' laboratory was supported by the Swiss National Science Foundation and by the Forschungskommission der ETH Zürich.

References

Abrahams JP, Leslie AG, Lutter R, Walker JE (1994) Structure at 2.8 Å resolution of F_1-ATPase from bovine heart mitochondria. Nature 370:621–628
Aksimentiev A, Balabin IA, Fillingame RH, Schulten K (2004) Insight into the molecular mechanism of rotation in the F_0 sector of ATP synthase. Biophys J 86:1332–1344
Angevine CM, Fillingame RH (2003) Aqueous access channels in subunit a of rotary ATP synthase. J Biol Chem 278:6066–6074
Angevine CM, Herold KA, Fillingame RH (2003) Aqueous access pathways in subunit a of rotary ATP synthase extend to both sides of the membrane. Proc Natl Acad Sci USA 100:13179–13183
Berg M, Dimroth P (1998) The biotin protein MadF of the malonate decarboxylase from *Malonomonas rubra*. Arch Microbiol 170:464–468
Berg M, Hilbi H, Dimroth P (1996) The acyl carrier protein of malonate decarboxylase of *Malonomonas rubra* contains 2'-(5''-phosphoribosyl)-3'-dephosphocoenzyme A as a prosthetic group. Biochemistry 35:4689–4696
Berg M, Hilbi H, Dimroth P (1997) Sequence of a gene cluster from *Malonomonas rubra* encoding components of the malonate decarboxylase Na^+ pump and evidence for their function. Eur J Biochem 245:103–115
Bott M (1997) Anaerobic citrate metabolism and its regulation in enterobacteria. Arch Microbiol 167:78–88
Bott M, Dimroth P (1994) *Klebsiella pneumoniae* genes for citrate lyase and citrate lyase ligase: localization, sequencing and expression. Mol Microbiol 14:347–356
Bott M, Meyer M, Dimroth P (1995) Regulation of anaerobic citrate metabolism in *Klebsiella pneumoniae*. Mol Microbiol 18:533–546
Bott M, Pfister K, Burda P, Kalbermatter O, Woehlke G, Dimroth P (1997) Methylmalonyl-CoA decarboxylase from *Propionigenium modestum*: cloning and sequencing of the structural genes and purification of the enzyme complex. Eur J Biochem 250:590–599
Boyer PD (1993) The binding change mechanism for ATP synthase—some probabilities and possibilities. Biochim Biophys Acta 1140:215–250
Buckel W (2001). Sodium ion-translocating decarboxylases. Biochim Biophys Acta 1505:15–27
Buckel W, Semmler R (1983) Purification, characterisation and reconstitution of glutaconyl-CoA decarboxylase, a biotin-dependent sodium pump from anaerobic bacteria. Eur J Biochem 136:427–434
Busch W, Saier MH Jr (2002) The transporter classification (TC) system, 2002. Crit Rev Biochem Mol Biol 37:287–337

Capaldi RA, Aggeler R (2002) Mechanism of the F_1F_0-type ATP synthase, a biological rotary motor. Trends Biochem Sci 27:154–160

Capaldi RA, Schulenberg B, Murray J, Aggeler R (2000) Cross-linking and electron microscopy studies of the structure and functioning of the *Escherichia coli* ATP synthase. J Exp Biol 203(Pt1):29–33

Dahinden P, Dimroth P (2004) Oxaloacetate decarboxylase of *Archaeoglobus fulgidus*: cloning of genes and expression in *Escherichia coli*. Arch Microbiol 182:414–420

Dahinden P, Auchli Y, Granjon T, Taralczak M, Wild M, Dimroth P (2005a) Oxaloacetate decarboxylase of *Vibrio cholerae*: purification, characterization, and expression of the genes in *Escherichia coli*. Arch Microbiol 183:121–129

Dahinden P, Pos KM, Taralczak M, Dimroth P (2005b) Identification of a domain in the α-subunit of the oxaloacetate decarboxylase Na^+ pump that accomplishes complex formation with the γ-subunit. FEBS J 272:846–855

Dehning I, Schink B (1989) *Malonomonas rubra* gen. nov., sp. nov., a microaerotolerant anaerobic bacterium growing by decarboxylation of malonate. Arch Microbiol 151:427–433

Denger K, Schink B (1982) Energy conservation by succinate decarboxylation in *Veillonella parvula*. J Gen Microbiol 138:967–971

DiBerardino M, Dimroth P (1995) Synthesis of the oxaloacetate decarboxylase Na^+ pump and its individual subunits in *Escherichia coli* and analysis of their function. Eur J Biochem 231:790–801

DiBerardino M, Dimroth P (1996) Aspartate 203 of the oxaloacetate decarboxylase β-subunit catalyses both the chemical and vectorial reaction of the Na^+ pump. EMBO J 15:1842–1849

Diez M, Zimmermann B, Börsch M, König M, Schweinberger E, Steigmiller S, Reuter R, Felekyan S, Kudryavtsev V, Seidel CA, Gräber P (2004) Proton-powered subunit rotation in single membrane-bound F_0F_1-ATP synthase. Nat Struct Mol Biol 11:135–141

Dimroth P (1980) A new sodium-transport system energized by the decarboxylation of oxaloacetate. FEBS Lett 122:234–236

Dimroth P (1982a) The role of biotin and sodium in the decarboxylation of oxaloacetate by the membrane-bound oxaloacetate decarboxylase from *Klebsiella aerogenes*. Eur J Biochem 121:435–441

Dimroth P (1982b) The generation of an electrochemical gradient of sodium ions upon decarboxylation of oxaloacetate by the membrane-bound and Na^+-activated oxaloacetate decarboxylase from *Klebsiella aerogenes*. Eur J Biochem 121:443–449

Dimroth P (1997) Primary sodium ion translocating enzymes. Biochim Biophys Acta 1318:11–51

Dimroth P (2004) Molecular basis for bacterial growth on citrate or malonate. In: Curtiss R (ed) EcoSal—*Escherichia coli* and *Salmonella*: cellular and molecular biology. ASM, Washington, DC, chap. 3.4.6

Dimroth P, Cook GM (2004) Bacterial Na^+- or H^+-coupled ATP synthases operating at low electrochemical potential. Adv Microb Physiol 49:175–218

Dimroth P, Hilbi H (1997) Enzymic and genetic basis for bacterial growth on malonate. Mol Microbiol 25:3–10

Dimroth P, Schink B (1998) Energy conservation in the decarboxylation of dicarboxylic acids by fermenting bacteria. Arch Microbiol 170:69–77

Dimroth P, Thomer A (1983) Subunit composition of oxaloacetate decarboxylase and characterization of the alpha chain as carboxyltransferase. Eur J Biochem 137:107–112

Dimroth P, Thomer A (1993) On the mechanism of sodium ion translocation by oxaloacetate decarboxylase of *Klebsiella pneumoniae*. Biochemistry 32:1734–1739

Dimroth P, Wang H, Grabe M, Oster G (1999) Energy transduction in the sodium F-ATPase of *Propionigenium modestum*. Proc Natl Acad Sci USA 96:4924–4929

Dimroth P, von Ballmoos C, Meier T, Kaim G (2003) Electrical power fuels rotary ATP synthase. Structure (Camb) 11:1469–1473

Dimroth P, von Ballmoos C, Meier T (2006) Catalytic and mechanical cycles in F-ATP synthases. Fourth in the cycles review series. EMBO Rep 7:276–282

Elston T, Wang H, Oster G (1998) Energy transduction in ATP synthase. Nature 391:510–513

Feniouk BA, Kozlova MA, Knorre DA, Cherepanov DA, Mulkidjanian AY, Junge W (2004) The proton-driven rotor of ATP synthase: ohmic conductance (10 fS) and absence of voltage gating. Biophys J 86:4094–4109

Ferguson SA, Keis S, Cook GM (2006) Biochemical and molecular characterization of a Na^+-translocating F_1F_0 ATPase from the thermoalkaliphilic bacterium *Clostridium paradoxum*. J Bacteriol 188:5045–5054

Fillingame RH, Angevine CM, Dmitriev OY (2003) Mechanics of coupling proton movements to c-ring rotation in ATP synthase. FEBS Lett 555:29–34

Gay NJ (1984) Construction and characterization of an *Escherichia coli* strain with a *uncI* mutation. J Bacteriol 158:820–825

Girvin ME, Rastogi VK, Abildgaard F, Markley JL, Fillingame RH (1998) Solution structure of the transmembrane H^+-transporting subunit c of the F_1F_0 ATP synthase. Biochemistry 37:8817–8824

Hall PR, Zheng R, Lizamma A, Pusztai-Carey M, Carey PR, Lee VC (2004) Transcarboxylase 5 S structures: assembly and catalytic mechanism of a multienzyme complex subunit. EMBO J 23:3621–3631

Hermolin J, Dmitriev OY, Zhang Y, Fillingame RH (1999) Defining the domain of binding of F_1 subunit ε with the polar loop of F_0 subunit c in the *Escherichia coli* ATP synthase. J Biol Chem 274:17011–17016

Hilbi H, Dehning I, Schink B, Dimroth P (1992) Malonate decarboxylase of *Malonomonas rubra*, a novel type of biotin-containing acetyl enzyme. Eur J Biochem 207:117–123

Hilpert W, Dimroth P (1982) Conversion of the chemical energy of methylmalonyl-CoA decarboxylation into a Na^+ gradient. Nature 296:584–585

Hilpert W, Dimroth P (1983) Purification and characterization of a new sodium-transport decarboxylase. Methylmalonyl-CoA decarboxylase from *Veillonella alcalescens*. Eur J Biochem 132:579–587

Hilpert W, Dimroth P (1991) On the mechanism of sodium ion translocation by methylmalonyl-CoA decarboxylase from *Veillonella alcalescens*. Eur J Biochem 19:79–86

Hilpert W, Schink B, Dimroth P (1984) Life by a new decarboxylation-dependent energy conservation mechanism with Na^+ as coupling ion. EMBO J 3:1665–1670

Hoppe J, Scheirer HU, Friedl P, Sebald W (1982) An Asp–Asn substitution in the proteolipid subunit of the ATP synthase from *Escherichia coli* leads to a non-functional proton channel. FEBS Lett 145:21–29

Hoppe J, Brunner J, Joergensen BB (1984) Structure of the membrane-embedded F_0 part of F_1F_0 ATP synthase from *Escherichia coli* as inferred from labeling with 3-(trifluoromethyl)-3-(m-[^{125}I]iodophenyl)diazirine. Biochemistry 23:5610–5616

Huder JB, Dimroth P (1993) Sequence of the sodium ion pump methylmalonyl-CoA decarboxylase from *Veillonella parvula*. J Biol Chem 268:24564–24571

Itoh H, Takahashi A, Adachi K, Noji H, Yasuda R, Yoshida M, Kinosita K Jr (2004) Mechanically driven ATP synthesis by F_1-ATPase. Nature 427:465–468

Jäger H, Birkenhäger R, Stalz W-D, Altendorf K, Deckers-Hebestreit G (1998) Topology of subunit a of the *Escherichia coli* ATP synthase. Eur J Biochem 251:122–132

Jiang W, Fillingame RH (1988) Interacting helical faces of subunit a and c in the F_1F_0 ATP synthase of *Escherichia coli* defined by disulfide cross-linking. Proc Natl Acad Sci USA 95:6607–6612

Jiang W, Hermolin J, Fillingame RH (2001) The preferred stoichiometry of c subunits in the rotary motor sector of *Escherichia coli* ATP synthase is 10. Proc Natl Acad Sci USA 98:4966–4971

Jockel P, Di Berardino M, Dimroth P (1999) Membrane topology of the β subunit of the oxaloacetate decarboxylase Na^+ pump from *Klebsiella pneumoniae*. Biochemistry 38:13461–13472

Jockel P, Schmid M, Choinowski T, Dimroth P (2000a) Essential role of tyrosine 229 of the oxaloacetate decarboxylase beta-subunit in the energy coupling mechanism of the Na^+ pump. Biochemistry 39:4320–4326

Jockel P, Schmid M, Steuber J, Dimroth P (2000b) A molecular coupling mechanism for the oxaloacetate decarboxylase Na^+ pump as inferred from mutational analysis. Biochemistry 39:2307–2315

Junge W, Lill H, Engelbrecht S (1997) ATP synthase: an electrochemical transducer with rotatory mechanics. Trends Biochem Sci 22:420–423

Junge W, Pänke O, Cherepanov DA, Gumbiowski K, Müller M, Engelbrecht S (2001) Intersubunit rotation and elastic power transmission in F_0F_1 ATPase. FEBS Lett 504:152–160

Kaim G, Dimroth P (1993) Formation of a functionally active sodium-translocating F_1F_0 ATPase in *Escherichia coli* by homologous recombination. Eur J Biochem 218:937–944

Kaim G, Dimroth P (1994) Construction, expression and characterization of a plasmid-encoded Na^+-specific ATPase hybrid consisting of *Propionigenium modestum* F_0-ATPase and *Escherichia coli* F_1-ATPase. Eur J Biochem 222:615–623

Kaim G, Dimroth P (1995) A double mutation in subunit c of the Na^+-specific F_1F_0 ATPase of *Propionigenium modestum* results in a switch from Na^+- to H^+-coupled ATP synthesis in the *Escherichia coli* host cells. J Mol Biol 253:726–738

Kaim G, Dimroth P (1998a) A triple mutation in the a subunit of the *Escherichia coli/Propionigenium modestum* F_1F_0 ATPase hybrid causes a switch from Na^+ stimulation to Na^+ inhibition. Biochemistry 37:4626–4634

Kaim G, Dimroth P (1998b) Voltage-generated torque drives the motor of the ATP synthase. EMBO J 17:5887–5895

Kaim G, Dimroth P (1999) ATP synthesis by F-type ATP synthase is obligatorily dependent on the transmembrane voltage. EMBO J 18:4118–4127

Kaim G, Ludwig W, Dimroth P, Schleifer KH (1992) Cloning, sequencing and in vivo expression of genes encoding the F_0 part of the sodium-ion-dependent ATP synthase of *Propionigenium modestum* in *Escherichia coli*. Eur J Biochem 207:463–470

Kaim G, Wehrle F, Gerike U, Dimroth P (1997) Molecular basis for the coupling ion selectivity of F_1F_0 ATP synthases: probing the liganding groups for Na^+ and Li^+ in the c subunit of the ATP synthase from *Propionigenium modestum*. Biochemistry 36:9185–9194

Kaim G, Matthey U, Dimroth P (1998) Mode of interaction of the single a subunit with the multimeric c subunits during the translocation of the coupling ions by F_1F_0 ATPases. EMBO J 17:688–695

Kaim G, Prummer M, Sick B, Zumofen G, Renn A, Wild UP, Dimroth P (2002) Coupled rotation within single F_0F_1 enzyme complexes during ATP synthesis or hydrolysis. FEBS Lett 525:156–163

Kato-Yamada Y, Noji H, Yasuda R, Kinosita K, Yoshida M (1998) Direct observation of the rotation of the ε subunit in F_1-ATPase. J Biol Chem 273:19375–19377

Kluge C, Dimroth P (1992) Studies on Na^+ and H^+ translocation through the F_0 part of the Na^+-translocating F_1F_0 ATPase from *Propionigenium modestum*: discovery of a membrane potential dependent step. Biochemistry 31:12665–12672

Kluge C, Dimroth P (1993a) Kinetics of inactivation of the F_1F_0 ATPase of *Propionigenium modestum* by dicyclohexylcarbodiimide in relationship to H^+ and Na^+ concentration: probing the binding site for the coupling ions. Biochemistry 32:10378–10386

Kluge C, Dimroth P (1993b) Specific protection by Na^+ and Li^+ of the F_1F_0 ATPase of *Propionigenium modestum* from the reaction with dicyclohexylcarbodiimide. J Biol Chem 268:14557–14560

Kluge C, Dimroth P (1994) Modification of isolated subunit c of the F_1F_0-ATPase from *Propionigenium modestum* by dicyclohexylcarbodiimide. FEBS Lett 340:245–248

Krumholz LR, Esser U, Simoni RD (1992) Characterization of the genes coding for the F_1F_0 subunits of the sodium dependent ATPase of *Propionigenium modestum*. FEMS Microbiol Lett 91:37–42

Laubinger W, Dimroth P (1987) Characterization of the Na^+-stimulated ATPase of *Propionigenium modestum* as an enzyme of the F_1F_0 type. Eur J Biochem 168:475–480

Laubinger W, Dimroth P (1988) Characterization of the ATP synthase of *Propionigenium modestum* as a primary sodium pump. Biochemistry 27:7531–7537

Laubinger W, Dimroth P (1989) The sodium ion translocating adenosinetriphosphatase of *Propionigenium modestum* pumps protons at low sodium ion concentrations. Biochemistry 28:7194–7198

Lolkema JS, Boekema EJ (2003) The A-type ATP synthase subunit K of *Methanopyrus kandleri* is deduced from its sequence to form a monomeric rotor comprising 13 hairpin domains. FEBS Lett 543:47–50

Long JC, Wang S, Vik SB (1998) Membrane topology of subunit a of the F_1F_0 ATP synthase as determined by labeling of unique cysteine residues. J Biol Chem 273:16235–16240

Matthey U, Kaim G, Braun D, Wüthrich K, Dimroth P (1999) NMR studies of subunit c of the ATP synthase from *Propionigenium modestum* in dodecylsulfate micelles. Eur J Biochem 261:459–467

Matthey U, Braun D, Dimroth P (2002) NMR investigations of subunit c of the ATP synthase from *Propionigenium modestum* in chloroform/methanol/water (4:4:1). Eur J Biochem 269:1942–1946

Matthies C, Schink B (1992a) Energy conservation in fermentative glutarate degradation by the bacterial strain WoGl3. FEMS Microbiol Lett 79:221–225

Matthies C, Schink B (1992b) Reciprocal isomerization of butyrate and isobutyrate by the strictly anaerobic bacterium strain WoGl3 and methanogenic isobutyrate degradation by a defined triculture. Appl Environ Microbiol 58:1435–1439

Matthies C, Springer N, Ludwig W, Schink B (2000) *Pelospora glutarica* gen. nov., sp. nov., a glutarate-fermenting, strictly anaerobic, spore-forming bacterium. Int J Syst Evol Microbiol 50(Pt2):645–648

Meier T, Dimroth P (2002) Intersubunit bridging by Na^+ ions as a rationale for the unusual stability of the c-rings of Na^+-translocating F_1F_0 ATP synthases. EMBO Rep 3:1094–1098

Meier T, Matthey U, von Ballmoos C, Vonck J, Krug von Nidda T, Kühlbrandt W, Dimroth P (2003) Evidence for structural integrity in the undecameric c-rings isolated from sodium ATP synthases. J Mol Biol 325:389–397

Meier T, Polzer P, Diederichs K, Welte W, Dimroth P (2005) Structure of the rotor ring of F-type Na^+-ATPase from *Ilyobacter tartaricus*. Science 308:659–662

Mellwig C, Böttcher B (2003) A unique resting position of the ATP synthase from chloroplasts. J Biol Chem 278:18544–18549

Mitome N, Suzuki T, Hayashi S, Yoshida M (2004) Thermophilic ATP synthase has a decamer c ring: indication of noninteger 10 : 3 H$^+$/ATP ratio and permissive elastic coupling. Proc Natl Acad Sci USA 101:12159–12164

Müller V, Aufurth S, Rahlfs S (2001) The Na$^+$ cycle in *Acetobacterium woodii*: identification and characterization of a Na$^+$-translocating F_1F_0-ATPase with a mixed oligomer of 8 and 16 kDa proteolipids. Biochim Biophys Acta 1505:108–120

Murata T, Yamato I, Kakinuma Y, Leslie AG, Walker JE (2005) Structure of the rotor of the V-type Na$^+$-ATPase from *Enterococcus hirae*. Science 308:654–659

Nakanishi-Matsui M, Kashiwagi S, Hosokawa H, Cipriano DJ, Dunn SD, Wada Y, Futai M (2006) Stochastic high-speed rotation of *Escherichia coli* ATP synthase F_1 sector: the epsilon subunit sensitive rotation. J Biol Chem 281:4126–4131

Neumann S, Matthey U, Kaim G, Dimroth P (1998) Purification and properties of the F_1F_0 ATPase of *Ilyobacter tartaricus*, a sodium ion pump. J Bacteriol 180:3312–3316

Nishizaka T, Oiwa K, Noji H, Kimura S, Muneyuki E, Yoshida M, Kinosita K Jr (2004) Chemomechanical coupling in F_1-ATPase revealed by simultaneous observation of nucleotide kinetics and rotation. Nat Struct Mol Biol 11:142–148

Noji H, Yasuda R, Yoshida M, Kinosita K (1997) Direct observation of the rotation of F_1 ATPase. Nature 386:299–302

Oberfeld B, Brunner J, Dimroth P (2006) Phospholipids occupy the internal lumen of the c ring of the ATP synthase of *Escherichia coli*. Biochemistry 45:1841–1851

Pänke O, Gumbiowski K, Junge W, Engelbrecht S (2000) F-ATPase: specific observation of the rotating c subunit oligomer of EF_0EF_1. FEBS Lett 472:34–38

Pogoryelov D, Yu J, Meier T, Vonck J, Dimroth P, Müller DJ (2005) The c_{15} ring of the *Spirulina platensis* F-ATP synthase: F_1F_0 symmetry mismatch is not obligatory. EMBO Rep 6:5474–5483

Pos KM, Dimroth P (1996) Functional properties of the purified Na$^+$-dependent citrate carrier of *Klebsiella pneumoniae*: evidence for asymmetric orientation of the carrier in proteoliposomes. Biochemistry 35:1018–1026

Rastogi VK, Girvin ME (1999) Structural changes linked to proton translocation by subunit c of the ATP synthase. Nature 402:263–268

Rondelez Y, Tresset G, Nakashima T, Kato-Yamada Y, Fujita H, Takeuchi S, Noji H (2005) Highly coupled ATP synthesis by F_1 single molecules. Nature 433:773–777

Rubinstein JL, Walker JE, Henderson R (2003) Structure of the mitochondrial ATP synthase by electron cryomicroscopy. EMBO J 22:6182–6192

Sambongi Y, Iko Y, Tanabe M, Omote H, Iwamoto-Kihara A, Ueda I, Yanagida T, Wada Y, Futai M (1999) Mechanical rotation of the c subunit oligomer in ATP synthase F_1F_0: direct observation. Science 286:1722–1724

Schaffitzel C, Berg M, Dimroth P, Pos KM (1998) Identification of an Na$^+$-dependent malonate transporter of *Malonomonas rubra* and its dependence on two seperate genes. J Bacteriol 180:2689–2693

Schink B, Pfennig N (1982) *Propionigenium modestum* gen. nov. sp. nov., a new strictly anaerobic, nonsporing bacterium growing on succinate. Arch Microbiol 133:209–216

Schmid M, Wild M, Dahinden P, Dimroth P (2002a) Subunit γ of the oxaloacetate decarboxylase Na$^+$ pump: interaction with other subunits/domains of the complex and binding site for the Zn^{2+} metal ion. Biochemistry 41:1285–1292

Schmid M, Vorburger T, Pos KM, Dimroth P (2002b) Role of conserved residues within helices IV and VIII of the oxaloacetate decarboxylase β subunit in the energy coupling mechanism of the Na$^+$ pump. Eur J Biochem 269:2997–3004

Schneider K, Dimroth P, Bott M (2000a) Biosynthesis of the prosthetic group of citrate lyase. Biochemistry 39:9438–9450

Schneider K, Dimroth P, Bott M (2000b) Identification of triphosphoribosyl-dephospho-CoA as precursor of the citrate lyase prosthetic group. FEBS Lett 483:165–168

Schneider K, Kästner CN, Meyer M, Wessel M, Dimroth P, Bott M (2002) Identification of a gene cluster in *Klebsiella pneumoniae* which includes *citX*, a gene required for biosynthesis of the citrate lyase prosthetic group. J Bacteriol 184:2439–2446

Sebald W, Machleidt W, Wachter E (1980) N,N'-dicyclohexylcarbodiimide binds specifically to a single glutamyl residue of the proteolipid subunit of the mitochondrial adenosinetriphosphatase from *Neurospora crassa* and *Saccharomyces cerevisiae*. Proc Natl Acad Sci USA 77:785–789

Seelert H, Poetsch A, Dencher NA, Engel A, Stahlberg H, Müller DJ (2000) Proton-powered turbine of a plant motor. Nature 405:418–419

Shimabukuro K, Yasuda R, Muneyuki E, Hara KY, Kinosita K Jr, Yoshida M (2003) Catalysis and rotation of F_1 motor: cleavage of ATP at the catalytic site occurs in 1 ms before 40 degree substep rotation. Proc Natl Acad Sci USA 100:13731–13736

Stahlberg H, Müller DJ, Suda K, Fotiadis D, Engel A, Meier T, Matthey U, Dimroth P (2001) Bacterial Na^+-ATP synthase has an undecameric rotor. EMBO Rep 2:229–233

Steuber J, Krebs W, Bott M, Dimroth P (1999) A membrane-bound $NAD(P)^+$-reducing hydrogenase provides reduced pyridine nucleotides during citrate fermentation by *Klebsiella pneumoniae*. J Bacteriol 181:241–245

Stock D, Leslie AG, Walker JE (1999) Molecular architecture of the rotary motor in ATP synthase. Science 286:1700–1705

Studer R, Dahinden P, Wang W-W, Auchli Y, Li X-D, Dimroth P (2007) Crystal structure of the carboxyltransferase domain of the oxaloacetate decarboxylase Na^+ pump from *Vibrio cholerae*. J Mol Biol 367:547–557

Thauer RK, Jungermann K, Decker K (1977) Energy conservation in chemotrophic anaerobic bacteria. Bacteriol Rev 41:100–180

Tsunoda SP, Aggeler R, Yoshida M, Capaldi RA (2001) Rotation of the c subunit oligomer in fully functional F_1F_0 ATP synthase. Proc Natl Acad Sci USA 98:898–902

Ueno H, Suzuki T, Kinosita K Jr, Yoshida M (2005) ATP-driven stepwise rotation of F_0F_1-ATP synthase. Proc Natl Acad Sci USA 102:1333–1338

Valiyaveetil FI, Fillingame RH (1997) On the role of Arg-210 and Glu-219 on subunit a in proton translocation by the *Escherichia coli* F_1F_0 ATP synthase. J Biol Chem 272:32635–32641

Valiyaveetil FI, Fillingame RH (1998) Transmembrane topography of subunit a in the *Escherichia coli* F_1F_0 ATP synthase. J Biol Chem 273:16241–16247

Vik SB, Antonio BJ (1994) A mechanism of proton translocation by F_1F_0 ATP synthase suggested by double mutants of the a subunit. J Biol Chem 269:30364–30369

von Ballmoos C, Dimroth P (2004) A continuous fluorescent method for measuring Na^+ transport. Anal Biochem 335:334–337

von Ballmoos C, Appoldt Y, Brunner J, Granier T, Vasella A, Dimroth P (2002a) Membrane topography of the coupling ion binding site in Na^+-translocating F_1F_0 ATP synthase. J Biol Chem 277:3504–3510

von Ballmoos C, Meier T, Dimroth P (2002b) Membrane embedded location of Na^+ or H^+ binding sites on the rotor ring of F_1F_0 ATP synthases. Eur J Biochem 269:5581–5589

von Ballmoos C, Brunner J, Dimroth P (2004) The ion channel of F-ATP synthase is the target of toxic organotin compounds. Proc Natl Acad Sci USA 101:11239–11244

Vonck J, von Nidda TK, Meier T, Matthey U, Mills DJ, Kühlbrandt W, Dimroth P (2002) Molecular architecture of the undecameric rotor of a bacterial Na^+-ATP synthase. J Mol Biol 321:307–316

Watts SD, Zhang Y, Fillingame RH, Capaldi RA (1995) The gamma subunit in the *Escherichia coli* ATP synthase complex (ECF_1F_0) extends through the stalk and contacts the c subunit of the F_0 part. FEBS Lett 368:235–238

Wehrle F, Appoldt Y, Kaim G, Dimroth P (2002a) Reconstitution of F_0 of the sodium ion translocating ATP synthase of *Propionigenium modestum* from its heterologously expressed and purified subunits. Eur J Biochem 269:2567–2573

Wehrle F, Kaim G, Dimroth P (2002b) Molecular mechanism of the ATP synthase's F_0 motor probed by mutational analyses of subunit a. J Mol Biol 322:369–381

Wendt KS, Schall I, Huber R, Buckel W, Jacob U (2003) Crystal structure of the carboxyltransferase subunit of the bacterial sodium ion pump glutaconyl-coenzyme A decarboxylase. EMBO J 22:3493–3502

Wifling K, Dimroth P (1989) Isolation and characterization of oxaloacetate decarboxylase of *Salmonella typhimurium*, a sodium pump. Arch Microbiol 152:584–588

Wild MR, Pos KM, Dimroth P (2003) Site-directed sulfhydryl labeling of the oxaloacetate decarboxylase Na^+ pump of *Klebsiella pneumoniae*: helix VIII comprises a portion of the sodium ion channel. Biochemistry 42:11615–11624

Woehlke G, Dimroth P (1994) Anaerobic growth of *Salmonella typhimurium* on L(+)- and D(−)-tartrate involves an oxaloacetate decarboxylase Na^+ pump. Arch Microbiol 162:233–237

Woehlke G, Wifling K, Dimroth P (1992) Sequence of the sodium ion pump oxaloacetate decarboxylase from *Salmonella typhimurium*. J Biol Chem 267:22798–22803

Xing J, Wang H, von Ballmoos C, Dimroth P, Oster G (2004) Torque generation by the F_0 motor of the sodium ATPase. Biophys J 87:2148–2163

Yasuda R, Noji H, Yoshida M, Kinosita K Jr, Itoh H (2001) Resolution of distinct rotational substeps by submillisecond kinetic analysis of F_1-ATPase. Nature 410:898–904

The Three Families of Respiratory NADH Dehydrogenases

Stefan Kerscher · Stefan Dröse · Volker Zickermann · Ulrich Brandt (✉)

Molecular Bioenergetics Group, Centre of Excellence "Macromolecular Complexes",
Johann Wolfgang Goethe-Universität, 60590 Frankfurt am Main, Germany
brandt@zbc.kgu.de

Abstract Most reducing equivalents extracted from foodstuffs during oxidative metabolism are fed into the respiratory chains of aerobic bacteria and mitochondria by NADH:quinone oxidoreductases. Three families of enzymes can perform this task and differ remarkably in their complexity and role in energy conversion. Alternative or NDH-2-type NADH dehydrogenases are simple one subunit flavoenzymes that completely dissipate the redox energy of the NADH/quinone couple. Sodium-pumping NADH dehydrogenases (Nqr) that are only found in procaryotes contain several flavins and are integral membrane protein complexes composed of six different subunits. Proton-pumping NADH dehydrogenases (NDH-1 or complex I) are highly complicated membrane protein complexes, composed of up to 45 different subunits, that are found in bacteria and mitochondria. This review gives an overview of the origin, structural and functional properties and physiological significance of these three types of NADH dehydrogenase.

1
Introduction

One of the universal principles of oxidative metabolism is that reducing equivalents extracted from foodstuffs are transferred onto the pyridine nucleotides NAD^+ or $NADP^+$ and are then fed into the quinone "pool" of the respiratory electron transfer chain located in the cell membrane of archaebacteria and eubacteria, or in the inner membrane of mitochondria. In various phyla, this reaction may be catalysed by three fundamentally different types of respiratory NADH dehydrogenases: non-pumping "alternative" NAD(P)H-dehydrogenase (also called NDH-2), sodium-pumping NADH dehydrogenase (also called Nqr) and proton-pumping NADH dehydrogenase (also known as complex I or NDH-1).

NDH-2 is found in eubacteria, archaebacteria, yeasts, fungi, and plants (Kerscher 2000; Melo et al. 2004). Some bacteria like *Sinorhizobium meliloti* 1021 and *Bradyrhizobium japonicum* USDA 110, some lower eukaryotes like *Saccharomyces cerevisiae* or *Neurospora crassa*, and plant mitochondria may possess multiple alternative NAD(P)H-dehydrogenases. Nqr is found in various eubacteria, including many marine species, but also in some well-known human pathogens (Häse et al. 2001; Kogure 1998). Complex I (Brandt 2006) is found in eubacteria and the vast majority of eucaryotic organisms, with few notable exceptions that include fermentative yeasts like *S. cerevisiae*. Eubacterial genomes may encode any possible combination of the three types of NADH:quinone oxidoreductases (Melo et al. 2004). A few examples of the distribution of NADH dehydrogenases in characteristic respiratory chains are depicted in Fig. 1. Note that multiple versions of NDH-2 may be present in fungi and plants on both sides of the inner mitochondrial membrane.

The three families of NADH:quinone oxidoreductases are of completely independent evolutionary origin and have remarkably different catalytic mechanisms and levels of complexity. The most prominent difference is that the free energy gap of around 0.4 V between NADH and ubiquinone is completely

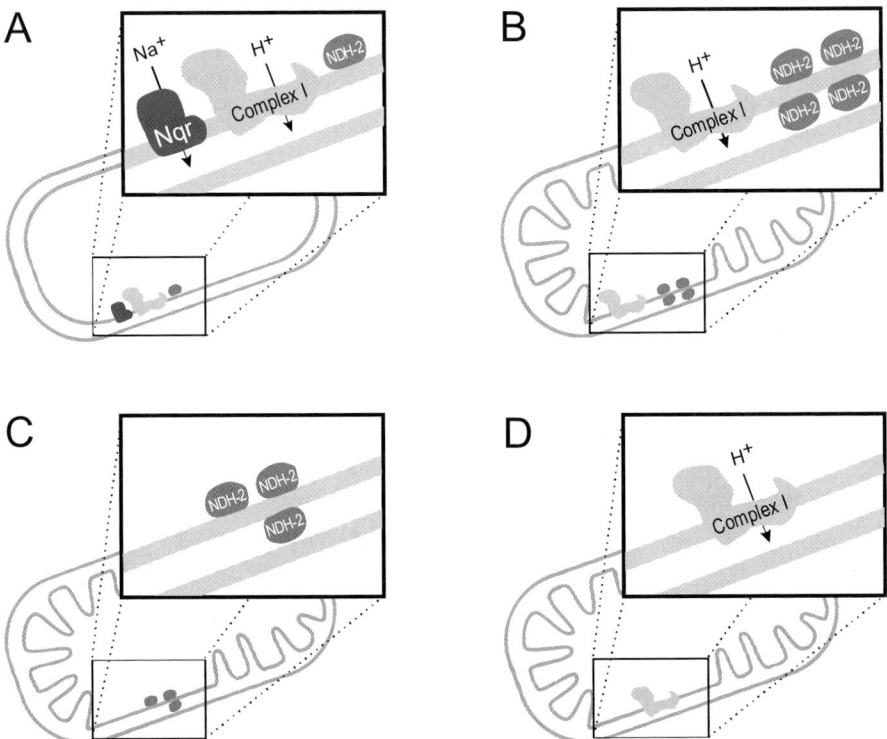

Fig. 1 Occurrence of the three families of respiratory NADH dehydrogenases. **A** Prokaryotes may contain all three types of NADH dehydrogenases. **B** Mitochondria from fungi and strictly aerobic yeast may contain complex I and up to four different NDH-2-type dehydrogenases that may be oriented to both faces of the inner mitochondrial membrane. **C** Mitochondria from fermentative yeasts do not contain complex I, but up to three NDH-2-type dehydrogenases. **D** Mitochondria from higher eukaryotes only contain complex I. *Nqr* sodium pumping NADH dehydrogenase, *NDH-2* alternative NADH dehydrogenase, *Complex I* proton-pumping NADH dehydrogenase

dissipated by the alternative NADH dehydrogenases, while Nqr uses it to pump sodium ions and complex I to pump protons. Thus the presence of different types of NADH dehydrogenases in the same cell allows for adjustments of the flow of reducing equivalents to ensure e.g. ATP supply by oxidative phosphorylation and at the same time sufficient regeneration of NAD^+. However, little is known about the details of such regulatory relationships between the different types of NADH dehydrogenases.

The purpose of this review is to provide a comprehensive overview of the three types of respiratory NADH dehydrogenases to highlight differences in subunit composition, structure and reaction mechanism.

2
"Alternative" or NDH-2-Type NADH Dehydrogenases

2.1
Characteristics

In contrast to the two other families of respiratory NADH dehydrogenases, NDH-2-type enzymes do not contribute to the formation of a membrane potential across the respiratory membrane. Procaryotic alternative NADH dehydrogenases are associated with the cytoplasmic face of the cell membrane, but eucaryotic alternative NADH dehydrogenases can have either an external or an internal orientation, and in many cases, several isoenzymes are found (Fig. 1).

Only very few potent inhibitors are known for NDH-2-type enzymes. Platanetin (6-dimethylallyl-3,5,7,8-tetrahydroxyflavone) inhibits the external alternative NADH:ubiquinone oxidoreductase activity of intact potato tuber mitochondria with an I_{50} of $2\,\mu M$ (Ravanel et al. 1990). In unsealed mitochondrial membrane preparations from the yeast *Yarrowia lipolytica*, HDQ (1-hydroxy-2-dodecyl-4(1H)quinolone) inhibits NDH2 with an I_{50} of 200 nM (Eschemann et al. 2005).

Typically, NDH-2-type enzymes consist of a single polypeptide chain with a molecular mass of 50–60 kDa and contain a non-covalently bound molecule of FAD as redox prosthetic group (Fig. 2). In several alternative NADH dehydrogenases from thermophilic archaebacteria, covalently attached FMN is found instead of FAD (Bandeiras et al. 2002; Gomes et al. 2001). Non-covalently bound FMN was reported to occur in the alternative NADH dehydrogenase from the parasite *Trypanosoma brucei* (Fang and Beattie 2002).

The flavine cofactor and the substrate NADH are bound to dinucleotide binding motifs (Wierenga et al. 1985) consisting of β-strand-α-helix-β-strand structures with GX(X)GXXG motifs at the junction between β-strand-1 and α-helix-2. Typically, at the end of β-strand-2, there is an acidic or hydroxyl amino acid residue that makes hydrogen bonds to the 2′ and 3′ OH groups

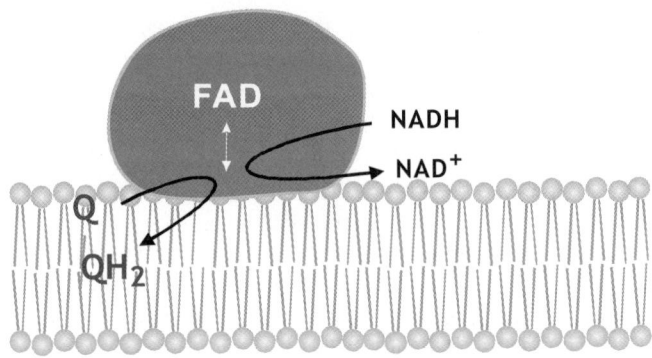

Fig. 2 Cartoon of alternative NADH dehydrogenase (NDH-2)

of the adenine ribose of NADH (Lesk 1995). Some NADH dehydrogenases are able to react with NADPH as well, but it is unclear how this difference correlates with specific features of the primary sequence. Several alternative NADH dehydrogenase sequences contain a Ca^{2+}-binding EF hand motif (Melo et al. 2001; Rasmusson et al. 1999). On the basis of these sequence features, alternative NADH dehydrogenases have been classified into three groups (Melo et al. 2004): enzymes from groups A and B possess two GX(X)GXXG motifs, with a calcium binding motif present in group B only. All group C enzymes identified so far either on the genome or the proteome level occur in thermophilic archaebacteria. They are characterized by the presence of a single, N-terminal GX(X)GXXG motif. From this observation, it may seem clear that the first GX(X)GXXG motif in groups A and B represents the binding site for the substrate NADH, while the second GX(X)GXXG motif represents the FAD binding site. This view, however, is challenged by the known X-ray structures of a large number of related enzymes, in which the FAD binding site corresponds to the first GX(X)GXXG motif (see below).

2.2
Physiological Roles

There is no definite answer as to why the respiratory chain of many plants, yeasts, fungi and bacteria is branched at the NADH:ubiquinone oxidoreductase step. Functions that have been tentatively proposed for alternative NADH dehydrogenases include elimination of excess reducing equivalents (Raghavendra and Padmasree 2003), and prevention of formation of reactive oxygen species (ROS) (Møller 2001). Nonetheless, there are isolated reports indicating that alternative NADH dehydrogenases may promote ROS production (Fang and Beattie 2003a). Only in a few cases have specific metabolic roles of alternative NADH dehydrogenases been identified, some of which are described in the following.

In *Azotobacter vinelandii*, NDH-2 is a vital component of the respiratory protection mechanism for the nitrogenase complex (Bertsova et al. 2001). For *Escherichia coli*, it is known that while complex I is the major respiratory chain NADH dehydrogenase under anaerobic conditions, the alternative enzyme prevails under aerobic or nitrate respiratory conditions. Under anaerobic conditions, *ndh-2* is repressed by FNR, the regulator of fumarate and nitrate reduction (Unden 1998; Unden and Bongaerts 1997). During anaerobic growth in rich medium, in the absence of FNR, *ndh-2* expression is activated by the amino acid response regulator (Green et al. 1997). Also, the *ndh-1* promoter is positively regulated by FIS during the transition from the lag to the early exponential phase, the late exponential and the stationary phase (Wackwitz et al. 1999). In addition to NADH:quinone oxidoreductase activity, NDH-2 of *E. coli* has NADH-linked Cu^{2+} reductase activity (Rapisarda et al. 1999). This enzyme, and a group of closely related alternative NADH dehydrogenases contain a HMA (heavy metal associated) consensus pattern (Rapisarda et al. 2002), corresponding to pattern PS01047 in the PROSITE database. The Cu^{2+} reductase activity of *E. coli* NDH-2 was reported to improve bacterial growth in extreme copper concentrations and increase the resistance to copper and hydrogen peroxide (Rodríguez-Montelongo et al. 2006).

In baker's yeast, *S. cerevisiae*, ethanolic fermentation is the preferred mode of glucose utilization even under aerobic conditions (Lagunas 1986). *S. cerevisiae* is a Crabtree effect positive yeast, i.e. it shows inhibition of oxygen consumption at high concentrations of glucose. *S. cerevisiae* lacks respiratory chain complex I but does possess three alternative NADH dehydrogenases, two external (SCNDE1 and SCNDE2) that are involved in the reoxidation of cytosolic NADH and an internal one (SCNDI1) that acts as substitute for the electron transfer function of complex I. In shake flasks, a mutant lacking the major external alternative NADH dehydrogenase (*nde1Δ*) and the double deletion mutant (*nde1Δ, nde2Δ*) exhibited reduced specific growth rates on ethanol and galactose but not on glucose (Luttik et al. 1998). Glucose metabolism in aerated, glucose-limited chemostat cultures of these deletion mutants was essentially aerobic. No ethanol and only traces of glycerol, which are the secreted metabolites typically formed during anaerobic reoxidation of cytosolic NADH, could be detected in culture supernatants. These findings indicate that *S. cerevisiae* has additional ways to feed electrons from cytosolic NADH into the mitochondrial respiratory chain. Candidate pathways are the glycerol-3-phosphate dehydrogenase system and the ethanol-acetaldehyde shuttle (Bakker et al. 2000; Luttik et al. 1998). Such shuttle mechanisms may also explain why deletion of the internal enzyme NDI1 is still compatible with respiratory growth on carbon sources that lead to NADH production not only in the cytosol (through the operation of the glyoxylate cycle) but even more so (through the operation of the citric acid cycle) in the mitochondrial matrix (Bakker et al. 2000; Marres et al. 1991).

Kluyveromyces lactis, like *S. cerevisiae*, is a facultatively anaerobic yeast, but does not show a Crabtree effect. Activity of the pentose phosphate cycle is higher in *K. lactis*, leading to higher cytosolic NADPH concentrations (Tarrío et al. 2006a). Recent results indicate that the two external NDH-2-type enzymes of *K. lactis* have an important function in the reoxidation of this cofactor. Expression of KLNDE1 and KLNDE2 in the *S. cerevisiae* phosphoglucose isomerase mutant, *pgi1*, restored growth in glucose, with KLNDE1 being more efficient. It was proposed that the lower K_M for NADPH of KLNDE1 may be a major determinant of this difference (Tarrío et al. 2006b).

External alternative NADH dehydrogenase activity of mitochondria from the obligate aerobic yeast *Y. lipolytica* depends on a single enzyme only, termed YLNDH2 (Kerscher et al. 1999). A second locus encoding an NDH-2-type enzyme (YALI0E05599g), which contains a Ca^{2+}-binding EF hand motif and is predicted to have an external localization, is present in the *Y. lipolytica* genome as deposited at the *genolevures* website (http://cbi.labri.fr/Genolevures/), but its metabolic role remains to be determined. Deletion of YLNDH2 did not produce any appreciable growth defect, but deletions of nuclear genes for essential subunits of complex I were not viable in standard laboratory strains of this yeast. Complex I mutants could only be generated after a truncated version of YLNDH2 had been redirected to the mitochondrial matrix by the N-terminal attachment of a mitochondrial import sequence (Kerscher et al. 2001a). This strict requirement of a mitochondrial matrix NADH dehydrogenase, which is in contrast to the situation in *S. cerevisiae*, implies that shuttle systems for the translocation of reducing equivalents from the matrix to the cytosol are either absent or inefficient in *Y. lipolytica*.

T. brucei, the parasite that causes African sleeping sickness has a dual life cycle in the bloodstream of the mammalian host and in the insect vector. Trypanosomes in mammalian bloodstream lack well-developed mitochondria and meet their energy requirements by anaerobic glycolysis, while trypanosomes in the insect host possess a large single mitochondrion with a cyanide-sensitive, cytochrome-containing electron transport chain (Hajduk et al. 1992). Mitochondrial complex I has been partially purified from these so-called procyclic forms (Fang et al. 2001), but alternative NADH dehydrogenase(s) are present as well. A novel type of internal alternative NADH dehydrogenase has been described that exists as a dimer consisting of two 33 kDa subunits with non-covalently bound FMN as a cofactor and employs a ping-pong mechanism (Fang and Beattie 2002). However, no sequence data are available so far. A gene for a canonical alternative 54 kDa NADH dehydrogenase (AY125472) has been detected in the *T. brucei* genome and catalytic activity was detected following heterologous expression in *E. coli* (Fang and Beattie 2003b). However, the orientation in *T. brucei* mitochondria and the relationship to the 33 kDa subunit enzyme are unknown.

Plant mitochondria possess complex I and several alternative NADH dehydrogenases. Two cDNA clones from potato (*Solanum tuberosum*), termed

*St-nda*1 and *St-ndb*1, were found to encode proteins localized to the internal and external sides of the inner mitochondrial membrane, respectively (Rasmusson et al. 1999). NDB1 contains an insertion carrying an EF hand motif for calcium binding, at the same sequence position as in the external calcium-dependent NADPH dehydrogenase, NDE1, in *N. crassa* (Melo et al. 1999, 2001). Seven alternative NADH dehydrogenase genes have been identified in the genome of *Arabidopsis thaliana*. Two and four of them are related to *St-nda*1 and *St-ndb*1, respectively, while the seventh one, *At-ndc1*, affiliates with cyanobacterial NDH-2-type genes, suggesting that this gene entered the eukaryotic cell via the chloroplast progenitor (note that *At-ndc1* belongs to group A, as defined in (Melo et al. 2004)). All of these genes are expressed in several organs of the plant and expression of one of them (*At-nda*1) is under circadian rhythm control, with peak expression early in the morning (Michalecka et al. 2003).

Although mammals do not have NDH-2-type enzymes, expression of SC-NDI1, the internal alternative NADH dehydrogenase from *S. cerevisiae* could restore NADH:ubiquinone oxidoreductase activity in several mammalian cells lines with defects in nuclear or mitochondrially coded complex I subunits (Bai et al. 2001; Seo et al. 1999; Yagi et al. 2006). In a rat dopaminergic cell line (PC12) and a human neuroblastoma cell line (SK-N-MC), SCNDI1 expression efficiently reduced the oxidative stress caused by complex I inhibitors such as rotenone (Seo et al. 2006a). Following unilateral injection of SCNDI1-recombinant adeno-associated virus particles into the substantia nigra of mice, resistance to MPTP-induced neuronal injury was observed (Seo et al. 2006b). Taken together, SCNDI1 holds the promise of a molecular remedy for hereditary and acquired complex I defects.

2.3
Other Enzymes Related to Alternative NADH Dehydrogenases

Alternative NADH dehydrogenases belong to the "FAD/NAD-linked reductases" family as defined in the "Structural Classification of Proteins" (SCOP) database accessible at http://scop.mrc-lmb.cam.ac.uk/scop/index.html. Members of this family catalyse redox reactions with NADH and a variety of electron acceptors. Multiple family members are present in all genomes, and as an example, a T-COFFEE (Notredame et al. 2000) alignment of several FAD/NAD-linked reductases from baker's yeast, *S. cerevisiae*, is shown in Fig. 3

- LPD1, GLR1 and TRR2 are members of an important subfamily of FAD/NAD-linked reductases family, namely the flavoprotein disulfide reductases. In addition to FAD, these enzymes use at least one non-flavin redox centre, most commonly an enzymic disulfide (Argyrou and Blanchard 2004).

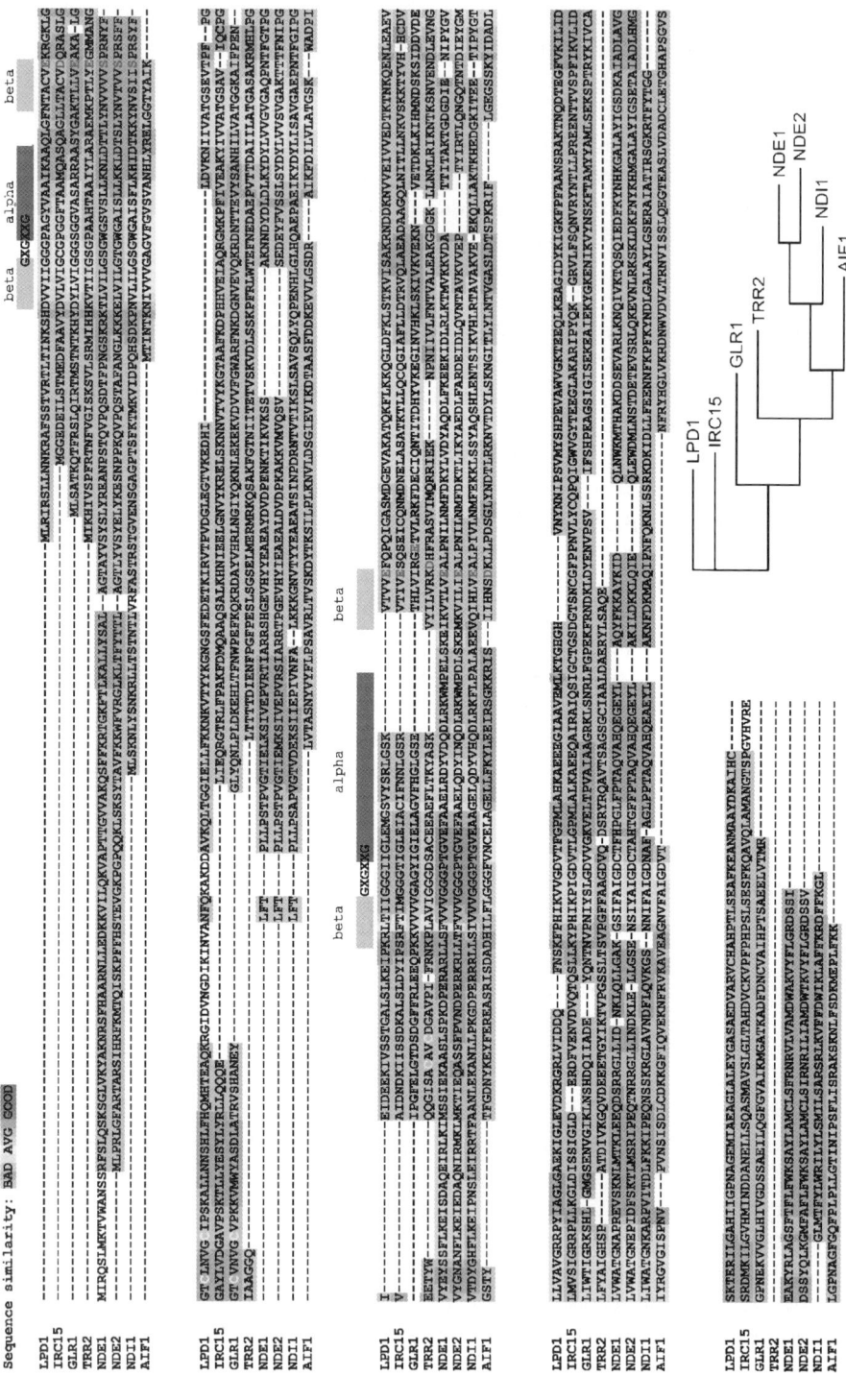

Fig. 3 Alignment of several FAD/NAD-linked reductases from *S. cerevisiae*. *LPD1* lipoamide dehydrogenase; *IRC15* putative S-adenosylmethionine-dependent methyltransferase; *GLR1* glutathione reductase; *TRR2* mitochondrial thioredoxin reductase; *NDE1, NDE2, NDI1* NDH-2-type enzymes; *AIF* apoptosis inducing factor. Dinucleotide binding sites, identified by GXGXXG motifs and β-α-β structural folds, are indicated above the alignment. Acidic or hydroxyl residues at the ends of β-strands-2 are indicated in *red*, reactive cysteine pairs (in LPD1, GLR1 and TRR2) in *yellow*. Secondary structure prediction for SCNDE1 was done using the PROF (Ouali and King 2000) server at http://www.aber.ac.uk/~phiwww/prof/; the phylogram was drawn using PAUP (Swofford 1993)

- Dihydrolipoamide dehydrogenase (LPD) is the lipoamide dehydrogenase component (E3) of the pyruvate dehydrogenase and 2-oxoglutarate dehydrogenase multi-enzyme complexes. Glutathione reductase (GLR) carries out the NADPH-dependent reduction of oxidized glutathione (GSSG) to yield free glutathione (GSH), a key antioxidant molecule.
- Thioredoxin reductases (TRR) are important antioxidant enzymes that determine the redox state of thioredoxins (TRX), which act as disulfide reductases on cytoplasmic proteins. *S. cerevisiae* has two different thioredoxin systems, a cytosolic (TRR1, TRX1 TRX2) and a mitochondrial (TRR2, TRX3) one. Mutagenesis studies have revealed that GLR1 and TRR2 have an overlapping function in the mitochondria (Trotter and Grant 2005). In comparison with LPD1 and GLR1, the reactive cysteine pair in TRR2 resides much further downstream in the primary sequence, indicating that similar reaction mechanisms of flavoprotein disulfide reductases are the result of convergent evolution.
- Human apoptosis inducing factor (AIF) was initially discovered by its ability to induce signs of apoptotic cell death (chromatin condensation, DNA fragmentation) in isolated nuclei. It is a flavoprotein with a molecular mass of 57 kDa, which, together with other proapoptotic factors like cytochrome *c* (Kluck et al. 1997; Liu et al. 1996), is released from the mitochondrial intermembrane space and translocates to the nucleus to induce the degradation phase of apoptosis (Susin et al. 1999). AIF appears to have an additional role as an assembly or maintenance factor for mitochondrial complex I: AIF depletion in mammalian cells, induced by homologous recombination or RNA interference, leads to high lactate production and enhanced dependency on glycolytic ATP generation, due to severe reduction of respiratory chain complex I activity. Similarly, Harlequin mice with reduced AIF expression due to a retroviral insertion display reduced expression of complex I subunits and reduced oxidative phosphorylation in retina and in brain, leading to retinal degeneration and neuronal defects (Vahsen et al. 2004). The molecular mechanisms underlying the vital and lethal function of AIF are poorly understood but none of the proposed models assumes that AIF can feed electrons into the mitochondrial respiratory chain (Modjtahedi et al. 2006). The AIF family has two addi-

tional members in humans, called AMID and AIFL. An X-ray structure of human AIF has been solved and structural models for human AMID, AIFL and *S. cerevisiae* AIF were constructed (Modjtahedi et al. 2006). AIF from *S. cerevisiae* is phylogenetically equidistant to human AIF, AMID and AIFL. The death effector function appears to be evolutionary conserved: purified AIF was able to induce DNA degradation in isolated yeast nuclei. Gene deletion and overexpression studies demonstrated that *S. cerevisiae* AIF plays a role in cell death induced by oxidative stress and chronological aging (Wissing et al. 2004).

- IRC15 is annotated as a putative *S*-adenosylmethionine-dependent methyltransferase in the *Saccharomyces* genome database at http://www.yeast genome.org/.

2.4
Structural Model and Mechanistic Implications

No X-ray structure for any alternative NADH dehydrogenase has been solved, but crystallization and preliminary structure determination of NADH:quinone oxidoreductase from the extremophile *Acidianus ambivalens* has been reported, and model building is in progress (Brito et al. 2006). A structural model has been created for NDH2 from *E. coli* through comparative modelling, using NADH peroxidase from *Enterococcus faecalis* (PDB entry: 2NPX) as a template (Schmid and Gerloff 2004). Both proteins share structural features typical for members of the SCOP family "FAD/NAD-linked reductases". The N-terminal FAD and the C-terminal NADH binding domain each contain a five-stranded parallel β-sheet, topped by a three-stranded antiparallel β-sheet. The C-terminal domain is less well conserved and structure prediction was not possible for the last 48 amino acids. The isoalloxazine ring system of FAD and the nicotinamide moiety of NADH are in planar apposition in the structural model of *E. coli* NDH-2 and it has been proposed that during catalysis, a triple sandwich is formed to allow electron transfer from NADH via FAD to the quinone headgroup (Schmid and Gerloff 2004).

Kinetic studies with NDH-2 from *Y. lipolytica*, however, have demonstrated that the reaction with NADH and *n*-decylubiquinone follows a ping-pong mechanism (Eschemann et al. 2005), and the same mechanism was observed for the reaction of partially purified SCNDI1 with NADH and the artificial electron acceptor dichlorophenol-indophenol (Velázques and Pardo 2001). These findings exclude the formation of a ternary complex during catalysis. A ping-pong mechanism may indicate either anti-cooperative binding of the substrates to two separate sites, an assumption that is implausible for a simple redox reaction, or mutually exclusive binding to the same site.

At first sight, it may seem difficult to envision that a hydrophilic substrate like NADH and a hydrophobic substrate like ubiquinone could bind to the same pocket, but two lines of indirect evidence support the implications of

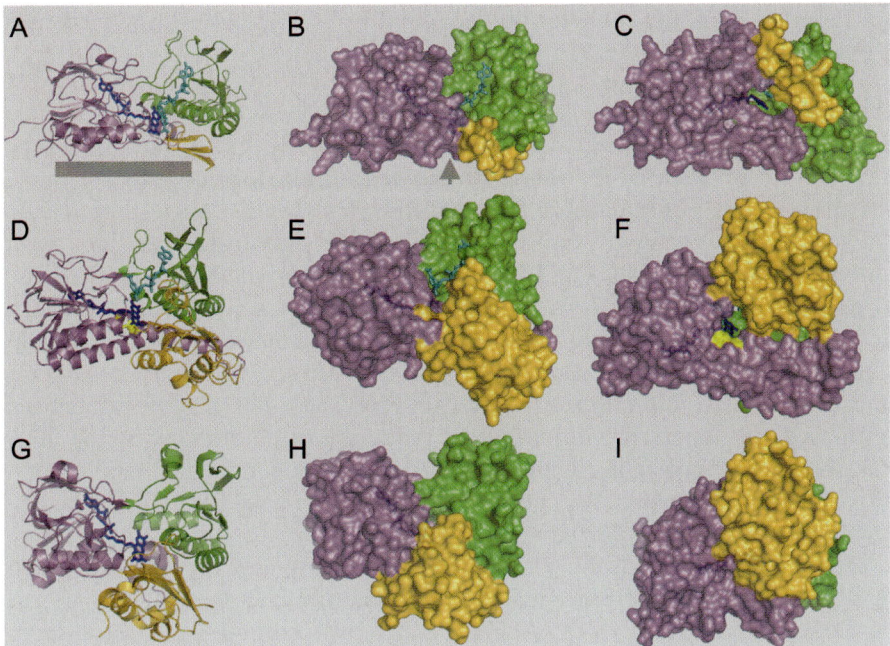

Fig. 4 Structural models of different enzymes related to NDH-2. Homology model of NDH-2 structure from *E. coli*, PDB: 1OZK (**A**, **B**, **C**), and X-ray structures of lipoamide dehydrogenase from *P. putida*, PDB: 1LVL (**D**, **E**, **F**) and *Homo sapiens* AIF, PDB: 1M6I (**G**, **H**, **I**), are shown as ribbon diagrams (**A**, **D**, **G**) and surface views (others). **B**, **E**, **G** have the same orientation as **A**, **D**, **G**, while **C**, **F**, **I** are rotated along the vertical axis to reveal the view indicated by the *arrow* in **B**. The *grey bar* in **A** indicates the approximate position of the respiratory membrane. In all proteins, the FAD-binding domain is in *magenta*, the NADH binding domain in *green* and the C-terminal domain in *light orange*. FAD and NADH are shown as *blue* and *light blue* stick diagrams, respectively. In the lipoamide dehydrogenase structures, the reactive cysteine pair is indicated in *yellow*. Please note that the nicotinamide ring in 1LVL does not seem to be in a catalytically relevant orientation. The figure was prepared using PyMOL v0.99

the kinetic data: (i) while the ubiquinone "tail" is hydrophobic, the headgroup is not. It can be assumed that while the ubiquinone tail is confined to the hydrophobic part of the phospholipid bilayer, the ubiquinone headgroup has the freedom to diffuse into the hydrophilic phosphodiester or the surrounding aqueous phase of the membrane. In fact, indications for a diffusion barrier, which can be modelled according to the concept of external diffusional control (Goldstein 1976), have been observed for the reaction of *Y. lipolytica* NDH2 with NADH and *n*-decylubiquinone (Eschemann et al., unpublished); (ii) in the structural model of *E. coli* NDH, the pocket that is occupied by the nicotinamide moiety of NADH on the top side of the view shown in models A and B of Fig. 4 is also accessible from the bottom side of the same view, i.e. from the face of the enzyme that is in contact with the phospholipid bilayer.

A channel that connects the top and bottom sides of the enzyme and the isoalloxazine ring system of FAD is clearly seen when the enzyme is rotated by about 90 degrees along its horizontal axis (model C). It may be argued that the significance of the model structure is limited, since the C-terminal 48 amino acids might occlude this channel. However, a similar channel can be seen in the X-ray structure of the lipoamide dehydrogenase from *Pseudomonas putida* (models D–F). *Mycobacterium tuberculosis* lipoamide dehydrogenase can catalyse electron transfer from NADH to lipoamide and a variety of quinones, including 2,6-dimethyl-1,4-benzoquinone and 5-hydroxy-1,4-naphthoquinone with similar efficiencies and seems to utilize a ping-pong mechanism with all these substrates (Argyrou et al. 2003). Also, it has been proposed that the extra-mitochondrial reduction of ubiquinone by the three flavoenzymes lipoamide dehydrogenase, glutathione reductase and thioredoxin reductase, makes an important contribution to the regeneration of this powerful antioxidant (Nordman et al. 2003).

Planar apposition of the FAD isoalloxazine ring and the quinone headgroup is not without precedent. X-ray structures of human and mouse NADH:quinone oxidoreductase (QR1), which protects cells from the deleterious and carcinogenic effects of quinoic compounds, have been solved both for the apo- and the duroquinone bound form (Faig et al. 2000). In QR1, which utilizes a ping-pong mechanism, the same binding pocket can accommodate the two substrates NADH and duroquinone in a mutually exclusive fashion. It is tempting to speculate that catalysis by NDH-2 proceeds via a similar mechanism, but eventually, X-ray structures of NDH-2, co-crystallized with each of the two substrates will be needed to clarify this issue.

3
Sodium-Pumping NADH Dehydrogenases (Nqr)

3.1
Occurrence, Cellular Functions and Physiological Significance in Bacteria

Some marine and pathogenic bacteria, as well as certain extremophilic bacteria, have a special energy-transducing mechanism that is different from those found in *E. coli* and most of the freshwater and terrestrial bacteria (Häse et al. 2001; Kogure 1998). These bacteria require sodium for their growth and utilize a Na^+ gradient across their cytoplasma membrane that serves as a driving force for various cellular functions like ATP synthesis, Na^+-dependent symport of nutrients or Na^+-dependent motility (flagellar rotation). The use of Na^+ as a coupling ion can be explained by the difficulty in generating a proton-motive force (pmf) under certain environmental conditions, e.g. in seawater where the external pH of about 8.0 would result in an "opposite" proton gradient (ΔpH). Among the three types of primary sodium pumps identified is a unique sodium motive, NADH quinone reductase (Nqr) that is not related to proton-pumping NADH quinone oxidoreductases (NDH-1; complex I) or NDH-2-type enzymes (Bogachev and Verkhovsky 2005; Häse and Barquera 2001; Rich et al. 1995).

This primary sodium pump was first discovered in the halophilic marine bacterium *Vibrio alginolyticus* (Tokuda and Unemoto 1981, 1984). Nqr purified from *V. alginolyticus* was first considered to be composed of three major subunits α, β and γ but later six structural genes (*nqrA* to *nqrF*; sometimes entitled *nqr1* to *nqr6* in older literature) were found in the *nqr* operon (Beattie et al. 1994; Hayashi et al. 1995) that is required to build a functional sodium pump (Hayashi et al. 2001a; Nakayama et al. 1998). Subsequently, an *nqr* operon was found in many marine and pathogenic bacteria including *Vibrio harveyi*, *V. cholerae*, *Haemophilus influenzae*, *Neisseria gonorrhoeae*, and *N. meningitidis*, *Yersinia pestis*, *Shewanella putrefaciens* and *Pseudomonas aeruginosa* (Zhou et al. 1999). Most of these bacteria belong to the proteobacteria gamma division (Mrazek et al. 2006), but the operon can also be found in genomes of bacteria that belong to other branches of the phylogenetic tree like *Chlamydia trachomatis*, *C. pneumoniae*, and *Actinobacillus actinomycetemcomitans*. The latter three bacteria contain orthologues for all six subunits of Nqr, but the genes are not organized in a single operon (Häse et al. 2001; Zhou et al. 1999). Database searches also reveal several more distant homologies, e.g. to nitrogen fixation proteins of *Rhodobacter capsulatus* (*rnf* operon), to proteins of *E. coli* (*ydg* operon or *nqrEGABCD*) and *Klebsiella pneumoniae* (*nqrEGABCD*) (see chapter by Vignais, in this volume). Homologous gene products (not always a complete operon) were also identified in proteobacteria alpha (*nqrD* and *nqrE* – *Silicibacter* sp. TM1040) and such "exotic" bacteria as Thermotogae (*nqr A* to *nqrE* – *Thermotoga maritima*; *nqrD* – *Fervidobacterium nodosum*) and Planctomycetes (*nqrE* – *Rhodopirellula baltica* SH 1), (Häse et al. 2001; Zhou et al. 1999 and BLAST search). In *Th. maritima*, the *nqrF* gene is replaced by a gene encoding a different iron–sulfur cluster-containing protein *nqrG*, (Häse et al. 2001) that is homologous to *rnfB* of *Rh. capsulatus*.

While the utilization of a sodium-motive force in marine bacteria is evident, its existence in blood-born pathogenic bacteria is less obvious, especially since most of them also encode primary H^+ pumps (Häse et al. 2001). While in *V. cholerae* Nqr is not essential for survival of the pathogen, it was implicated in the spreading of the free-living bacterium in its environmental phase (Häse and Barquera 2001).*V. cholerae* is found primarily in inland coastal areas and estuaries, but the bacterium thrives in seawater as well. Nqr might not only be involved in the adaptation to different environmental conditions. It was reported that when the *nqr* operon in *V. cholerae* is interrupted by transposon mutagenesis, virulence factors are expressed at higher levels. Therefore, the involvement of Nqr in sodium homeostasis appears to be linked to pathogenesis (Häse and Mekalanos 1999). Moreover, *V. cholerae* relies on Na^+-dependent polar flagella for its motility. The presence of *nqr* genes in genomes of obligate anaerobes like *Th. maritima* imply that Nqr can also work in the reverse direction, using the energy of a Na^+-gradient to reduce NAD^+ (Häse et al. 2001).

3.2
Subunit Composition and Cofactors

The Nqr complex is an integral membrane enzyme that is composed of the six subunits NqrA–NqrF encoded by the *nqr* operon (Fig. 5). Three of these subunits (NqrB, NqrD and NqrE) are very hydrophobic, whereas the others are rather hydrophilic. Table 1 summarizes the latest consensus model of their membrane topologies, which was predicted from a combination of different computer-based topology algorithms and gene fusion studies with subunits of the *V. cholerae* enzyme (Duffy and Barquera 2006), and also indicates the known binding sites of the binuclear iron–sulfur cluster, NADH, FAD, two FMNs and probably ubiquinone-8.

Among the six subunits, only NqrF shows significant homology to other proteins. Rich et al. (1995) identified three motifs in its sequence which they proposed to be: (1) the NADH binding site, (2) the binding site for the noncovalently bound FAD cofactor and (3) the iron–sulfur cluster binding site. The alignment of sequences from different bacterial genomes clearly reveals

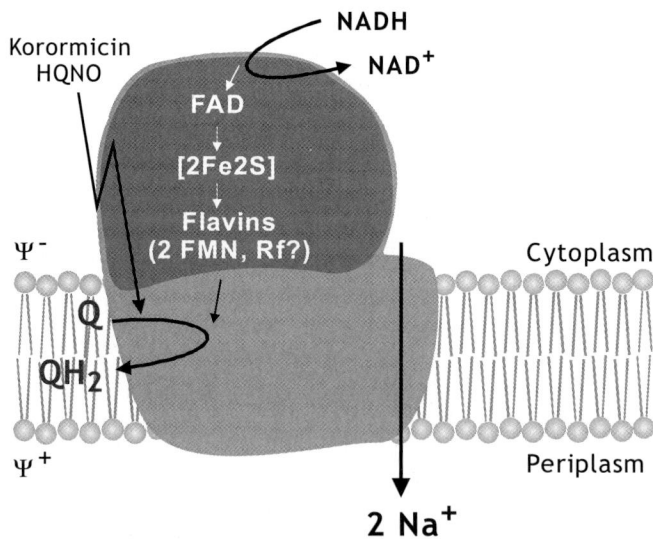

Fig. 5 Cartoon of sodium-pumping NADH dehydrogenase (Nqr)

Table 1 Na$^+$-NQR subunit composition and cofactor binding exemplified by the enzyme of *V. cholerae*

Subunit	Residues	Molecular weight	Predicted membrane topology Consensus/experimental[a]	Bound cofactor(s)
NqrA	446	48.6	0/0	?
NqrB	415	45.4	9C/9C	Covalently bound FMN(T236); Q binding?[b]
NqrC	257	27.6	1P/2P	Covalently bound FMN(T225)
NqrD	210	22.8	5C/6C	?
NqrE	198	21.5	6P/6P	?
NqrF	408	45.1	2P/1P	Binding site for NADH, contains A [2Fe2S] cluster and non-covalently bound FAD
Binding site unknown:				Riboflavin

[a] Predicted consensus model and experimentally supported model presented by Duffy & Barquera (2006). P and C denote predicted periplasmic and cytoplasmic C-terminal locations

[b] NADH:Q$_1$ activity of a Na$^+$-NOR complex from *V. alginolyticus* with a point mutation in *nqrB* (resulting in the conversion G140V) showed reduced sensitivity to the inhibitors korormicin and HQNO (Hayashi et al. 2002)

that these three motifs are conserved across all Nqr complexes (Barquera et al. 2004). The iron–sulfur binding motif is marked by four conserved cysteines that were initially proposed to ligate a tetranuclear cluster based on the homology of the N-terminal part of NqrF to known proteins containing tetranuclear iron–sulfur clusters like ferredoxins (Rich et al. 1995). However, EPR, visible and circular dichroism spectroscopy indicated the presence of a binuclear cluster in NqrF that is related to two-iron ferredoxins of the vertebrate type (Barquera et al. 2004; Bogachev et al. 1997; Lin et al. 2005; Pfenninger-Li et al. 1996; Türk et al. 2004). The alternate mutation of the four conserved cysteine residues (C70, C76, C79 and C111 in *V. cholerae* NqrF) of the canonical binding motif resulted in the loss of the iron–sulfur cluster and abolished specific NADH:ubiquinone-1 oxidoreductase activity (Barquera et al. 2004). The C-terminal part of NqrF shows local regions of homology that can be aligned with FAD and NADH binding sites in NADH-oxidizing enzymes belonging to a large family of flavoenzymes related to ferredoxin reductases (Rich et al. 1995). Mutation of three conserved residues (R210, Y212 and S246 in *V. cholerae* NqrF) in the FAD binding domain resulted in the elimination of one extractable flavin per complex, indicating that this motif indeed binds the cofactor (Barquera et al. 2004).

These results and the presence of an additional characteristic NADH binding site indicate that the NqrF subunit is the entry point of electrons into the Nqr complex. This view is supported by a high NADH oxidase activity of a purified soluble form of NqrF (Türk et al. 2004). PhoA fusions of the *V. cholerae* NqrF indicated that its cofactor binding regions are localized on the cytoplasmic side of the membrane (Duffy and Barquera 2006). Most computer-based topology algorithms predicted two transmembrane helices in the N-terminal part of NqrF (Duffy and Barquera 2006; Rich et al. 1995). However, the presence of the second predicted transmembrane α-helix is unlikely, since this would place one of the four cysteines ligating the binuclear iron–sulfur cluster in the middle of the membrane. Moreover, Türk et al. (2004) were able to produce a soluble variant of *V. cholerae* NqrF by removing residues 3–25, which include the first predicted transmembrane α-helix.

Two subunits contain covalently bound FMN and are: (i) the hydrophobic subunit NqrB (Nakayama et al. 2000) that is predicted to contain nine transmembrane α-helices (Duffy and Barquera 2006) and (ii) the relatively hydrophilic subunit C (Barquera et al. 2001; Nakayama et al. 2000; Zhou et al. 1999) featuring an N-terminal and a C-terminal α-helix (Duffy and Barquera 2006). In both subunits, FMN is attached by an ester bond between its phosphate group and a threonine residue of the protein (T235 in NqrB and T223 in NqrC; *V. alginolyticus* numbering) (Hayashi et al. 2001b). This mode of covalent flavin binding is unique to Nqr. Mutagenesis of the corresponding Thr-residues in NqrC (T225) and NqrB (T236) of *V. cholerae* resulted in the loss of the covalently bound FMNs (Barquera et al. 2001, 2006). A point mutation (G140V in *V. alginolyticus*) in NqrB resulted in a large shift in the affinity for the highly specific inhibitor korormicin (Hayashi et al. 2002). Since korormicin is a non-competitive inhibitor for Q_1, this suggests that NqrB might contain a (part of a) quinone binding site. Subunits NqrD and NqrE of *V. cholerae* both have six transmembrane α-helices (Duffy and Barquera 2006), which is in agreement with the topology of their homologues in *E. coli* (YdgQ and YdgL). Both subunits contain negatively charged amino acid residues within their transmembrane helices, and several of these residues are conserved in the sequences of other bacteria. For NqrA, no transmembrane helices were predicted by most algorithms, and protein fusions to PhoA and GFP confirm cytoplasmic localization (Duffy and Barquera 2006). Therefore, it can be assumed that NqrA is a soluble protein tightly attached to the complex by non-covalent interactions.

The existence of a fourth flavin as an intrinsic cofactor of Nqr is still a controversial matter in the field. Spectroscopic determinations of extractable (non-covalently bound) flavins and covalently bound flavins gave a ratio of around 1 : 1 (Barquera et al. 2002a; Bogachev et al. 2006). Barquera et al. (2002a) calculated a content of four flavins per Nqr complex, which would suggest the presence of a second non-covalently bound cofactor. Subsequently, the same group extracted riboflavin in addition to FAD from purified

His-tagged Nqr of *V. cholerae* (Barquera et al. 2002b, 2004). Since the ratio of the extractable riboflavin to FAD was close to 1 : 1, they concluded that the riboflavin is a bona fide component of the Nqr preparation. While the Nqr flavin cofactors FAD and FMN are both riboflavin derivatives, riboflavin itself has never been reported to be an enzyme-bound cofactor (Barquera et al. 2002b).

In contrast, Bogachev et al. (2006), who extracted tightly but non-covalently bound flavins from His-tagged Nqr from *V. harveyi* (which was expressed in a $nqr\Delta$ deletion strain from *V. cholerae*; Barquera et al. 2002a), found variable riboflavin to FAD ratios of 0.4–0.6 in addition to some free "soluble" FMN. These authors suggested that the riboflavin and the FMN found following acid extraction were the result of partial hydrolysis of the phosphoester bond of covalently bound FMN. They also argued that their kinetic and thermodynamic analysis showed that there are only three flavin cofactors in the active enzyme (Bogachev et al. 2002, 2006). However, Barquera et al. (2006) recently showed that in different Nqr mutants, where the known bound flavins were eliminated alternately, there is still a flavin radical present that cannot be assigned to non-covalently bound FAD or the two covalently bound FMNs. They concluded that this radical must arise from non-covalently bound riboflavin (see below). So far, there are no indications for a riboflavin binding site. From studies of Duffy and Barquera (2006) it seems clear that all the redox-active cofactors of Nqr are located on the cytoplasmic (negative) side of the membrane, with the possible exception of riboflavin, whose location in the protein has not been established. This topological organization bears some analogy to that of complex I.

In addition to the flavins and the iron–sulfur cluster, Nqr of *V. alginolyticus*, *V. harveyi* and *V. cholerae* contains tightly bound ubiquinone-8 (Barquera et al. 2002a; Pfenninger-Li et al. 1996; Zhou et al. 1999). However, the detergent used in the Nqr preparation has some influence on the ubiquinone content (Barquera et al. 2002a): enzyme prepared using dodecylmaltoside contained approximately one mole of ubiquinone-8 per mole of enzyme, whereas enzyme prepared using lauryldimethylamine-oxide (LDAO) was completely devoid of quinone.

3.3
Functional Properties and Catalytic Mechanism

Nqr complex has been purified from the three *Vibrio* species, *V. alginolyticus*, *V. harveyi* and *V. cholerae* (Barquera et al. 2002a; Pfenninger-Li et al. 1996; Zhou et al. 1999). The enzyme oxidizes NADH and deamino NADH and transfers electrons to its natural substrate ubiquinone-8, but can also use hydrophilic ubiquinones (like Q_1), menaquinone or ferricyanide as electron acceptors. The purified His-tagged Nqr complex of *V. cholerae* exhibits a specific NADH oxidation activity of $\sim 100\,\mu\text{mol min}^{-1}\,\text{mg}^{-1}$ (or $\sim 700\,\text{electrons s}^{-1}$)

and a specific ubiquinone-1 reductase activity of ~ 77 μmol min^{-1} mg^{-1} (Barquera et al. 2002a). The reported near 1 : 1 stoichiometry of NADH oxidation and quinone reduction contrasts with some previous reports of preparations of the enzyme from other sources (Pfenninger-Li et al. 1996) in which the rate of NADH oxidation was substantially greater than the rate of quinone reduction. Apparently, these preparations are able to donate electrons to alternate electron acceptors. It was proposed by two independent laboratories (Barquera et al. 2002a; Bogachev et al. 2001) that the detergent LDAO used in those "early" preparations was responsible for this discrepancy, because it might serve as an electron acceptor by itself. Between the preparations with LDAO and those with the more suitable dodecylmaltoside there is also a discrepancy in the rates of menaquinone reduction. The Nqr complex of *V. cholerae* purified in dodecylmaltoside has a NADH:quinone-1 oxidoreductase activity that is stimulated fivefold by 200 mM NaCl and that is highly sensitive to the inhibitor heptylquinoline-*N*-oxide (HQNO) (Barquera et al. 2002a). In contrast, the LDAO preparation is insensitive to HQNO, suggesting a short-circuit of the coupled redox reaction.

Sodium ions are indispensable for the catalytic activity of Nqr (Bogachev and Verkhovsky 2005). For example, the ubiquinone-1 reductase activity of the *V. harveyi* complex is stimulated by Na$^+$ with a K_m of ~ 3 mM (Bogachev et al. 2001). This activity is sensitive to the inhibitors korormicin ($K_i \sim$ 0.1 nM; Yoshikawa et al. 1999) and HQNO (K_i = 0.3 – 0.4 μM; Yoshikawa et al. 1997; Zhou et al. 1999). Both compounds act as non-competitive inhibitors for Q_1. Inhibition of the Nqr complex by korormicin and HQNO is mutually exclusive, suggesting partially overlapping binding sites (Hayashi et al. 2002). In contrast to the quinone reductase reaction, the so-called NADH dehydrogenase reaction with artificial electron acceptors is Na$^+$-independent and insensitive to the inhibitors korormicin and HQNO (Bogachev and Verkhovsky 2005).

Transport of ^{22}Na$^+$ was directly shown with (partially) purified and reconstituted Nqr from *V. alginolyticus* (Pfenninger-Li et al. 1996). Ionophores (valinomycin, FCCP) stimulated the rate of Na$^+$ transport, indicating that the transport is a primary event. A stoichiometry of 0.5 Na$^+$ per oxidized NADH was calculated, but this value is probably underestimated, considering the low coupling of the proteoliposomes used and the low rate of quinone reduction of the reconstituted complex. Subsequently, reconstituted Nqrs from *V. alginolyticus*, *V. harveyi* and *V. cholerae* were shown to act as primary sodium pumps by indirect methods (Barquera et al. 2002a; Bogachev et al. 1997; Zhou et al. 1999). The formation of $\Delta\Psi$ (monitored by the dye oxonol VI) was dependent on the presence of Na$^+$, stimulated by monensin, slightly inhibited by CCCP and entirely arrested by the sodium ionophore ETH-157. These data indicate that Nqr is an electrogenic sodium pump and that the enzyme is not able to pump protons. A Na$^+$/e$^-$ stoichiometry of ≈ 1 (experimentally 0.71 $\pm$ 0.06) was indirectly determined with *V. alginolyticus* cells, following the alkalization of the medium (influx of protons) in the presence of electrogenic protonophores (Bogachev et al. 1997).

The molecular mechanism of Nqr that couples the electrogenic transport of Na^+ ions to the electron transfer from NADH through several redox-active cofactors to ubiquinone is still elusive. However, redox titrations, time-resolved measurements of the reduction of redox-active cofactors and EPR and ENDOR measurements not only deciphered some of the events that occur during the enzyme's catalytic cycle, but also continuously changed the interpretation of previous results since "new" and "hidden" cofactors and intermediates were identified. EPR measurement of reduced Nqr detected a signal that was assigned to the iron–sulfur cluster and one radical signal that occurred at a spin concentration of 1 : 1 with respect to the iron–sulfur cluster (Barquera et al. 2002a; Bogachev et al. 2001). The signal of the iron-sulfur cluster disappeared in the air-oxidized complex, but the radical signal can be still detected with almost unchanged intensity (Barquera et al. 2002a; Bogachev et al. 2001, 2002). Subsequently, this radical signal was shown to originate from a neutral flavin-semiquinone in the oxidized enzyme whereas an anionic flavin-semiquinone has been reported for the reduced enzyme (Barquera et al. 2003; Bogachev et al. 2002). Both flavin radicals exhibit an unusually high stability. The occurrence and disappearance of the flavin radicals were also monitored by optical spectroscopy.

The time course of Nqr reduction by NADH showed three distinct phases corresponding to the reduction of three different flavin species (Bogachev et al. 2002). The first phase is fast both in the presence and absence of sodium, and is assigned to reduction of FAD to $FADH_2$. The rates of the other two phases are strongly dependent on sodium concentration, and these phases were attributed to the reduction of the two covalently bound FMNs (but see below). These results place the coupling site for sodium transport between the sodium-independent and sodium-dependent redox reactions. Combination of data from optical and EPR spectroscopy suggested that a neutral flavosemiquinone may become reduced to the fully reduced flavine by NADH. The other FMN moiety is initially oxidized, and becomes reduced to anionic flavosemiquinone. This would imply that the two radicals found in Nqr originate from different flavin cofactors of the enzyme (Bogachev et al. 2002, 2006; Bogachev and Verkhovsky 2005). Bogachev and colleagues concluded that at least one of these one-electron transitions is coupled to the transmembrane sodium translocation. However, subsequent investigations revealed that there is no sodium dependence of the midpoint potential for any of the detectable Nqr cofactors, which would be expected for an electron transition that is coupled to the sodium translocation (Bogachev et al. 2006).

EPR spectra of FAD-deficient mutants in the oxidized and reduced forms exhibit neutral and anionic flavosemiquinone radical signals, respectively. This confirms that FAD in NqrF is not the source of either radical signal (Barquera et al. 2004). Furthermore, in both FAD and iron–sulfur cluster mutants the line widths of the neutral and anionic flavosemiquinone EPR signals are unchanged as compared to the wild-type enzyme, indicating that neither of these centres is coupled to the radicals. Recently, Bogachev et al. (2006) studied the reduction of Nqr from *V. harveyi* in more detail by spectro-electrochemistry using full spectrum absorbance detection. They confirmed three redox transitions involving flavins: (1) two-electron reduction of a flavoquinone; (2) one-electron reduction of a flavoquinone to form an anionic flavosemiquinone; (3) one-electron reduction of a neutral flavosemiquinone. However, recent data from Barquera and colleagues (Barquera et al. 2006) suggest that the mechanism of Nqr might be even more complicated: EPR and ENDOR analysis of mutants in NqrB and NqrC of the *V. cholerae* enzyme replacing the threonine ligands for the two FMNs by other amino acids revealed two distinct forms of the anionic radical (anionic radical I and II). When the enzyme becomes partially reduced, a mixture of anionic radicals I and II is observed, whereas in the fully reduced enzyme, only anion radical I can be detected. The authors concluded that the sodium-translocating NADH:quinone oxidoreductase forms three spectroscopically distinct flavin radicals: (1) a neutral radical in the oxidized enzyme that most likely arises from riboflavin; (2) an anionic radical observed in the fully reduced enzyme, which is present in wild-type and a NqrC mutant but not the NqrB mutant; (3) a second anionic radical, seen primarily under weakly reducing conditions, which is present in wild-type and a NqrB mutant but not a NqrC mutant. On this basis the first anionic radical was tentatively assigned to the FMN in subunit B and the second to the FMN in subunit C (Barquera et al. 2006). Mechanistically this implies that reduction of the enzyme would involve at least four one-electron flavin redox transitions and one two-electron flavin transition as follows:

1. One-electron reduction of neutral riboflavin semiquinone to the corresponding hydroquinone
2. One-electron reduction of FMN in NqrC to form anionic semiquinone (anionic radical II)
3. Further one-electron reduction of this FMN semiquinone to the corresponding hydroquinone
4. One-electron reduction of FMN in NqrB to form anionic semiquinone (anionic radical I)
5. Two-electron reduction of FAD in NqrF

In addition, the reaction should also include redox transitions of the iron–sulfur cluster, and possibly of tightly bound quinone.

The inconsistency in the identified Nqr cofactors and in the number of one-electron flavin redox transitions and some surprising recent results suggest that all working models proposed so far for the mechanism of Nqr (Bogachev et al. 2002; Bogachev and Verkhovsky 2005) have to be revised. For example, it will be critical to establish whether there is an effect of the sodium concentration on the recently discovered second anionic flavin radical – if its existence is confirmed by independent investigations – because this could identify or exclude another possible coupling site. So far, only the initial part of the electron transport pathway seems well established (Barquera et al. 2004; Bogachev et al. 2006): non-covalently bound FAD in the NqrF subunit is the first cofactor to accept electrons from the substrate NADH; the electrons then go to the iron–sulfur cluster before they are transferred to the flavins.

4
Proton-Pumping NADH:Ubiquinone Oxidoreductase (Complex I)

4.1
Occurrence and General Features

Proton-pumping complex I is by far the largest and most complicated representative of the NADH:ubiquinone oxidoreductases. Complex I is the major entry point for electrons of the respiratory chain in mitochondria from most eucaryotes and in prokaryotes, where this enzyme is frequently termed NDH-1. Several archaebacterial genomes encode complex I-like enzymes, in which essential subunits from the NADH dehydrogenase module (see below) are replaced by other proteins. The *Archaeoglobus fulgidus* enzyme functions as a $F_{420}H_2$:quinone oxidoreductase (Brüggemann et al. 2000) and the enzyme from *Methanosarcina mazei* as an $F_{420}H_2$:methanophenazine oxidoreductase (Bäumer et al. 2000). A similar situation is seen in cyanobacteria (Prommeenate et al. 2004) and chloroplasts (Rumeau et al. 2005) and the human pathogens *Helicobacter pylori* and *Campylobacter jejuni* (Smith et al. 2000). For all these enzymes, the nature of the direct electron donor is speculative.

A growing body of evidence indicates that complex I is involved in neurodegenerative processes like Parkinson's disease (Greenamyre et al. 2001) and it was shown that the enzyme is a major source of reactive oxygen species (ROS) generated in the respiratory chain (Galkin and Brandt 2005; Kussmaul and Hirst 2006).

NDH-1 is regarded as a minimal form of the enzyme as it is able to carry out the complete bioenergetic function (NADH oxidation, quinone reduction, proton pumping). It consists of 14 subunits with a mass of about 550 kDa (Friedrich 1998). The eukaryotic enzyme is significantly more complicated with up to 45 different subunits in bovine (Carroll et al. 2003, 2006a) and at least different 40 subunits in fungi like the strictly aerobic yeast *Y. lipolytica* (Abdrakhmanova et al. 2004). The function of these up to 31 "accessory" subunits is largely unknown (Brandt 2006) and will not be further discussed here. All 14 bacterial subunits are present in the eukaryotic enzyme. These "central" subunits can be divided into seven hydrophilic and seven hydrophobic proteins. In most eukaryotes the latter are encoded by the mitochondrial genome. Unfortunately, the nomenclature of the subunits is somewhat confusing for historic reasons. In the best-characterized bovine system, subunits are either named by their N-termini (e.g. PSST) or by their molecular mass (e.g. 49-kDa or B15 in the case where the N-terminus is blocked). The seven central hydrophobic subunits coded by the mitochondrial genome are designated ND1 to ND6 and ND4L. However, for other organisms different designations are used. For clarity we will use the bovine system here also for orthologues from other species.

The enzyme contains FMN as the primary electron acceptor and eight iron–sulfur clusters designated N1a, N1b, N2, N3, N4, N5, N6a and N6b (Ohnishi 1998; Rasmussen et al. 2001). In NDH-1 of some bacteria, one additional cluster (N7) is found (Uhlmann and Friedrich 2005). A plethora of inhibitors are known to block ubiquinone reduction by complex I. The most prominent examples are rotenone, piericidine, capsaicin and the acetogenous annonins (Degli Esposti 1998). Despite their diverse chemical nature they were all shown to bind to a large binding pocket with overlapping binding sites (Okun et al. 1999). The redox chemistry is coupled to the translocation of protons across the membrane with a stoichiometry of $4H^+/2e^-$ (Galkin et al. 2006; Wikström 1984) contributing about 40% of total $\Delta\mu H$ in mammalian mitochondria. It has been proposed that bacterial complex I also pumps sodium ions (Gemperli et al. 2003; Krebs et al. 1999). If true, this would have far-reaching mechanistic consequences. Therefore, this critical issue is discussed in detail in the following section.

4.2
Can Bacterial Complex I Act as a Sodium Pump?

It has been demonstrated that complex I from *E. coli* (Bogachev et al. 1996; Stolpe and Friedrich 2004) and *K. pneumoniae* (Bertsova and Bogachev 2004) pumps protons. In contrast, Steuber and colleagues have presented evidence suggesting that complex I from *Klebsiella pneumoniae* (Gemperli et al. 2003; Krebs et al. 1999) and *E. coli* (Steuber et al. 2000) pumps sodium ions at a stoichiometry of $2Na^+/2e^-$ (Gemperli et al. 2002). This is an intriguing proposal since the membrane integral subunits ND2/ND4/ND5 of complex I exhibit remarkable similarity to bacterial Na^+/H^+ antiporters antiporter (Fearnley and Walker 1992; Mathiesen and Hägerhäll 2002), and the activities of complex I and these antiporters are inhibited by the same amiloride derivatives (Nakamaru-Ogiso et al. 2003).

From a thermodynamic point of view however, it seems rather unlikely that $2Na^+/2e^-$ could be pumped in addition to $4H^+/2e^-$ and in fact, both activities have not been observed in the same experiment. Alternatively, it could be envisioned that complex I from these organisms exhibits variable ion specificities, depending on the experimental conditions. However, the sodium pumping capacity itself of complex I has been questioned (Bertsova and Bogachev 2004; Stolpe and Friedrich 2004). Thus, before discussing a mechanism of how the ion specificity could change and what the physiological implications may be, it seems necessary to scrutinize the evidence that has been presented to show pumping of either of the cations.

Obtaining quantitative measurements of proton pumping to determine H^+/e^- stoichiometries is not a simple task. However, it is rather straightforward to qualitatively demonstrate proton pumping by using pH-sensitive dyes to detect formation of a proton gradient across the membrane and applying uncouplers to selectively dissipate this gradient. For reconstituted *E. coli* complex I it was shown that the formation of a proton gradient and a membrane potential was completely abolished by the addition of the specific complex I inhibitor piericidin A, demonstrating that the observed pumping activity was in fact due to complex I (Stolpe and Friedrich 2004). Uncoupler-sensitive proton pumping by *K. pneumoniae* complex I was demonstrated in whole cells and sub-bacterial particles and inhibition of proton pumping by capsaicin, another specific inhibitor of complex I, was reported (Bertsova and Bogachev 2004). In this study, it was also demonstrated that proton pumping and piericidin A-sensitive deaminoNADH oxidase activity were absent in a strain of *K. pneumonia* lacking complex I. Thus, there is compelling evidence that the complexes from *E. coli* and *K. pneumoniae*, like those from all other organisms tested so far, operate as proton pumps.

There are no dyes or electrodes available that would allow following the changes in sodium concentration in real time with sufficient specificity to measure the formation of a Na^+ gradient. Therefore, sodium ions can only be determined after quantitative separation of the two compartments. As an alternative to radioactive $^{22}Na^+$, Steuber's group used atomic absorption spectroscopy to measure Na^+ uptake into sub-bacterial particles or proteoliposomes (Dimroth 1986), after removing external sodium by rapidly passing samples that were taken at different times through a Dowex 50 cation-exchange cartridge (Krebs et al. 1999). The only ionophor available that exhibits sufficient specificity for sodium ions is the Na^+/H^+ exchanger monensin.

Another problem when studying sodium pumping in bacteria is the presence of endogenous Na^+/H^+ antiporters. In addition, *K. pneumoniae* contains three types of NADH:ubiquinone oxidoreductases, namely complex I (NDH-1; (Krebs et al. 1999)), NDH-2 (Krebs et al. 1999) and Nqr (Bertsova and Bogachev 2004; Hayashi et al. 2001a). Only the "alternative" NADH dehydrogenase NDH-2 cannot use deaminoNADH as a substrate (Matsushita et al. 1987). Therefore, the only way to discriminate between the NADH:ubiquinone oxidoreductase activities of complex I and Nqr is the use of specific inhibitors. For complex I, numerous specific high affinity inhibitors are available (Degli Esposti 1998) and Nqr from *V. alginolyticus* was shown to be selectively inhibited by Ag^+ (Steuber et al. 1997) and korormicin (Nakayama et al. 1999). HQNO also inhibits Nqr and other quinone-dependent enzymes but not complex I (Bertsova and Bogachev 2004; Nakayama et al. 1999).

The presence of sodium-pumping NADH:ubiquinone oxidoreductase in *K. pneumoniae* was first demonstrated by Dimroth and Thomer (Dimroth and Thomer 1989). Later, complex I was identified in this organism and partially purified (Krebs et al. 1999). In the same study, membrane vesicles were shown to take up sodium ions at a rate of $0.18\,\mu\mathrm{mol\,min^{-1}\,mg^{-1}}$. Based on the observation that a complex I inhibitor blocked this activity by 80–90%, it was proposed that complex I was the Na^+-pumping NADH dehydrogenase in *K. pneumoniae*. However, both Nqr and complex I are present in *K. pneumoniae*, which was not known at the time. Moreover, the membranes contained Na^+/H^+ antiporters, which may have converted a primary proton gradient formed by complex I into a sodium gradient explaining the inhibitory effect of rotenone that is only shown in the presence of uncoupler. Along these lines, the transiently increased sodium uptake that was seen in the presence of uncoupler (Krebs et al. 1999; Steuber 2001) may reflect an initial stimulation of Nqr resulting from the absence of a pH gradient that is otherwise formed in parallel by complex I, while the subsequent breakdown of the sodium gradient reflected a delayed response of the much slower antiporters. Overall, it seems impossible to draw any firm conclusions from the data presented in this study.

Using an improved preparation of complex I from *K. pneumoniae* that was reconstituted into proteoliposomes, the same group published a pumping stoichiometry of $2Na^+/2e^-$ (Gemperli et al. 2002). The major problem with this study is that even very high concentrations (50 μM) of the potent complex I inhibitor annonin IV had only a very limited effect on sodium pumping and 80–90% inhibition of the ubiquinone-1 reductase activity by rotenone was shown only for membrane vesicles. This raises the question of whether the observed Na^+-uptake was in fact due to a contamination of the complex I preparation by Nqr. The authors mention that the Q-reductase activity was not sensitive to Ag^+, which had been shown previously to efficiently inhibit Nqr from *V. alginolyticus* (Steuber et al. 1997). It remains unclear whether this cation inhibits *K. pneumoniae* Nqr as well. Subunit NqrF, the subunit that has been shown to be the binding site for Ag^+ (Nakayama et al. 1999), is only 65% homologous between the two organisms (Hayashi et al. 2001a) and therefore the binding site for Ag^+ may be absent in *K. pneumoniae*.

The second argument used to exclude the presence of Nqr is the claim that FAD was not found in the complex I preparation. However, Gemperli and co-workers show that after a sucrose gradient, a complex I sample did contain about $0.05\,\mathrm{nmol\,mg^{-1}}$ of FAD (Table II in Gemperli et al. 2002). If taken as a measure for Nqr, compared to $0.7\,\mathrm{nmol\,mg^{-1}}$ of complex I (assuming that Nqr contains two FMN), this would mean that the molar ratio between the two enzymes was 1 : 14. This is significant because specific ubiquinone-1 reductase activities of 40 µmol min^{-1} mg^{-1} were reported for purified Nqr from *Vibrio harveyi* (Bertsova et al. 2001), which is more than 15-times higher than the rate reported for the complex I preparation from *K. pneumoniae* (Gemperli et al. 2002) or the specific activity of purified complex I from *E. coli* after relipidation (Stolpe and Friedrich 2004). It follows that the presence of the small amount of Nqr indicated by the FAD content may have been responsible for a large portion of the observed activity. If one takes the different molecular masses of the two enzyme complexes into account, it can also be calculated that the contaminating Nqr would only correspond to a few percent of the total protein, an amount that may easily escape detection by Edman degradation or mass-spectrometry. It should be noted that Bertsova and Bogachev observed a stimulation of deaminoNADH oxidase activity by Na$^+$ only in *K. pneumoniae* membranes that contained Nqr (Bertsova and Bogachev 2004), suggesting that this effect seen with the preparation from Steuber's group was also due to the presence of Nqr (Gemperli et al. 2002). At any rate, together with the possibility of additional contamination by Na$^+$/H$^+$ antiporters, it seems impossible to reliably interpret the reported results. The authors used the same preparation in a study to show sodium ion cycling with the sodium-dependent ATP synthase from *Ilyobacter tartaricus* (Gemperli et al. 2003). Therefore, the same arguments hold for this work that also presents no inhibitor controls.

Steuber et al. also claimed that complex I from *E. coli* is a Na$^+$ pump (Steuber et al. 2000). In this study strain EP432, lacking Na$^+$/H$^+$ antiporters, was compared to strain ANN021 lacking complex I. The problem here is that the reported maximal rates of sodium uptake were only 30 nmol min^{-1} mg^{-1}, which is only a few percent of the deaminoNADH oxidation rate of the membrane vesicles used. Moreover, the sodium uptake was again only partially blocked by 5 µM rotenone. To explain these data, Stolpe et al. reasoned that some electrons may become distributed from complex I to other enzymes that may act as primary sodium pumps, thereby explaining a partial rotenone sensitivity of the sodium uptake even though complex I was not a sodium ion pump (Stolpe and Friedrich 2004). These authors also considered that *E. coli* complex I may act as a Na$^+$/H$^+$ antiporter under certain conditions.

In summary, convincing evidence of sodium ion pumping, like for example a demonstration of efficient and fully inhibitor-sensitive Na$^+$ uptake by complex I vesicles from a strain lacking Nqr, has not been presented. Therefore, it seems unlikely that complex I can operate as a primary sodium pump.

4.3
The Functional Modules of Complex I

The central subunits of the minimal form can be grouped into three functional modules (Brandt 2006). Serving as the electron input domain, the N-module binds NADH, consists of the 75-kDa, 51-kDa and 24-kDa subunits, and comprises iron–sulfur clusters N1a, N1b and N3, N4 and N5. The latter two subunits form the major part of the FP fragment that can be dissociated from eukaryotic complex I by chaotropic agents (Finel et al. 1992; Sazanov et al. 2000). Transferring the electrons onto ubiquinone, the Q-module consists of the 49-kDa, 30-kDa, the TYKY and PSST subunits and comprises the remaining three iron–sulfur clusters. Site-directed mutagenesis (Darrouzet et al. 1998; Kerscher et al. 2001b) as well as EPR (Magnitsky et al. 2002) and inhibitor labelling studies (Schuler et al. 1999) indicate that the ubiquinone reactive domain is located near the interface between the 49-kDa subunit and the PSST subunit where cluster N2 is bound to the PSST subunit. The 30-kDa subunit is fused in some bacteria to the 49-kDa subunit. The homology of the major part of the N-module (51-kDa, 24-kDa and the N-terminal part of the 75-kDa subunit) to an NAD^+ reducing hydrogenase and the evolutionary relationship of the Q-module to [NiFe] hydrogenases have been extensively discussed (Brandt 2006; Walker 1992). The membrane integral P-module carries out the proton pumping function. Critical residues have been identified by site-directed mutagenesis in the ND4L subunit (Kao et al. 2005; Kervinen et al. 2004) and homology to Na^+/H^+ antiporters antiporter suggests a function in proton translocation for subunits ND2, ND4 and ND5 (Mathiesen and Hägerhäll 2002). A scheme for composing complex I during evolution from pre-existing functional entities has been proposed (Friedrich and Weiss 1997).

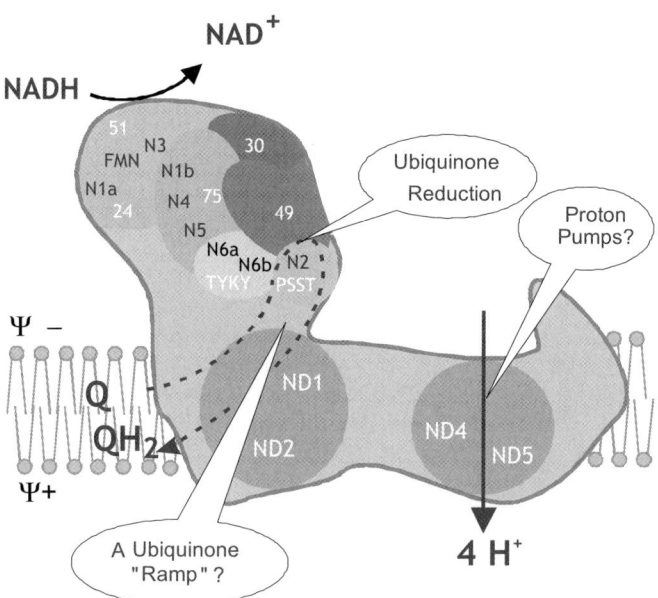

Fig. 6 Cartoon of proton-pumping NADH dehydrogenase (complex I)

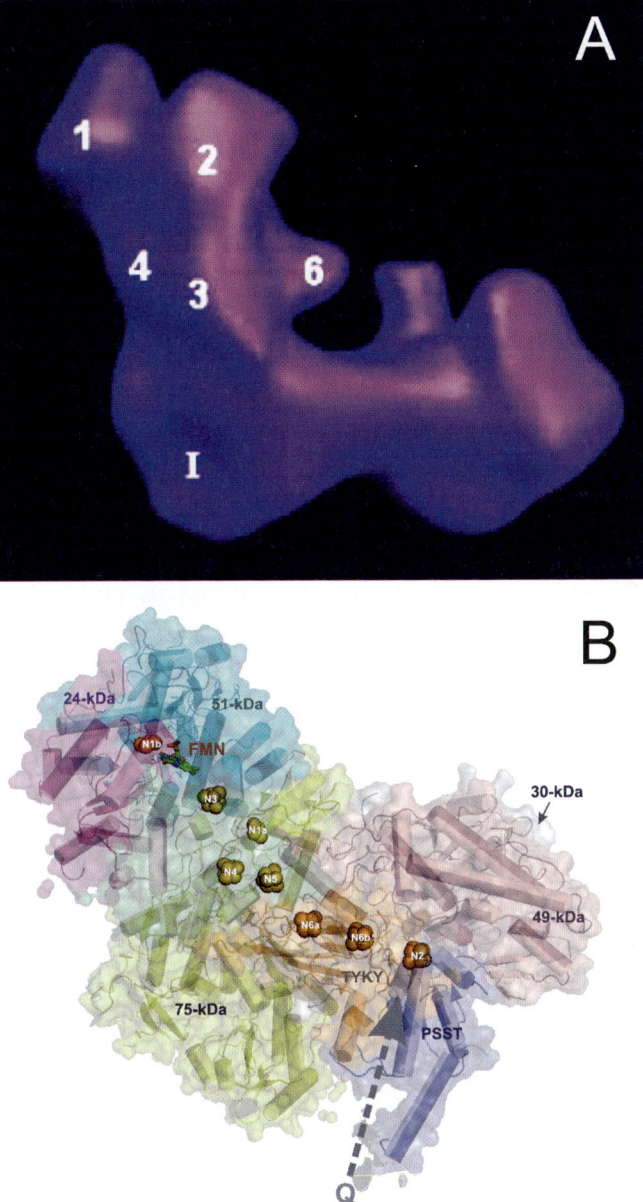

Fig. 7 Structure of complex I. **A** 3D model of holo-complex I from *Y. lipolytica* (Fig. 6C from Radermacher et al. 2006, with permission). The *numbers* indicate five of the six subdomains that were identified in the peripheral arm. Subdomain 5 is in the back of the peripheral arm and therefore not seen in this orientation. *I* marks the indentation in the membrane arm. **B** Structure of the peripheral arm of complex I from *Th. thermophilus*. The *hatched arrow* indicates the proposed path of entry for the substrate ubiquinone. The figure was prepared using PyMOL v0.99

Up to now, structural information on the holoenzyme is based exclusively on electron microscopy. The enzymes from *N. crassa* (Guenebaut et al. 1997), bovine (Grigorieff 1998), *Y. lipolytica* (Radermacher et al. 2006), *A. thaliana* (Dudkina et al. 2005), *E. coli* (Böttcher et al. 2002; Sazanov et al. 2003) and *Aquifex aeolicus* (Peng et al. 2003) were investigated. Overall, the enzyme exhibits an L-shaped arrangement of a peripheral arm comprising the N- and the Q-modules and the membrane arm corresponding to the P-module (Fig. 6). A comparison of eukaryotic and prokaryotic enzymes reveals a largely consistent structure with the most significant differences in the connecting region between both arms (Guenebaut et al. 1998). Both arms have a length of about 200 Å in procaryotes and eucaryotes. The 3D reconstruction of *Y. lipolytica* complex I has the highest resolution of an electron microscopic complex I structure so far (Radermacher et al. 2006). In the peripheral arm six subdomains can be distinguished (Fig. 7b). Labelling with antibodies identified the approximate position of the 30-kDa and 49-kDa subunits (Zickermann et al. 2003). The membrane arm shows two major protrusions on the matrix side and a flexible structure was observed that seems to connect these protrusions with the peripheral arm.

The X-ray structure of the peripheral arm from *Th. thermophilus* NDH-1 was determined recently (Sazanov and Hinchliffe 2006). Overall this complex I fragment has a Y-shaped appearance and a length of about 140 Å. A chain of seven iron–sulfur clusters connects the electron input and output domains that are arranged at the distal ends of the structure. The spatial arrangement indicates the following electron transfer path: NADH → FMN → N3 → N1b → N4 → N5 → N6a → N6b → N2 → quinone. Cluster N1a is located in the N-module on the other side of the NADH binding domain and probably serves a special function due to its very negative redox potential. Where present, cluster N7 is probably not involved in electron transfer because it is too far away from all other redox centres (Fig. 7a).

It is controversially discussed how the crystallized peripheral arm fragment is oriented with respect to the membrane arm. Sazanov suggested that the 49-kDa subunit is directly attached to the membrane part, allowing immediate access for hydrophobic ubiquinone to its binding site (Sazanov and Hinchliffe 2006). This view is not in line with antibody labelling studies, which indicate a position of the 49-kDa subunit more distant to the membrane part (Zickermann et al. 2003). The latter finding is supported by a recent 3D reconstruction of a subcomplex lacking the 51-kDa and 24-kDa subunits and fitting of the *Th. thermophilus* fragment into the 3D *Y. lipolytica* complex I structure (M. Radermacher, personal communication). It has been proposed that a hydrophobic ubiquinone "ramp" within complex I (Zickermann et al. 2003) allows the substrate to get from the membrane to the quinone reduction centre in the membrane arm (Fig. 6).

4.4
Mechanism of Proton Translocation

It has been shown that subunits ND4 and ND5, which were implicated to be involved in proton translocation, reside in the distal end of the membrane arm (Holt et al. 2003). On the other hand it is now clear that all redox groups of complex I are located in the peripheral arm (Carroll et al. 2006b; Sazanov and Hinchliffe 2006) and that the ubiquinone reduction site is not membrane intrinsic. Therefore, the architecture of complex I raises the question of how the redox energy is transferred to the site of proton translocation. At this stage it is straightforward to hypothesize that energy conversion is exerted predominantly by conformational coupling. In fact, experimental evidence for this notion is growing: As early as 1994, it was reported that the pattern of chemical crosslinks (Belogrudov and Hatefi 1994) and trypsin-fragmentation (Yamaguchi et al. 1998) of complex I subunits are dependent on its redox state and the binding of substrates. More recently, similar studies have been combined with 2D electron microscopic analysis showing global structural changes of *E. coli* complex I (Mamedova et al. 2004). The remarkable structural flexibility reported for complex I from *Y. lipolytica* (Radermacher et al. 2006) may also be considered as an indication that long-range conformational changes play a significant role for the mechanism of complex I.

Tunnelling simulations based on the X-ray coordinates of *Th. thermophilus* complex I have shown that there is probably no thermodynamic barrier between NADH and ubiquinone and that electrons arrive rapidly at the ubiquinone-reducing iron–sulfur cluster N2 (Moser et al. 2006). We have shown that in a mutant of *Y. lipolytica* complex I, in which the redox potential of cluster N2 is shifted to become essentially isopotential with the other clusters forming the chain from NADH to ubiquinone, complex I pumps protons with unchanged stoichiometry as the wild-type (Zwicker et al. 2006). This strongly suggests that the conversion between redox and conformational energy occurs at the level of ubiquinone redox intermediates. Obtaining a high-resolution structure of holo-complex I will be prerequisite to identifying the parts of the device that has to be postulated to transmit the conformational energy from the Q- to the P-module.

References

Abdrakhmanova A, Zickermann V, Bostina M, Radermacher M, Schägger H, Kerscher S, Brandt U (2004) Subunit composition of mitochondrial complex I from the yeast *Yarrowia lipolytica*. Biochim Biophys Acta 1658:148–156

Argyrou A, Blanchard JS (2004) Flavoprotein disulfide reductases: advances in chemistry and function. Progr Nuc Ac Res Mol Biol 78:89–142

Argyrou A, Sun G, Palfey BA, Blanchard JS (2003) Catalysis of diaphorase reactions by *Mycobacterium tuberculosis* lipoamide dehydrogenase occurs at the EH$_4$ level. Biochem 42:2218–2228

Bai Y, Hájek P, Chomyn A, Seo BB, Matsuno-Yagi A, Yagi T, Attardi G (2001) Lack of complex I activity in human cells carrying a mutation in MtDNA-encoded ND4 subunit is corrected by the *Saccharomyces cerevisiae* NADH-quinone oxidoreductase (NDI1) gene. J Biol Chem 276:38808–38813

Bakker B, Bro C, Kötter P, Luttik MAH, van Dijken JP, Pronk JT (2000) The mitochondrial alcohol dehydrogenase Adh3p is involved in a redox shuttle in *Saccharomyces cerevisiae*. J Bacteriol 182:4730–4737

Bandeiras TM, Salgueiro C, Kletzin A, Gomes CM, Teixeira M (2002) *Acidianus ambivalens* type-II NADH dehydrogenase: genetic characterisation and identification of the flavin moiety as FMN. FEBS Lett 531:273–277

Barquera B, Häse CC, Gennis RB (2001) Expression and mutagenesis of the NqrC subunit of the NQR respiratory Na$^+$ pump from *Vibrio cholerae* with covalently attached FMN. FEBS Lett 492:45–49

Barquera B, Hellwig P, Zhou WD, Morgan JE, Häse CC, Gosink KK, Nilges M, Bruesehoff PJ, Roth A, Lancaster CRD, Gennis RB (2002a) Purification and characterization of the recombinant Na$^+$-translocating NADH:quinone oxidoreductase from *Vibrio cholerae*. Biochem 41:3781–3789

Barquera B, Morgan JE, Lukoyanov D, Scholes CP, Gennis RB, Nilges MJ (2003) X- and W-band EPR and Q-band ENDOR studies of the flavin radical in the Na$^+$-translocating NADH:quinone oxidoreductase from *Vibrio cholerae*. J Am Chem Soc 125:265–275

Barquera B, Nilges MJ, Morgan JE, Ramirez-Silva L, Zhou WD, Gennis RB (2004) Mutagenesis study of the 2Fe-2S center and the FAD binding site of the Na$^+$-translocating NADH:ubiquinone oxidoreductase from *Vibrio cholerae*. Biochem 43:12322–12330

Barquera B, Ramirez-Silva L, Morgan JE, Nilges MJ (2006) A new flavin radical signal in the Na$^+$-pumping NADH:quinone oxidoreductase from *Vibrio cholerae* – An EPR/electron nuclear double resonance investigation of the role of the covalently bound flavins in subunits B and C. J Biol Chem 281:36482–36491

Barquera B, Zhou WD, Morgan JE, Gennis RB (2002b) Riboflavin is a component of the Na$^+$-pumping NADH:quinone oxidoreductase from *Vibrio cholerae*. Proc Natl Acad Sci USA 99:10322–10324

Bäumer S, Ide T, Jacobi C, Johann A, Gottschalk G, Deppenmeier U (2000) The F$_{420}$H$_2$ dehydrogenase from *Methanosarcina mazei* is a redox-driven proton pump closely related to NADH dehydrogenases. J Biol Chem 275:17968–17973

Beattie P, Tan K, Bourne RM, Leach D, Rich PR, Ward FB (1994) Cloning and sequencing of four structural genes for the Na$^+$-translocating NADH-ubiquinone oxidoreductase of *Vibrio alginolyticus*. FEBS Lett 356:333–338

Belogrudov G, Hatefi Y (1994) Catalytic sector of complex I (NADH:ubiquinone oxidoreductase): subunit stoichiometry and substrate-induced conformation changes. Biochem 33:4571–4576

Bertsova YV, Bogachev AV (2004) The origin of the sodium-dependent NADH oxidation by the respiratory chain of *Klebsiella pneumoniae*. FEBS Lett 563:207–212

Bertsova YV, Bogachev AV, Skulachev VP (2001) Noncoupled NADH:ubiquinone oxidoreductase of *Azotobacter vinelandii* is required for diazotrophic growth at high oxygen concentrations. J Bacteriol 183:6869–6874

Bogachev AV, Bertsova YV, Barquera B, Verkhovsky MI (2001) Sodium-dependent steps in the redox reactions of the Na$^+$-motive NADH:quinone oxidoreductase from *Vibrio harveyi*. Biochem 40:7318–7323

Bogachev AV, Bertsova YV, Bloch DA, Verkhovsky MI (2006) Thermodynamic properties of the redox centers of Na$^+$-translocating NADH:quinone oxidoreductase. Biochem 45:3421–3428

Bogachev AV, Bertsova YV, Ruuge EK, Wikstrom M, Verkhovsky MI (2002) Kinetics of the spectral changes during reduction of the Na$^+$-motive NADH:quinone oxidoreductase from *Vibrio harveyi*. Biochim Biophys Acta 1556:113–120

Bogachev AV, Murtazina RA, Skulachev VP (1996) H$^+$/e$^-$ stoichiometry for NADH dehydrogenase I and dimethyl sulfoxide reductase in anaerobically grown *Escherichia coli* cells. J Bacteriol 178:6233–6237

Bogachev AV, Murtazina RA, Skulachev VP (1997) The Na$^+$/e$^-$ stoichiometry of the Na$^+$-motive NADH:quinone oxidoreductase in *Vibrio alginolyticus*. FEBS Lett 409:475–477

Bogachev AV, Verkhovsky MI (2005) Na$^+$-translocating NADH:quinone oxidoreductase: Progress achieved and prospects of investigations. Biochemistry-Moscow 70:143–149

Böttcher B, Scheide D, Hesterberg M, Nagel-Steger L, Friedrich T (2002) A novel, enzymatically active conformation of the *Escherichia coli* NADH:ubiquinone oxidoreductase (complex I). J Biol Chem 277:17970–17977

Brandt U (2006) Energy converting NADH:quinone oxidoreductases. Annu Rev Biochem 75:69–92

Brito JA, Bandeiras TM, Teixeira M, Vonrhein C, Archer M (2006) Crystallisation and preliminary structure determination of a NADH:quinone oxidoreductase from the extremophile *Acidianus ambivalens*. Biochim Biophys Acta 1764:842–845

Brüggemann H, Falinski F, Deppenmeier U (2000) Structure of the F420H2:quinone oxidoreductase of *Archaeoglobus fulgidus* identification and overproduction of the $F_{420}H_2$-oxidizing subunit. Eur J Biochem 267:5810–5814

Carroll J, Fearnley IM, Shannon RJ, Hirst J, Walker JE (2003) Analysis of the subunit composition of complex I from bovine heart mitochondria. Mol Cell Proteomics 2:117–126

Carroll J, Fearnley IM, Skehel JM, Shannon RJ, Hirst J, Walker JE (2006a) Bovine complex I is a complex of 45 different subunits. J Biol Chem 281:32724–32727

Carroll J, Fearnley IM, Walker JE (2006b) Definition of the mitochondrial proteome by measurement of molecular masses of membrane proteins. Proc Natl Acad Sci USA 103:16170–16175

Darrouzet E, Issartel JP, Lunardi J, Dupuis A (1998) The 49-kDa subunit of NADH-ubiquinone oxidoreductase (complex I) is involved in the binding of piericidin and rotenone, two quinone-related inhibitors. FEBS Lett 431:34–38

Degli Esposti M (1998) Inhibitors of NADH-ubiquinone reductase: an overview. Biochim Biophys Acta 1364:222–235

Dimroth P (1986) Preparation, characterization, and reconstitution of oxaloacetate decarboxylase from *Klebsiella aerogenes*, a sodium pump. Methods Enzymol 125:530–540

Dimroth P, Thomer A (1989) A primary respiratory Na$^+$ pump of an anaerobic bacterium: the Na$^+$-dependent NADH:quinone oxidoreductase of *Klebsiella pneumononiae*. Arch Microbiol 151:439–444

Dudkina NV, Eubel H, Keegstra W, Boekema EJ, Braun HP (2005) Structure of a mitochondrial supercomplex formed by respiratory-chain complexes I and III. Proc Natl Acad Sci USA 102:3225–3229

Duffy EB, Barquera B (2006) Membrane topology mapping of the Na$^+$-pumping NADH:quinone oxidoreductase from *Vibrio cholerae* by PhoA-green fluorescent protein fusion analysis. J Bacteriol 188:8343–8351

Eschemann A, Galkin A, Oettmeier W, Brandt U, Kerscher S (2005) HDQ (1-hydroxy-2-dodecyl-4(1H)quinolone), a high affinity inhibitor for mitochondrial alternative NADH dehydrogenase. J Biol Chem 280:3138–3142

Faig M, Bianchet MA, Talalay P, Chen S, Winski S, Ross D, Amzel LM (2000) Structures of recombinant human and mouse NAD(P)H:quinone oxidoreductases: Species comparison and structural changes with substrate binding and release. Proc Natl Acad Sci USA 97:3177–3182

Fang J, Beattie DS (2002) Novel FMN-containing rotenone-insensitive NADH dehydrogenase from *Trypanosoma brucei* mitochondria: isolation and characterization. Biochem 41:3065–3072

Fang J, Beattie DS (2003a) External alternative NADH dehydrogenase of *Saccharomyces cerevisiae*: a potential source of superoxide. Free Radic Biol Med 34:478–488

Fang J, Beattie DS (2003b) Identification of a gene encoding a 54 kDa alternative NADH dehydrogenase in *Trypanosoma brucei*. Mol Biochem Parasitol 127:73–77

Fang J, Wang Y, Beattie DS (2001) Isolation and characterization of complex I, rotenone-sensitive NADH:ubiquinone oxidoreductase, from the procyclic forms of *Trypanosoma brucei*. Eur J Biochem 268:3075–3082

Fearnley IM, Walker JE (1992) Conservation of sequences of subunits of mitochondrial complex I and their relationships with other proteins. Biochim Biophys Acta 1140:105–134

Finel M, Skehel JM, Albracht SPJ, Fearnley IM, Walker JE (1992) Resolution of NADH: ubiquinone oxidoreductase from bovine heart mitochondria into two subcomplexes, one of which contains the redox centers of the enzyme. Biochem 31:11425–11434

Friedrich T (1998) The NADH:ubiquinone oxidoreductase (complex I) from *Escherichia coli*. Biochim Biophys Acta 1364:134–146

Friedrich T, Weiss H (1997) Modular evolution of the respiratory NADH:ubiquinone oxidoreductase and the origin of its modules. J Theor Biol 187:529–540

Galkin A, Brandt U (2005) Superoxide radical formation by pure complex I (NADH:ubiquinone oxidoreductase) from *Yarrowia lipolytica*. J Biol Chem 280:30129–30135

Galkin A, Dröse S, Brandt U (2006) The proton pumping stoichiometry of purified-mitochondrial complex I reconstituted into proteoliposomes. Biochim Biophys Acta 1757:1575–1581

Gemperli AC, Dimroth P, Steuber J (2002) The respiratory complex I (NDH-I) from *Klebsiella pneumoniae*, a sodium pump. J Biol Chem 277:33811–33817

Gemperli AC, Dimroth P, Steuber J (2003) Sodium ion cycling mediates energy coupling beetween complex I and ATP synthase. Proc Natl Acad Sci USA 100:839–844

Goldstein L (1976) Kinetic behavior of immobilized enzyme systems. Methods Enzymol 44:397–443

Gomes CM, Bandeiras TM, Teixeira M (2001) A new type-II NADH dehydrogenase from the archaeon *Acidianus ambivalens*: characterization and in vitro reconstitution of the respiratory chain. J Bioenerg Biomembr 33:1–8

Green J, Anjum MF, Guest JR (1997) Regulation of the ndh gene of *Escherichia coli* by integration host factor and a novel regulator, Arr. Microbiology 143:2865–2875

Greenamyre JT, Sherer TB, Betarbet R, Panov A (2001) Complex I and Parkinson's disease. IUBMB Life 52:135–141

Grigorieff N (1998) Three-dimensional structure of bovine NADH:ubiquinone oxidoreductase (complex I) at 22 Å in ice. J Mol Biol 277:1033–1046

Guenebaut V, Schlitt A, Weiss H, Leonard K, Friedrich T (1998) Consistent structure between bacterial and mitochondrial NADH:ubiquinone oxidoreductase (complex I). J Mol Biol 276:105–112

Guenebaut V, Vincentelli R, Mills D, Weiss H, Leonard K (1997) Three-dimensional structure of NADH dehydrogenase from *Neurospora crassa* by electron microscopy and conical tilt reconstruction. J Mol Biol 265:409–418

Hajduk S, Adler B, Bertrand K, Fearon K, Hager K, Hancock K, Harris M, Blanc AL, Moore R, Pollard V, Priest J, Wood Z (1992) Molecular biology of African trypanosomes: development of new strategies to combat an old disease. Am J Med Sci 303:258–270

Häse CC, Barquera B (2001) Role of sodium bioenergetics in *Vibrio cholerae*. Biochim Biophys Acta 1505:169–178

Häse CC, Fedorova ND, Galperin MY, Dibrov PA (2001) Sodium ion cycle in bacterial pathogens: evidence from cross-genome comparisons. Microbiol Mol Biol Rev 65:353–370

Häse CC, Mekalanos JJ (1999) Effects of changes in membrane sodium flux on virulence gene expression in *Vibrio cholerae*. Proc Natl Acad Sci USA 96:3183–3187

Hayashi M, Hirai K, Unemoto T (1995) Sequencing and the alignment of structural genes in the Nqr operon encoding the Na$^+$-translocating NADH-quinone reductase from *Vibrio alginolyticus*. FEBS Lett 363:75–77

Hayashi M, Nakayama Y, Unemoto T (2001a) Recent progress in the Na$^+$-translocating NADH-quinone reductase from the marine *Vibrio alginolyticus*. Biochim Biophys Acta 1505:37–44

Hayashi M, Nakayama Y, Yasui M, Maeda M, Furuishi K, Unemoto T (2001b) FMN is covalently attached to a threonine residue in the NqrB and NqrC subunits of Na$^+$-translocating NADH-quinone reductase from *Vibrio alginolyticus*. FEBS Lett 488:5–8

Hayashi M, Shibata N, Nakayama Y, Yoshikawa K, Unemoto T (2002) Korormicin insensitivity in *Vibrio alginolyticus* is correlated with a single point mutation of Gly-140 in the NqrB subunit of the Na$^+$-translocating NADH-quinone reductase. Arch Biochem Biophys 401:173–177

Holt PJ, Morgan DJ, Sazanov LA (2003) The location of NuoL and NuoM subunits in the membrane domain of the *Escherichia coli* complex I – implications for the mechanism of proton pumping. J Biol Chem 278:43114–43120

Kao MC, Nakamaru-Ogiso E, Matsuno-Yagi A, Yagi T (2005) Characterization of the membrane domain subunit NuoK (ND4L) of the NADH-quinone oxidoreductase from *Escherichia coli*. Biochem 44:9545–9554

Kerscher S (2000) Diversity and origin of alternative NADH:ubiquinone oxidoreductases. Biochim Biophys Acta 1459:274–283

Kerscher S, Eschemann A, Okun PM, Brandt U (2001a) External alternative NADH:ubiquinone oxidoreductase redirected to the internal face of the mitochondrial inner membrane rescues complex I deficiency in *Yarrowia lipolytica*. J Cell Sci 114:3915–3921

Kerscher S, Kashani-Poor N, Zwicker K, Zickermann V, Brandt U (2001b) Exploring the catalytic core of complex I by *Yarrowia lipolytica* yeast genetics. J Bioenerg Biomembr 33:187–196

Kerscher S, Okun JG, Brandt U (1999) A single external enzyme confers alternative NADH:ubiquinone oxidoreductase activity in *Yarrowia lipolytica*. J Cell Sci 112:2347–2354

Kervinen M, Patsi J, Finel M, Hassinen IE (2004) A pair of membrane-embedded acidic residues in the NuoK subunit of *Escherichia coli* NDH-1, a counterpart of the ND4L subunit of the mitochondrial complex I, are required for high ubiquinone reductase activity. Biochem 43:773–781

Kluck RM, Bossy-Wetzel E, Green DR, Newmeyer DD (1997) The release of cytochrome *c* from mitochondria: A primary site for Bcl-2 regulation of apoptosis. Science 275:1132–1136

Kogure K (1998) Bioenergetics of marine bacteria. Curr Opin Biotechnol 9:278–282

Krebs W, Steuber J, Gemperli AC, Dimroth P (1999) Na$^+$ translocation by the NADH:ubiquinone oxidoreductase (complex I) from *Klebsiella pneumoniae*. Mol Microbiol 33:590–598

Kussmaul L, Hirst J (2006) The mechanism of superoxide production by NADH:ubiquinone oxidoreductase (complex I) from bovine heart mitochondria. Proc Natl Acad Sci USA 103:7607–7612

Lagunas R (1986) Misconceptions about the energy metabolism of *Saccharomyces cerevisiae*. Yeast 2:221–228

Lesk AM (1995) NAD-binding domains of dehydrogenases. Curr Opin Struct Biol 5:775–783

Lin PC, Puhar A, Turk K, Piligkos S, Bill E, Neese F, Steuber J (2005) A vertebrate-type ferredoxin domain in the Na^+-translocating NADH dehydrogenase from *Vibrio cholerae*. J Biol Chem 280:22560–22563

Liu X, Kim CN, Yang J, Jemmerson R, Wang X (1996) Induction of apoptotic program in cell-free extracts: requirement for dATP and cytochrome *c*. Cell 86:147–157

Luttik MAH, Overkamp KM, Kötter P, de Vries S, van Dijken P, Pronk JT (1998) The *Saccharomyces cerevisiae NDE1* and *NDE2* genes encode separate mitochondrial NADH dehydrogenases catalyzing the oxidation of cytosolic NADH. J Biol Chem 273:24529–24534

Magnitsky S, Toulokhonova L, Yano T, Sled VD, Hagerhall C, Grivennikova VG, Burbaev DS, Vinogradov AD, Ohnishi T (2002) EPR characterization of ubisemiquinones and iron–sulfur cluster N2, central components of the energy coupling in the NADH-ubiquinone oxidoreductase (complex I) in situ. J Bioenerg Biomembr 34:193–208

Mamedova AA, Holt PJ, Carroll J, Sazanov LA (2004) Substrate-induced conformational change in bacterial complex I. J Biol Chem 279:23830–23836

Marres CAM, de Vries S, Grivell LA (1991) Isolation and inactivation of the nuclear gene encoding the rotenone-insensitive internal NADH:ubiquinone oxidoreductase of mitochondria from *Saccharomyces cerevisiae*. Eur J Biochem 195:857–862

Mathiesen C, Hägerhäll C (2002) Transmembrane topology of the NuoL, M and N subunits of NADH:quinone oxidoreductase and their homologues among membrane-bound hydrogenases and bona fide antiporters. Biochim Biophys Acta 1556:121–132

Matsushita K, Ohnishi T, Kaback HR (1987) NADH-ubiquinone oxidoreductases of the *Escherichia coli* aerobic respiratory chain. Biochem 26:7732–7737

Melo AM, Bandeiras TM, Teixeira M (2004) New insights into type II NAD(P)H:quinone oxidoreductases. Microbiol Mol Biol Rev 68:603–616

Melo AM, Duarte M, Möllers IM, Prokisch H, Dolan PL, Pinto L, Nelson MA, Videira A (2001) The external calcium-dependent NADPH dehydrogenase from *Neurospora crassa* mitochondria. J Biol Chem 276:3947–3951

Melo AM, Duarte M, Videira A (1999) Primary structure and characterisation of a 64-kDa NADH dehydrogenase from the inner membrane of *Neurospora crassa* mitochondria. Biochim Biophys Acta 1412:282–287

Michalecka AM, Svensson AS, Johansson FI, Agius SC, Johanson U, Brennicke A, Binder S, Rasmusson AG (2003) Arabidopsis genes encoding mitochondrial type II NAD(P)H dehydrogenases have different evolutionary origin and show distinct responses to light. Plant Physiol 133:642–652

Modjtahedi N, Giordanetto F, Madeo F, Kroemer G (2006) Apoptosis-inducing factor: vital and lethal. Trends Cell Biol 16:264–272

Møller IM (2001) Plant mitochondria and oxidative stress: electron transport, NADPH turnover, and metabolism of reactive oxygen species. Annu Rev Plant Physiol 52:561–591

Moser CC, Farid TA, Chobot SE, Dutton PL (2006) Electron tunneling chains of mitochondria. Biochim Biophys Acta 1757:1096–1109

Mrazek J, Spormann AM, Karlin S (2006) Genomic comparisons among gamma-proteobacteria. Env Microbiol 8:273–288

Nakamaru-Ogiso E, Seo BB, Yagi T, Matsuno-Yagi A (2003) Amiloride inhibition of the proton-translocating NADH-quinone oxidoreductase of mammals and bacteria. FEBS Lett 14:43–46

Nakayama Y, Hayashi M, Unemoto T (1998) Identification of six subunits constituting Na^+-translocating NADH-quinone reductase from the marine *Vibrio alginolyticus*. FEBS Lett 422:240–242

Nakayama Y, Hayashi M, Unemoto T, Yoshikawa K, Mochida K (1999) Inhibitor studies of a new antibiotic, korormicin, 2-*n*-heptyl-4-hydroxyquinoline *N*-oxide and Ag^+ toward the Na^+-translocating NADH-quinone reductase from the marine *Vibrio alginolyticus*. Biol Pharm Bull 22:1064–1067

Nakayama Y, Yasui M, Sugahara K, Hayashi M, Unemoto T (2000) Covalently bound flavin in the NqrB and NqrC subunits of Na^+-translocating NADH-quinone reductase from *Vibrio alginolyticus*. FEBS Lett 474:165–168

Nordman T, Xia L, Björkhem-Bergman L, Damdimopoulos A, Nalvarte I, Arnér E, Spyrou G, Eriksson L, Björnstedt M, Olsson J (2003) Regeneration of the antioxidant ubiquinol by lipoamide dehydrogenase, thioredoxin reductase and glutathione reductase. BioFactors 18:45–50

Notredame C, Higgins D, Herringa J (2000) T-Coffee: a novel method for multiple sequence alignments. J Mol Biol 302:205–217

Ohnishi T (1998) Iron–sulfur clusters/semiquinones in complex I. Biochim Biophys Acta 1364:186–206

Okun JG, Lümmen P, Brandt U (1999) Three classes of inhibitors share a common binding domain in mitochondrial complex I (NADH:ubiquinone oxidoreductase). J Biol Chem 274:2625–2630

Ouali M, King RD (2000) Cascaded multiple classifiers for secondary structure prediction. Protein Sci 9:1162–1176

Peng G, Fritzsch G, Zickermann V, Schägger H, Mentele R, Lottspeich F, Bostina M, Radermacher M, Huber R, Stetter KO, Michel H (2003) Isolation, characterization and electron microscopic single particle analysis of the NADH:ubiquinone oxidoreductase (complex I) from the hyperthermophilic eubacterium aquifex aeolicus. Biochem 42:3032–3039

Pfenninger-Li XD, Albracht SPJ, van Belzen R, Dimroth P (1996) NADH:ubiquinone oxidoreductase of *Vibrio alginolyticus*: purification, properties, and reconstitution of the Na^+ pump. Biochem 35:6233–6242

Prommeenate P, Lennon AM, Markert C, Hippler M, Nixon PJ (2004) Subunit composition of NDH-1 complexes of *Synechocystis* sp PCC 6803 – identification of two new *ndh* gene products with nuclear-encoded homologues in the chloroplast Ndh complex. J Biol Chem 279:28165–28173

Radermacher M, Ruiz T, Clason T, Benjamin S, Brandt U, Zickermann V (2006) The three-dimensional structure of complex I from *Yarrowia lipolytica*: A highly dynamic enzyme. J Struct Biol 154:269–279

Raghavendra AS, Padmasree K (2003) Beneficial interactions of mitochondrial metabolism with photosynthetic carbon assimilation. Trends Plant Sci 8:546–553

Rapisarda VA, Chehín RN, De Las Rivas J, Rodriguez-Montelongo L, Farías RN, Massa EM (2002) Evidence for Cu(I)-thiolate ligation and prediction of a putative copper-binding site in the *Escherichia coli* NADH dehydrogenase-2. Arch Biochem Biophys 405: 87–94

Rapisarda VA, Rodríguez-Montelongo L, Farías RN, Massa EM (1999) Characterization of an NADH-linked cupric reductase activity from the *Escherichia coli* respiratory chain. Arch Biochem Biophys 370:143–150

Rasmussen T, Scheide D, Brors B, Kintscher L, Weiss H, Friedrich T (2001) Identification of two tetranuclear FeS clusters on the ferredoxin-type subunit of NADH:ubiquinone oxidoreductase (complex I). Biochem 40:6124-6131

Rasmusson AG, Svensson AS, Knoop V, Grohmann L, Brennicke A (1999) Homologues of yeast and bacterial rotenone-insensitive NADH dehydrogenases in higher eukaryotes: two enzymes are present in potato mitochondria. Plant J 20:79-87

Ravanel P, Creuzet S, Tissut M (1990) Inhibitory effect of hydroxyflavones on the exogenous NADH dehydrogenase of plant mitochondrial inner membranes. Phytochemistry 29:441-445

Rich PR, Meunier B, Ward FB (1995) Predicted structure and possible ionmotive mechanism of the sodium-linked NADH-ubiquinone oxidoreductase of *Vibrio alginolyticus*. FEBS Lett 375:5-10

Rodríguez-Montelongo L, Volentini SI, Fárias RN, Massa EM, Rapisarda VA (2006) The Cu(II)-reductase NADH dehydrogenase-2 of *Escherichia coli* improves the bacterial growth in extreme copper concentrations and increases the resistance to the damage caused by copper and hydroperoxide. Arch Biochem Biophys 451:1-7

Rumeau D, Bécuwe-Linka N, Beyly A, Louwagie M, Peltier G (2005) New subunits NDH-M, -N, and -O, encoded by nuclear genes, are essential for plastid Ndh complex functioning in higher plants. Plant Cell 17:219-232

Sazanov LA, Carroll J, Holt P, Toime L, Fearnley IM (2003) A role for native lipids in the stabilization and two-dimensional crystallization of the *Escherichia coli* NADH:ubiquinone oxidoreductase (complex I). J Biol Chem 278:19483-19491

Sazanov LA, Hinchliffe P (2006) Structure of the hydrophilic domain of respiratory complex I from *Thermus thermophilus*. Science 311:1430-1436

Sazanov LA, Peak-Chew SY, Fearnley IM, Walker JE (2000) Resolution of the membrane domain of bovine complex I into subcomplexes: implications for the structural organization of the enzyme. Biochem 39:7229-7235

Schmid R, Gerloff D (2004) Functional properties of the alternative NADH:ubiquinone oxidoreductase from *E. coli* through comparative 3-D modelling. FEBS Lett 578:163-168

Schuler F, Yano T, Di Bernardo S, Yagi T, Yankovskaya V, Singer TP, Casida JE (1999) NADH-quinone oxidoreductase: PSST subunit couples electron transfer from iron-sulfur cluster N2 to quinone. Proc Natl Acad Sci USA 96:4149-4153

Seo BB, Marella M, Yagi T, Matsuno-Yagi A (2006a) The single subunit NADH dehydrogenase reduces generation of reactive oxygen species from complex I. FEBS Lett 580:6105-6108

Seo BB, Matsuno-Yagi A, Yagi T (1999) Modulation of oxidative phosphorylation of human kidney 293 cells by transfection with the internal rotenone-insensitive NADH-quinone oxidoreductase (*NDI1*) gene of *Saccharomyces cerevisiae*. Biochim Biophys Acta 1412:56-65

Seo BB, Nakamaru-Ogiso E, Flotte TR, Matsuno-Yagi A, Yagi T (2006b) In vivo complementation of complex I by the yeast Ndi1 enzyme. J Biol Chem 281:14250-14255

Smith MA, Finel M, Korolik V, Mendz GL (2000) Characteristics of the aerobic respiratory chains of the microaerophiles *Campylobacter jejuni* and *Helicobacter pylori*. Arch Microbiol 174:1-10

Steuber J (2001) Na^+ translocation by bacterial NADH:quinone oxidoreductases: an extension to the complex-I family of primary redox pumps. Biochim Biophys Acta 1505:45-56

Steuber J, Krebs W, Dimroth P (1997) The Na^+-translocating NADH:ubiquinone oxidoreductase from *Vibrio alginolyticus*. Redox states of the FAD prosthetic group and mechanism of Ag^+ inhibition. Eur J Biochem 249:770-776

Steuber J, Schmid C, Rufibach M, Dimroth P (2000) Na$^+$ translocation by complex I (NADH:quinone oxidoreductase) of *Escherichia coli*. Mol Microbiol 35:428–434

Stolpe S, Friedrich T (2004) The *Escherichia coli* NADH:ubiquinone oxidoreductase (complex I) is a primary proton pump but may be capable of secondary sodium antiport. J Biol Chem 279:18377–18383

Susin SA, Lorenzo HK, Zamzami IM, Marzo I, Snow BE, Brothers GM, Mangion J, Jacotot E, Costantini P, Loeffler M, Larochette N, Goodlett DR, Aebersold R, Siderovski DP, Penninger JM, Kroemer G (1999) Molecular characterization of mitochondrial apoptosis-inducing factor. Nature 397:441–446

Swofford DL (1993) PAUP: phylogenetic analysis using parsimony version 3.1.1 (computer program). Illinois Natural History Survey, Champaign, Illinois

Tarrío N, Beccera M, Cerdán ME, González Siso MI (2006a) Reoxidation of cytosolic NADPH in *Kluyveromyces lactis*. FEMS Yeast Res 6:371–380

Tarrío N, Cerdan ME, González Siso MI (2006b) Characterization of the second external alternative dehydrogenase from mitochondria of the respiratory yeast *Kluyveromyces lactis*. Biochim Biophys Acta 1757:1476–1484

Tokuda H, Unemoto T (1981) A respiration-dependent primary sodium extrusion system functioning at alkaline pH in the marine bacterium *Vibrio alginolyticus*. Biochem Biophys Res Comm 102:265–271

Tokuda H, Unemoto T (1984) Na$^+$ is translocated at NADH-quinone oxidoreductase segment in the respiratory-chain of *Vibrio alginolyticus*. J Biol Chem 259:7785–7790

Trotter EW, Grant CM (2005) Overlapping roles of the cytoplasmic and mitochondrial redox regulatory systems in the yeast *Saccharomyces cerevisiae*. Eukar Cell 4:392–400

Türk K, Puhar A, Neese F, Bill E, Gunter F, Steuber J (2004) NADH oxidation by the Na$^+$-translocating NADH:quinone oxidoreductase from *Vibrio cholerae* – Functional role of the NqrF subunit. J Biol Chem 279:21349–21355

Uhlmann M, Friedrich T (2005) EPR signals assigned to Fe/S cluster N1c of the *Escherichia coli* NADH:ubiquinone oxidoreductase (complex I) derive from cluster N1a. Biochem 44:1653–1658

Unden G (1998) Transcriptional regulation and energetics of alternative respiratory pathways in facultatively anaerobic bacteria. Biochim Biophys Acta 1365:220–224

Unden G, Bongaerts J (1997) Alternative respiratory pathways of *Escherichia coli*: energetics and transcriptional regulation in response to electron acceptors. Biochim Biophys Acta 1320:217–234

Vahsen N, Cande C, Briere JJ, Bénit P, Joza N, Larochette N, Mastroberardino PG, Pequignot MO, Casares N, Lazar V, Feraud O, Debili N, Wissing S, Engelhardt S, Madeo F, Piacentini M, Penninger JM, Schägger H, Rustin P, Kroemer G (2004) AIF deficiency compromises oxidative phosphorylation. EMBO J 23:4679–4689

Velázques I, Pardo JP (2001) Kinetic characterization of the rotenone-insensitive internal NADH:ubiquinone oxidoreductase of mitochondria from *Saccharomyces cerevisiae*. Arch Biochem Biophys 389:7–14

Wackwitz B, Bongaerts J, Goodman SD, Unden G (1999) Growth phase-dependent regulation of nuoA-N expression in *Escherichia coli* K-12 by the Fis protein: upstream binding sites and bioenergetic significance. Mol Gen Genet 262:876–883

Walker JE (1992) The NADH:ubiquinone oxidoreductase (complex I) of respiratory chains. Q Rev Biophys 25:253–324

Wierenga RK, de Maeyer MCH, Hol WGJ (1985) Interaction of pyrophosphate moieties with α-helixes in dinucleotide binding proteins. Biochem 24:1346–1357

Wikström MKF (1984) Pumping of protons from the mitochondrial matrix by cytochrome oxidase. Nature 308:558–560

Wissing S, Ludovico P, Herker E, Büttner S, Engelhardt S, Decker T, Link A, Proksch A, Rodrigues F, Corte-Real M, Fröhlich K-U, Manns J, Cande C, Sigrist SJ, Kroemer G, Madeo F (2004) An AIF orthologue regulates apoptosis in yeast. J Cell Biol 166:969–974

Yagi T, Seo BB, Nakamaru-Ogiso E, Marella M, Barber-Singh J, Yamashita T, Matsuno-Yagi A (2006) Possibility of transkingdom gene therapy for complex I diseases. Biochim Biophys Acta 1757:708–714

Yamaguchi M, Belogrudov G, Hatefi Y (1998) Mitochondrial NADH-ubiquinone oxidoreductase (complex I). Effect of substrates on the fragmentation of subunits by trypsin. J Biol Chem 273:8094–8098

Yoshikawa K, Nakayama Y, Hayashi M, Unemoto T, Mochida K (1999) Korormicin, an antibiotic specific for gram-negative marine bacteria, strongly inhibits the respiratory chain-linked Na^+-translocating NADH:quinone reductase from the marine *Vibrio alginolyticus*. J Antibiotics 52:182–185

Yoshikawa K, Takadera T, Adachi K, Nishijima M, Sano H (1997) Korormicin, a novel antibiotic specifically active against marine gram-negative bacteria, produced by a marine bacterium. J Antibiotics 50:949–953

Zhou WD, Bertsova YV, Feng BT, Tsatsos P, Verkhovskaya ML, Gennis RB, Bogachev AV, Barquera B (1999) Sequencing and preliminary characterization of the Na^+-translocating NADH:ubiquinone oxidoreductase from *Vibrio harveyi*. Biochem 38:16246–16252

Zickermann V, Bostina M, Hunte C, Ruiz T, Radermacher M, Brandt U (2003) Functional implications from an unexpected position of the 49-kDa subunit of complex I. J Biol Chem 278:29072–29078

Zwicker K, Galkin A, Dröse S, Grgic L, Kerscher S, Brandt U (2006) The redox-Bohr group associated with iron–sulfur cluster N2 of complex I. J Biol Chem 218:23013–23017

Hydrogenases and H^+-Reduction in Primary Energy Conservation

Paulette M. Vignais

CEA Grenoble, Laboratoire de Biochimie et Biophysique des Systèmes Intégrés, UMR CEA/CNRS/UJF no. 5092, Institut de Recherches en Technologies et Sciences pour le Vivant (iRTSV), 17 rue des Martyrs, 38054 Grenoble cedex 9, France
p.vignais@wanadoo.fr

Abstract Hydrogenases are metalloenzymes subdivided into two classes that contain iron-sulfur clusters and catalyze the reversible oxidation of hydrogen gas ($H_2 \leftrightarrows 2H^+ + 2e^-$). Two metal atoms are present at their active center: either a Ni and an Fe atom in the [NiFe]hydrogenases, or two Fe atoms in the [FeFe]hydrogenases. They are phylogenetically distinct classes of proteins. The catalytic core of [NiFe]hydrogenases is a heterodimeric protein associated with additional subunits in many of these enzymes. The catalytic core of [FeFe]hydrogenases is a domain of about 350 residues that accommodates the active site (H cluster). Many [FeFe]hydrogenases are monomeric but possess additional domains that contain redox centers, mostly Fe–S clusters. A third class of hydrogenase, characterized by a specific iron-containing cofactor and by the absence of Fe–S cluster, is found in some methanogenic archaea; this Hmd hydrogenase has catalytic properties different from those of [NiFe]- and [FeFe]hydrogenases.

The [NiFe]hydrogenases can be subdivided into four subgroups: (1) the H_2 uptake [NiFe]hydrogenases (group 1); (2) the cyanobacterial uptake hydrogenases and the cytoplasmic H_2 sensors (group 2); (3) the bidirectional cytoplasmic hydrogenases able to bind soluble cofactors (group 3); and (4) the membrane-associated, energy-converting, H_2 evolving hydrogenases (group 4). Unlike the [NiFe]hydrogenases, the [FeFe]hydrogenases form a homogeneous group and are primarily involved in H_2 evolution.

This review recapitulates the classification of hydrogenases based on phylogenetic analysis and the correlation with hydrogenase function of the different phylogenetic groupings, discusses the possible role of the [FeFe]hydrogenases in the genesis of the eukaryotic cell, and emphasizes the structural and functional relationships of hydrogenase subunits with those of complex I of the respiratory electron transport chain.

1
Introduction

Hydrogen is the most abundant element in the Universe. Initially released by abiotic processes in the Earth's early reducing atmosphere, in which it predominated (Tian et al. 2005), it has been since then a major energy source for life. The prokaryotic world has the ability to use H_2 directly, by the activity of uptake hydrogenases, or to produce H_2 directly, by the activity of H_2-evolving hydrogenases.

The study of hydrogenase enzymes in extant microorganisms, present in particular in today's anaerobic ecosystems, may give insights into the earliest life on planet Earth. Besides, the existence of hydrogen-driven subsurface lithoautotrophic microbial ecosystems (SLIMEs), which can exist and persist independently of the products of photosynthesis (organic carbon and molecular oxygen) and probably appeared before chlorophyll-based photosynthesis was invented, may provide clues as to the nature of life in extraterrestrial worlds (Nealson et al. 2005).

If essential processes of all life, anabolic reactions via carbon and nitrogen fixation and catabolic energy metabolism via carbon oxidation and redox reaction, can be sustained by hydrogen metabolism, then the question is: how is energy conserved and converted during H_2 metabolism?

Hydrogenases catalyze the simplest chemical reaction: $2H^+ + 2\,e^- \leftrightarrows H_2$. The reaction is reversible and its direction depends on the redox potential of the components able to interact with the enzyme. In the presence of an electron acceptor, a hydrogenase will act as a H_2 uptake enzyme, while in the presence of an electron donor, the enzyme will produce H_2. About a thousand hydrogenase sequences have been identified (Vignais and Billoud, unpublished results), many by genome sequencing, and more than 100 have been characterized genetically and/or biochemically. By comparing their amino acid sequences, it has been possible to identify classes and subgroups of enzymes, to compare and correlate genetic, physiological and biochemical information relative to members of the subgroups, independently of their origin and their various roles in energy metabolism (Vignais et al. 2001). This review deals with the diversity of hydrogenases, their classification, and their various modes of energy conservation and conversion.

2
Diversity and Classification of Hydrogenases

Hydrogenases are generally Fe – S proteins with two metal atoms at their active site, either a Ni and an Fe atom (in [NiFe]hydrogenases) (Volbeda et al. 1995; Higuchi et al. 1997) or two Fe atoms (in [FeFe]hydrogenases) (Peters et al. 1998; Nicolet et al. 1999). A third type is the Fe – S cluster-free hydrogenase discovered in methanogenic archaea (Zirngibl et al. 1992), which functions as H_2-forming methylenetetrahydromethanopterin dehydrogenase, abbreviated Hmd. Hmd tightly binds an iron-containing light-sensitive cofactor (Lyon et al. 2004). The iron is coordinated by two CO molecules, one sulfur and a pyridone derivative linked via a phosphodiester bond to a guanosine base. The crystal structure of the apoenzyme of the Fe – S cluster-free hydrogenase has been published recently (Pilak et al. 2006). Evidence from amino acid sequences and structures indicates that the three types of hydrogenases are phylogenetically distinct classes of proteins (Vignais et al. 2001).

2.1
The [NiFe]hydrogenases

The [NiFe]hydrogenases are the most numerous and best studied class of hydrogenases. They are found in organisms belonging to the *Bacteria* and *Archaea* domains of life. The core enzyme consists of an $\alpha\beta$ heterodimer with the large subunit (α-subunit) of ca. 60 kDa hosting the bimetallic active site and the small subunit (β-subunit) of ca. 30 kDa, the Fe – S clusters. The latter conduct electrons between the H_2-activating center and the physiological electron acceptor/donor from/to hydrogenase. Crystal structures of *Desulfovibrio* hydrogenases have shown that the two subunits interact extensively through a large contact surface and form a globular heterodimer. The bimetallic NiFe center is deeply buried in the large subunit; it is coordinated to the protein by four cysteines (Volbeda et al. 1995; Higuchi et al. 1997; Garcin et al. 1999; Matias et al. 2001). Infrared spectroscopy studies have revealed the presence of three non-protein ligands, one CO and two CN^- bound to the Fe atom (Volbeda et al. 1996; Happe et al. 1997). The [4Fe – 4S] cluster that is proximal to the active site (within 14 Å) is "essential" to H_2 activation (Volbeda et al. 1995; Fontecilla-Camps et al. 1997). Gas access to the active site is facilitated by hydrophobic channels linking the active site to the surface of the molecule (Fontecilla-Camps et al. 1997; Montet et al. 1997). Alignments of the full amino acid sequences of the small and large subunits have shown that the two subunits of [NiFe]hydrogenases evolved conjointly. That analysis led to a classification of [NiFe]hydrogenases that is consistent with the functions of the enzymes (Vignais et al. 2001).

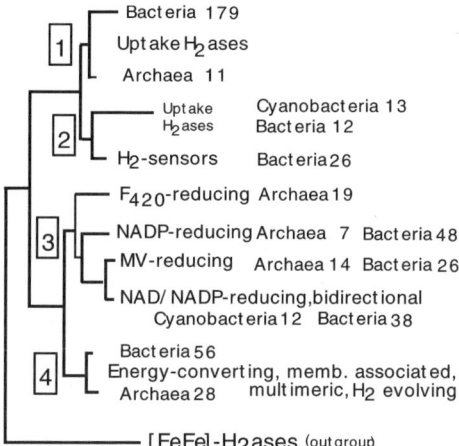

Fig. 1 Schematic representation of the phylogenetic tree of [NiFe]hydrogenases based on the complete sequences of the small and the large subunits (the same tree was obtained with each type of subunit) originally established by Vignais et al. 2001. The number of aligned hydrogenases in each subgroup is indicated. (Vignais and Billoud, unpublished results)

As shown in Fig. 1, the [NiFe]hydrogenases found in *Bacteria* and in *Archaea* cluster into four groups.

2.1.1
Group 1

In Group 1 are the membrane-bound enzymes, which perform respiratory hydrogen oxidation linked to quinone reduction. They allow the cells to use H_2 as an energy source and are called (H_2) uptake hydrogenases (generally termed Hup). The Hup hydrogenases and *Escherichia coli* hydrogenase-1 (Hya) are heterotrimeric enzymes consisting of a core heterodimer of an Fe – S cluster binding β-subunit (HupS, HyaA) and an α-subunit that binds the [NiFe] active site cofactor (HupL, HyaB). This associates with a third integral membrane cytochrome b γ-subunit (HupC, HyaC) to form the holoenzyme. The core hydrogenase dimer is anchored to the membrane by the di-heme cytochrome b

Fig. 2 Examples of electron transfer catalyzed by respiratory hydrogenases of group 1. ▶ Hypothetical mechanism of fumarate respiration with H_2 in *Wolinella succinogenes* (**a**) and in *Escherichia coli* (**b**). **a** Electron and proton transfer in the membrane of *W. succinogenes* according to the "E pathway hypothesis" of Lancaster et al. (2005), which proposes that transmembrane electron transfer via the heme groups of the di-hemic quinol:fumarate reductase is strictly coupled to cotransfer of protons via a transiently established pathway, where the side chain of residue Glu-C180 plays a prominent role. The two protons that are liberated upon oxidation of menaquinol (MKH_2) are released to the periplasm. In compensation, coupled to electron transfer via the two heme groups, protons are transferred from the periplasm to the cytoplasm (via the ring C propionate of the distal heme and the residue Glu-C180 of the membrane subunit of fumarate reductase), where they replace those protons that are bound during fumarate reduction (Kröger et al. 2002; Lancaster et al. 2005). The HydC protein of the hydrogenase forms four transmembrane helices; the heme b groups are shown as *diamonds*. The menaquinone binding site is close to the distal heme b group, near the cytoplasmic side of the membrane (Gross et al. 2004). [4Fe – 4S] and [3Fe – S] clusters are represented by *squares* and the [2Fe – 2S] cluster by a *rectangle*. **b** In *E. coli*, hydrogenase-2 donates electrons to heme-less fumarate reductase. Unlike trimeric uptake hydrogenases with a membrane integral cytochrome b as third subunit, *E. coli* hydrogenase-2 is heterotetrameric; besides the $\alpha\beta$ heterodimeric core, it includes a "16Fe" ferredoxin (HybA), most closely related to the periplasmically oriented HmcB protein from *Desulfovibrio vulgaris* (Hildenborough) (Dolla et al. 2000), and a large integral membrane protein (HybB), most closely related to the HmcC protein from *D. vulgaris* and predicted to comprise ten transmembrane helices (Dubini 2002).
c The trimeric F_{420}-nonreducing hydrogenase (Vho) from *Methanosarcina mazei* Göl, with a cytochrome b subunit that acts as the primary electron acceptor of the core hydrogenase, is shown to interact with the heterodisulfide reductase via methanophenazine (MP), the membrane integral electron carrier connecting protein complexes of the respiratory chain of *Ms. mazei*. The scheme shows that the membrane integral cytochrome b subunit accepts two protons from the cytoplasm for the reduction of MP and that the overall reaction leads to the production of two scalar protons (Ide et al. 1999), (adapted from Deppenmeier 2004)

(Dross et al. 1992; Bernhard et al. 1997), which connects it to the quinone pool of the respiratory chain in the membrane, and by the hydrophobic C-terminus of the small subunit (Cauvin et al. 1991; Dross et al. 1992). The prototype, the hydrogenase of *Wolinella succinogenes*, encoded by the *hydABC* genes (thoroughly studied by the group of the late Achim Kröger) is shown in (Fig. 2a). Other members of group 1, such as the Hyn enzyme from *Thiocapsa roseop-*

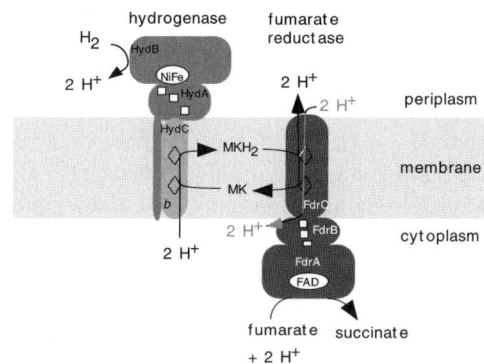

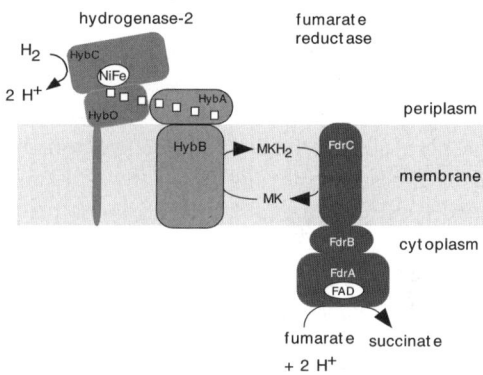

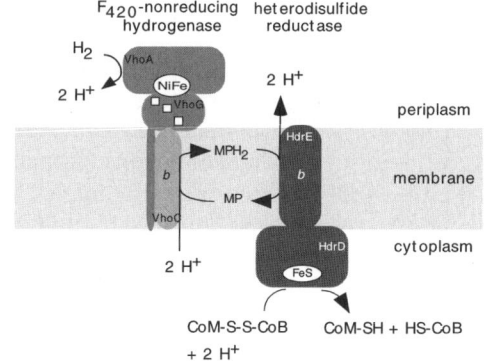

ersicina (Rákhely et al. 1998), the periplasmic *Desulfovibrio* hydrogenase able to interact with low-potential *c*-type cytochromes, and a transmembrane redox protein complex encoded by the *hmc* operon (Rossi et al. 1993) and *E. coli* hydrogenase-2 present a slightly different structure. *E. coli* hydrogenase-2 is predicted to be a large tetrameric complex consisting of the large (HybC) and the small (HybO) subunits associated to two other subunits, an Fe–S containing periplasmic subunit (HybA) and an integral membrane protein HybB (Dubini 2002) (Fig. 2b). Some *Desulfovibrio* species, e.g. *Desulfomicrobium baculatum* (formerly *Desulfovibrio baculatus*) contain a [NiFeSe]hydrogenase (HysSL). In this Se-containing hydrogenase, the carboxy-terminus of the gene encoding the large subunit contains a codon (TGA) for selenocysteine in a position homologous to a codon (TGC) for cysteine (Fauque et al. 1988). The SeCys in *Dm. baculatum* is a ligand to Ni (Garcin et al. 1999).

The uptake hydrogenases are characterized by the presence of a long signal peptide (30–50 amino acids residues) at the N-terminus of their small subunit. The signal peptide contains a conserved (S/T)RR×F×K motif recognized by a specific protein translocation pathway known as membrane targeting and translocation (Mtt) (Weiner et al. 1998) or twin-arginine translocation (Tat) (Sargent et al. 1998; Rodrigue et al. 1999) pathway, and serves as signal recognition to target fully folded mature heterodimer to the membrane and the periplasm (Wu et al. 2000; Sargent et al. 2002; Palmer et al. 2005). The twin-arginine motif has been shown to be required for successful assembly of the uptake hydrogenases from *Ralstonia eutropha* (Bernhard et al. 2000), and *W. succinogenes* (Gross et al. 1999). The Tat translocase transports fully folded proteins across the energy-transducing inner membrane using energy provided by the transmembrane Δp (Yahr and Wickner 2001). The Tat pathway is structurally and mechanistically similar to the ΔpH-dependent pathway used to import chloroplast proteins into the thylakoid (Mori and Cline 2001, 2002; Berks et al. 2003, 2005). Homologs of Tat proteins are found in many archaea, bacteria, chloroplasts, and mitochondria (Yen et al. 2002).

2.1.2
Group 2

Group 2 hydrogenases are not exported and remain in the cytoplasm. In accordance, their small subunit does not contain a signal peptide at its N-terminus. They are subdivided into (i) the cyanobacterial uptake hydrogenases induced under N_2 fixing conditions (Appel and Schulz 1998; Tamagnini et al. 2002) and (ii) the regulatory hydrogenases, which function as H_2 sensors in the regulatory cascade that controls the biosynthesis of some eubacterial uptake hydrogenases (Friedrich et al. 2005; Vignais et al. 2005). The third hydrogenase of *Aquifex aeolicus*, a soluble enzyme that clusters with group 2a hydrogenases has been proposed to provide reductant to the reductive TCA cycle for CO_2 fixation (Brugna-Guiral et al. 2003).

2.1.3
Group 3

In Group 3, the dimeric hydrogenase module is associated with other subunits able to bind soluble cofactors, such as cofactor 420 (F_{420}, 8-hydroxy-5-deazaflavin), NAD, or NADP. They are termed bidirectional hydrogenases for, physiologically, they function reversibly and can thus reoxidize the cofactors under anaerobic conditions by using the protons of water as electron acceptors. Many members of this group belong to the *Archaea* domain (Fig. 1).

Bidirectional NAD(P)-linked hydrogenases are also found in bacteria and in cyanobacteria. The first NAD-dependent [NiFe]hydrogenase was isolated from *R. eutropha* (formerly *Alcaligenes eutrophus* now renamed *Cupriavidus necator*) (Schneider and Schlegel 1976) in which it is encoded by genes located on a megaplasmid (Schwartz et al. 2003). Homologous NAD(P)-linked hydrogenases were later discovered in cyanobacteria and in the photosynthetic bacterium *T. roseopersicina* (Rákhely et al. 2004) (reviews by Appel and Schulz 1998; Vignais et al. 2001; Tamagnini et al. 2002). These bidirectional hydrogenases are composed of two moieties: the heterodimer [NiFe]hydrogenase encoded by the *hoxY* and *HoxH* genes and the diaphorase moiety, encoded by the *hoxU*, *hoxF* and *hoxE* genes, the products of which are homologous to subunits of complex I of the mitochondrial and bacterial respiratory chains and contain NAD(P), FMN, and Fe – S binding sites (Fig. 3, Table 1).

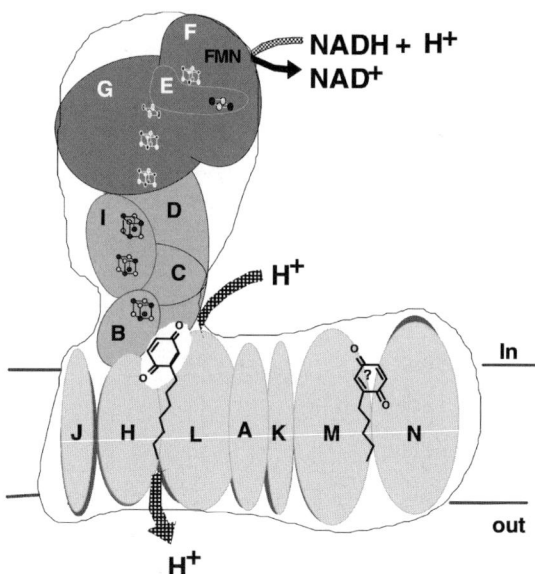

Fig. 3 Schematic representation of *Rhodobacter capsulatus* complex I. The [4Fe – 4S] and [2Fe – 2S] clusters are shown in the appropriate subunits (adapted from Dupuis et al. 2001; Holt et al. 2003; Sazanov and Hinchliffe 2006)

Table 1 Relationships between complex I and NDH-1 subunits and subunits of selected [NiFe]hydrogenases and of $F_{420}H_2$ dehydrogenase

	Bovine[1] Complex I	Synechocystis[2] NDH-1 H$_2$ase HoxEFUYH	E. coli[3] or R. capsulatus[4] NDH-1	P. denitrificans[5] NDH-1	E. coli[6] Hyc H$_2$ase	M. barkeri[7] Ech H$_2$ase	R. rubrum[8] Coo H$_2$ase	P. furiosus[9] Mbh H$_2$ase	Ms. Mazei[10] Fpo
Hydrophilic NADH-oxidizing module	9 kDa								
	24 kDa	HoxE	NuoE	Nqo2					FpoC
	51 kDa	HoxF	NuoF	Nqo1					FpoD
	75 kDa	HoxU[a]	NuoG	Nqo3					FpoB
									FpoI
Subunits of the connecting module	30 kDa	NdhJ		NuoCD (E.c.)[b] / NuoC (R.c.)	Nqo5	HycE / N-ter HycE	EchD			
	49 kDa	NdhH		NuoD (R.c.)	Nqo4	C-ter HycE	EchE	CooH	Mbh12	
	20 kDa (PSST)	NdhK	HoxH	NuoB	Nqo6	HycG	EchC	CooL	Mbh10	
	23 kDa (TYKY)	NdhI	HoxY	NuoI	Nqo9	HycF	EchF	CooX	Mbh14	
	39 kDa									
	18 kDa									
	13 kDa B				(Nqo15[c])					FpoO

Table 1 (continued)

	Bovine[1]	Synechocystis[2]	E. coli[3] or R. capsulatus[4]	P. denitrificans[5]	E. coli[6]	M. barkeri[7]	R. rubrum[8]	P. furiosus[9]	Ms. Mazei[10]
			HoxEFUYH						
	Complex I	NDH-1 H_2ase	NDH-1	NDH-1	Hyc H_2ase	Ech H_2ase	Coo H_2ase	Mbh H_2ase	Fpo
	39 kDa								
Intrinsic-	ND1	NdhA	NuoH	Nqo8	HycD	EchB	CooK	Mbh13	FpoH
membrane	ND2	NdhB	NuoN	Nqo14	HycC[d]	EchA[d]	N-ter CooM[d]	Mbh8	FpoN
hydrophobic	ND3	NdhC	NuoA	Nqo7					FpoA
subunits	ND4	NdhD	NuoM	Nqo13	HycC[d]	EchA[d]	N-ter CooM[d]		FpoM
	ND4L	NdhE	NuoK	Nqo11					FpoK
	ND5	NdhF	NuoL	Nqo12	HycC[d]	EchA[d]	N-ter CooM[d]		FpoL
	ND6	NdhG	NuoJ	Nqo10					FpoJ

References:
[1] Fearnley and Walker 1992; [2] Kaneko et al. 1996; Schmitz et al. 2002; [3] Weidner et al. 1993; [4] Dupuis et al. 1998; [5] Yagi 1993; [6] Sauter et al. 1992; [7] Künkel et al. 1998; [8] Fox et al. 1996a,b; [9] *P. furiosus* genome database (http://comb5-156.umbi.umd.edu/) and Sapra et al. 2000; [10] Bäumer et al. 2000

[a] Sequence similarities between HoxU and N-ter NuoG
[b] NuoC and NuoD are fused in *E. coli* (Blattner et al. 1997)
[c] Nqo15 in *Thermus thermophilus* (Hinchliffe et al. 2006)
[d] NuoL, NuoM, and NuoN are homologous to one particular class of Na^+/H^+ antiporters (Hamamoto et al. 1994)

2.1.4
Group 4

In Group 4 cluster multimeric (six subunits or more) membrane-bound hydrogenases, which comprise transmembrane subunits homologous to complex I subunits involved in proton pumping and energy coupling (Table 1, Fig. 4). They appear to be able to couple the oxidation of a carbonyl group (originating from formate, acetate, or carbon monoxide) with the reduction of protons to H_2 and form the group of *energy-converting*, H_2-*evolving hydrogenases*. The prototype of this group is *E. coli* hydrogenase-3, encoded by the *hyc* operon, part of the formate hydrogen lysase complex (FLH-1) (Böhm et al. 1990; Sauter et al. 1992), which metabolizes formate to H_2 and CO_2, the biosynthetic pathway of which has been deciphered by the group of A. Böck (Sawers et al. 2004). *E. coli* also contains the *hyf* operon, which can encode a putative 10-subunit hydrogenase complex (hydrogenase-4); seven of the *hyf* genes encode homologs of seven Hyc subunits of hydrogenase-3. Three additional genes (*hyfD*, *hyfE* and *hyfF*) have no counterpart in the Hyc complex

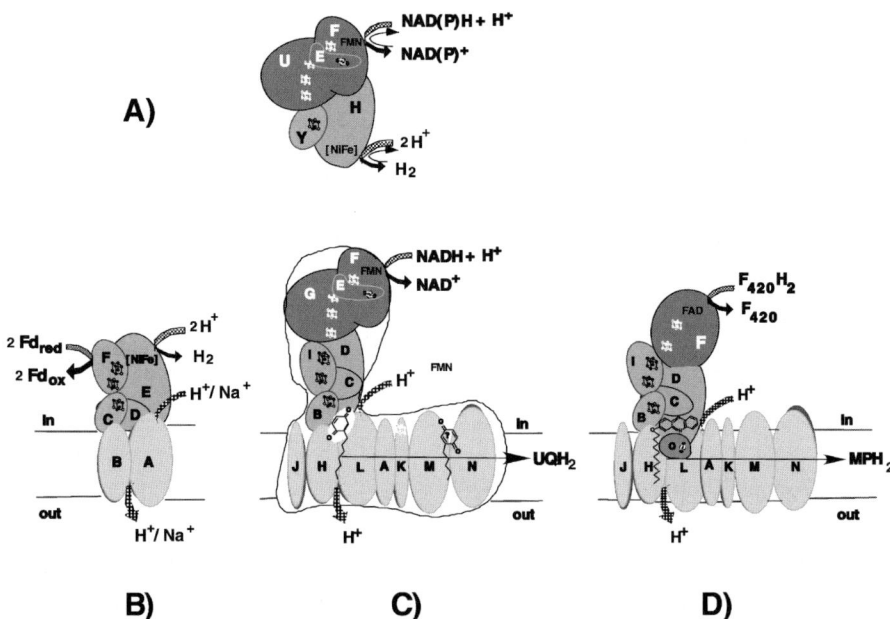

Fig. 4 Models of [NiFe]hydrogenases and $F_{420}H_2$ dehydrogenase compared with that of complex I from *Rhodobacter capsulatus* (**c**). **a** Bidirectional Hox hydrogenase from *Synechocystis* encoded by *hoxEFUYH*. **b** Ech hydrogenase from *Methanosarcina barkeri*, encoded by the *echABCDEF* genes (adapted from Hedderich 2004). **d** $F_{420}H_2$ dehydrogenase from *Methanosarcina mazei* encoded by the *fpoA-O* genes (adapted from Deppenmeier 2004)

and are capable of encoding integral membrane proteins, two of them sharing similarities with subunits that play a crucial role in proton translocation and energy coupling in the NADH:quinone oxidoreductase (complex I) (Andrews et al. 1997). Up to now, no Hyf-derived hydrogenase activity could be detected and no Ni-containing protein corresponding to HyfG, the large subunit of hydrogenase-4 has been observed (Skibinski et al. 2002).

The CO-induced hydrogenase of the photosynthetic bacterium *Rhodospirillum rubrum* is another member of group 4. It is part of the CO-oxidizing system that allows *R. rubrum* to grow in the dark with CO as sole energy source. CO dehydrogenase and the hydrogenase encoded by the *coo* operon oxidize CO to CO_2 with concomitant production of H_2. Since the CO dehydrogenase is a peripheral membrane protein, it was proposed that the hydrogenase component of the oxidizing system constitutes the energy coupling site (Fox et al. 1996a,b). A homologous CO-oxidizing complex has been isolated from the thermophilic Gram-positive bacterium *Carboxydothermus hydrogenoformans* (Soboh et al. 2002).

Group 4 hydrogenases were later isolated from *Archaea* and shown to be able to couple H_2 evolution and energy conservation. They include the EhA and Ehb hydrogenases from *Methanothermobacter* species (Tersteegen and Hedderich 1999), the Ech hydrogenase from *Methanosarcina barkeri* (Künkel et al. 1998; Meuer et al. 1999), and the Mbh hydrogenase from *Pyrococcus furiosus* (Sapra et al. 2000; Silva et al. 2000; recent reviews by Hedderich 2004; Hedderich and Forzi 2005; Vignais and Colbeau 2004). Some, found in present-day hyperthermophiles, were acquired from *Archaea* by horizontal gene transfer. According to Calteau et al. (2005) this would be the case for the 13-gene operon found in the genome of *Thermotoga maritima*, capable of encoding a Mbx hydrogenase, probably acquired by horizontal transfer from an archaebacterium belonging to the *Pyrococcus* group, and for the six-gene *ech* operon found in *Thermoanaerobacter tengcongensis* (Sobboh et al. 2004) and in *Desulfovibrio gigas* (Rodrigues et al. 2003), which was probably transferred independently from an archaebacterium belonging to the *Methanosarcina* clade.

2.2
The [FeFe]hydrogenases

[FeFe]hydrogenases are found in anaerobic prokaryotes, such as clostridia and sulfate reducers, and in eukaryotes (see reviews by Adams 1990; Atta and Meyer 2000; Vignais et al. 2001; Horner et al. 2000, 2002). [FeFe]hydrogenases are the only type of hydrogenases to have been found in eukaryotes, and they are located exclusively in membrane-limited organelles, i.e., in chloroplasts or in hydrogenosomes. They are usually involved in H_2 production.

Unlike [NiFe]hydrogenases composed of at least two subunits, many [FeFe]hydrogenases are monomeric and consist of the catalytic subunit only, although dimeric, trimeric, and tetrameric enzymes are also known (Vignais et al. 2001). The catalytic subunit of [FeFe]hydrogenases, in contrast to those of Ni-containing enzymes, vary considerably in size. Besides the conserved domains of ca. 350 residues containing the active site (H-cluster, Adams 1990), they often comprise additional domains, which accommodate Fe – S clusters (Fig. 5). The H-cluster consists of a binuclear iron subsite ([Fe_2S_3]) bound to a conventional [4Fe – 4S] cluster by a bridging cysteinyl sulfur. To each Fe atom a terminal carbon monoxide, a bridging carbon monoxide, and a cyanide ligand are bound. The Fe atoms also share two bridging sulfur ligands of a di(thiomethyl)amine molecule ($CH_3 - S^-)_2$ (Peters et al. 1998, 1999; Nicolet et al. 1999, 2000, 2002).

Although [FeFe]- and [NiFe]hydrogenases have completely different structures and are evolutionary unrelated, they share a common feature, namely the presence of endogenous CO and CN^- ligands bound to a Fe center in the

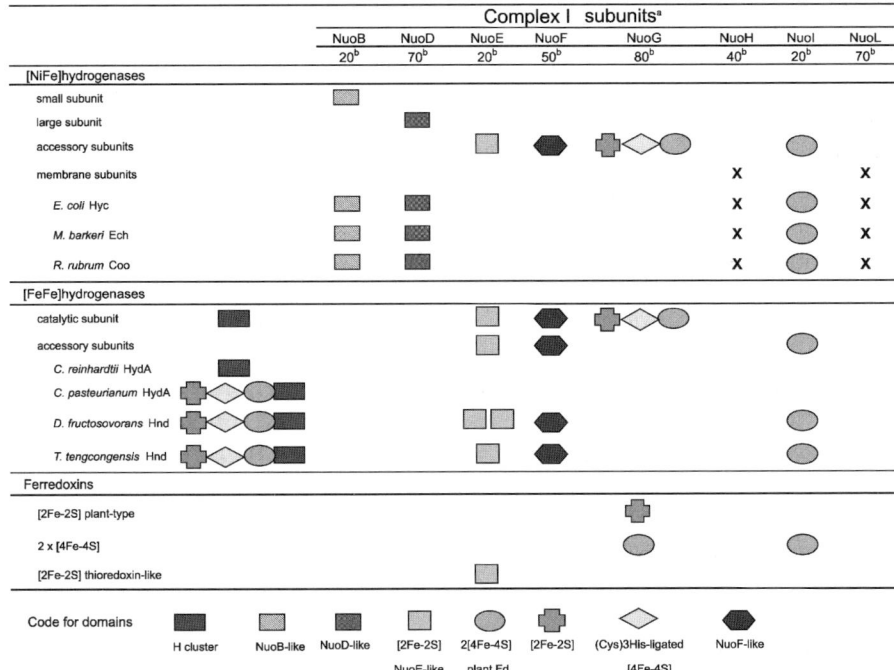

Fig. 5 Schematic representation of homologies between hydrogenases and complex I. The code for domains is indicated in the *lower part* of the figure. [a] Complex I subunits are designated by the *nuo* nomenclature used for *E. coli* and *Rb. capsulatus*. [b] Approximate masses (kDa) of subunits found in various bacteria. The H-cluster domain is included although it has no homolog in complex I only accessory subunits of some [FeFe]hydrogenases have (adapted from Fig. 10 of Vignais et al. (2001), where a detailed legend can be consulted)

active site. The presence of these ligands stabilizes iron in a low oxidation and spin state and makes it resemble transition metals (Ru, Pd, or Pt) known to be good catalysts for H_2 splitting (Adams and Stiefel 2000). Another common feature is the presence of an Fe – S cluster proximal to the dinuclear metallocenter, which is then wired to the surface for electron exchange with its partner redox proteins by a conduit of Fe – S clusters. Finally, both types of enzymes contain hydrophobic gas channel(s) that runs from the molecular surface to the buried active site (Nicolet et al. 2002).

A [FeFe]hydrogenase is proposed to have been a key enzyme at the origin of the eukaryotic cell. Two hypotheses posit that a metabolic symbiosis (syntrophy) between a methanogenic archaebacterium and a proteobacterium able to release H_2 in anaerobiosis was the first step in eukaryogenesis. The hydrogen hypothesis (Martin and Müller 1998) proposes that an anaerobic heterotrophic α-*Proteobacterium*, producing H_2 and CO_2 as waste products, formed a symbiotic metabolic association (syntrophy) with a strictly anaerobic, autotrophic archaebacterium, possibly a methanogen dependent on H_2. The intimate relationship over long periods of time allowed the symbiont and the host to co-evolve and become dependent on each other. In an anaerobic environment the symbiont was either lost, as in type I amitochondriate eukaryotes, or became a hydrogenosome (i.e., a hydrogen-generating and ATP-supplying organelle) as in type II amitochondriate eukaryotes (Müller 1993). By further evolution, the host lost its autotrophic pathway and its dependence on H_2 and the endosymbiont adopted a more efficient aerobic respiration to become the ancestral mitochondrion. Thus, the eukaryotic cell would have emerged as the result of endosymbiosis between two prokaryotes, an H_2-dependent, autotrophic archaebacterium (the host) and an H_2- and ATP-producing eubacterium (the symbiont), the common ancestor of mitochrondria and hydrogenosomes. The syntrophy hypothesis for the origin of eukaryotes, proposed at the same time and independently (Moreira and López-Garcia 1998; López-Garcia and Moreira 1999) is based on similar metabolic consideration (interspecies hydrogen transfer), but the latter authors speculated that the organisms involved in syntrophy with methanogenic archaea were δ-*Proteobacteria* (ancestral sulfate-reducing myxobacteria) (it was also suggested that a second anaerobic symbiont was involved in the origin of mitochondria). Thus, hydrogenosomes are either considered to be relics of ancestral endosymbiont and to share a common origin with mitochondria (Bui et al. 1996; Martin and Müller 1998) or to have evolved several times as adaptations of mitochondria to anaerobic environments (Hackstein 2005; Hackstein et al. 2001; Embley et al. 2003; Tjaden et al. 2004).

Eukaryotic organelles contain only [FeFe]hydrogenases. A phylogenetic analysis of eukaryotic [FeFe]hydrogenases (Horner et al. 2000, 2002) suggests a polyphyletic origin of these enzymes, implying an acquisition by lateral gene transfer from different prokaryotic sources. On the other hand, the [FeFe]hydrogenases from green algae emerge as a monophyletic group with hydrogenosomal [FeFe]hydrogenases from microaerophilic protists (Horner et al. 2002). The source of an ancestral [FeFe]hydrogenase is not resolved; its presence in eukaryotes may reflect an early lateral transfer from a eubacterium. The plastidial [FeFe]hydrogenases appear to have a non-cyanobacterial origin, since cyanobacteria, the progenitors of chloroplasts, contain only [NiFe]hydrogenases and no [FeFe]hydrogenases (Vignais et al. 2001; Tamagnini et al. 2002). Eukaryotes possess genes that encode proteins that are phylogenetically related to [FeFe]hydrogenases. Mitochondria do not contain [FeFe]hydrogenase but have kept a key enzyme, cysteine desulfurase (called IscS or Nfs1), which performs a crucial role in cellular Fe−S protein maturation (Mühlenhoff and Lill 2000; Lill and Mühlenhoff 2005) and appears to have originated from the ancestor endosymbiont.

2.3
Hydrogenases and Complex I

The energy-converting NADH-ubiquinone oxidoreductase is the main entry site of reducing equivalents into the mitochondrial and the bacterial respiratory chains (for a recent review see Brandt 2006). The mitochondrial enzyme is also called complex I, whereas the bacterial enzyme is more often referred to as type 1 NADH-dehydrogenase or NDH-1. The bovine mitochondrial complex I contains a total of 46 different subunits (Carroll et al. 2003; Hirst et al. 2003) whereas NDH-1 from the bacteria *Paracoccus denitrificans* (Yagi 1993; Yagi et al. 1998) and *Rhodobacter capsulatus* (Dupuis et al. 1998) contain 14 subunits, all of which have homologs in the bovine enzyme (Table 1). Both the mitochondrial and bacterial enzymes are L-shaped, with a membrane domain and a peripheral arm extending into the cytosol. The hydrophilic NADH-oxidizing module, distal from the membrane comprises three hydrophilic subunits containing FMN and five Fe−S clusters; a second hydrophilic module consisting of four subunits connects the NADH-oxidizing proteins to the membrane-bound hydrophobic subunits. The two extramembranous modules contain all the redox centers of the enzyme (Dupuis et al. 1998, 2001; Yagi et al. 1998; Friedrich 2001; Friedrich et al. 1998; Schultz and Chan 2001; Sazanov et al. 2000; Sazanov and Hinchliffe 2006) (Fig. 3). Sequence similarities between hydrogenases and complex I, first reported by Böhm et al. (1990) and Pilkington et al. (1991), have been emphasized in several subsequent reports (Friedrich and Weiss 1997; Friedrich and Scheide 2000; Albracht and Hedderich 2000; Dupuis et al. 2001; Friedrich 2001; Yano and Ohnishi 2001; Vignais et al. 2001, 2004; Hedderich 2004). Subunits NuoE, NuoF, NuoI, and the N-terminal Fe−S binding domain (ca. 220 residues) of NuoG have homologous counterparts in accessory subunits and domains of soluble [NiFe]hydrogenases of group 3 (Hox) and of [FeFe]hydrogenases (Fig. 5). In addition, three subunits located within the connecting module of complex I share similarities with subunits of the core [NiFe] enzyme, the NuoB subunit with the small hydrogenase subunit (the [4Fe–4S] cluster of NuoB, known as cluster N2 (Ohnishi et al.1998) and suggested to be a key component in redox-driven proton translocation (Flemming et al. 2005) is related to the hydrogenase proximal cluster), and the NuoC and NuoD subunits (fused as a single NuoCD protein in *E. coli*) with the large subunit. Furthermore, hydrophobic subunits of multimeric, membrane-bound [NiFe]hydrogenases belonging to group 4, e.g., *E. coli* Hyc and Hyf, *R. rubrum* Coo, *Ms. barkeri* Ech, *Methanothermobacter marburgensis* Eha and Ehb and *P. furiosus* Mbh are also homologous to transmembrane subunits of complex I (NuoH, NuoL, NuoM, NuoN). It should be noted that these hydrogenases of group 4 are also ion (H^+ or Na^+) pumps (the nature of the coupling ion used is still elusive). Thus, the presumed evolutionary links between hydrogenases and complex I concern not only the electron transferring subunits but also the ion pumping units, i.e., the coupling between electron transport and energy recovery by a chemiosmotic mechanism.

On the basis of the similarities between [NiFe]hydrogenases and the NuoB–NuoD dimer of the connecting module, Dupuis et al. (2001) have proposed (i) that the [NiFe] active site of hydrogenases was reorganized into a quinone-reduction site carried by the NuoB–NuoD dimer (Prieur et al. 2001) and a hydrophobic subunit such as NuoH (Fig. 3), and (ii) that NuoD might provide both the quinone gate and a potential proton channel entry for a minimal "proton pumping" module, composed of subunits NuoB, NuoD, NuoI, and NuoL (Friedrich and Scheide 2000; Dupuis et al. 2001; see also Kashani-Poor et al. 2001). Subunit NuoL (or NuoN or NuoM, considered up to recently to have evolved by triplication of an ancestral gene related to bacterial Na^+ or K^+/H^+ antiporters (Fearnley and Walker 1992; Friedrich and Weiss 1997)) would have provided the transmembrane channel required to complete the proton (or Na^+) pump (Dupuis et al. 2001). These membrane proteins are similar to an electrogenic Na^+/H^+ antiporter first identified in an alkalophilic *Bacillus* strain (Hamamoto et al. 1994). In *Bacillus subtilis* the corresponding proteins are encoded by a seven-gene operon, *mrp* (multiple resistance and pH), and are termed MrpA-G (Ito et al. 1999) (the *sha* nomenclature for "sodium hydrogen antiporter" is also used (Kosono et al. 2006)). The MrpA and the MrpD antiporters come in two subclasses, MrpA-type and MrpD-type, and it has been determined that NuoL is more closely related to MrpA and that NuoM and N are more closely related to the MrpD antiporter (Mathiesen and Hägerhäll 2002). NuoK has later been shown to be homologous to MrpC, suggesting that a multisubunit antiporter complex was recruited to the ancestral enzyme (Mathiesen and Hägerhäll 2003). The latter authors concluded that the last common ancestor of complex I and the membrane-bound [NiFe]hydrogenases of group 4 contained the NuoKLMN subunit module.

The nature of the ion translocated by complex I (Na^+ or H^+) is still a matter of debate. The mitochondrial enzyme of respiratory chains were shown to be proton pumps (Hinkle 2005) but some bacterial respiratory enzymes generate an Na^+ gradient (Dimroth and Cook 2004) and many marine and pathogenic bacteria have a sodium-translocating NADH:ubiquinone oxidoreductase, which generates an electrochemical Na^+ gradient during aerobic respiration (Barquera et al. 2004). Complex I from *Klebsiella pneumoniae* (Krebs et al. 1999) and from *E. coli* (Steuber 2001) have been proposed to work as an Na^+ pump. Since the membranous complex I subunits NuoL, NuoM, and NuoN are homologous to cation/proton antiporters (Friedrich and Scheide 2000; Mathiesen and Hägerhäll 2002, 2003) the question arises whether the complex is involved in primary proton translocation or is capable of secondary Na^+/H^+ antiport. Recently, Stolpe and Friedrich (2004) have shown that *E. coli* complex I is a primary proton pump but is capable of secondary sodium antiport.

3
Modes of Energy Conservation by Hydrogenases

3.1
Energy Conservation via Energy-Transducing Electron Transport Chains by Respiratory [NiFe]hydrogenases (Group 1)

The uptake hydrogenases link the oxidation of H_2 to the reduction of oxygen (aerobic respiration) or to the reduction of anaerobic electron acceptors such as NO_3^-, SO_4^{2-}, fumarate, and TMAO/DMSO (anaerobic respiration). Similarly to other substrate-specific dehydrogenases they feed electrons into a common quinone pool, from which electrons are transferred via specific quinol oxidases to terminal reductases, e.g., in the absence of oxygen, TMAO/DMSO, nitrate or fumarate reductase and, in the presence of oxygen, cytochrome oxidase. The oversimplified scheme of Fig. 6 is meant to emphasize the role of the quinone pool in respiration, the type of quinone (e.g., ubiquinone, UQ, menaquinone, MK) depending on the prevailing environmental conditions (Richardson 2000). The third subunit of the trimeric uptake hydrogenase, the di-heme cytochrome *b* encoded by *hupC* in *Rb. capsulatus* and *hoxZ* in *R. eutropha* and *Azotobacter vinelandii* is the necessary link for transfer of electrons from H_2 to the electron transport chain; furthermore, it plays a role in activating and maintaining the hydrogenase in a reduced, active state (Cauvin et al. 1991; Sayavedra-Soto and Arp 1992; Bernhard et al. 1997; Meek and Arp 2000). Electrons from H_2 are donated to the quinone pool (Henry and Vignais 1983) and the energy of H_2 ox-

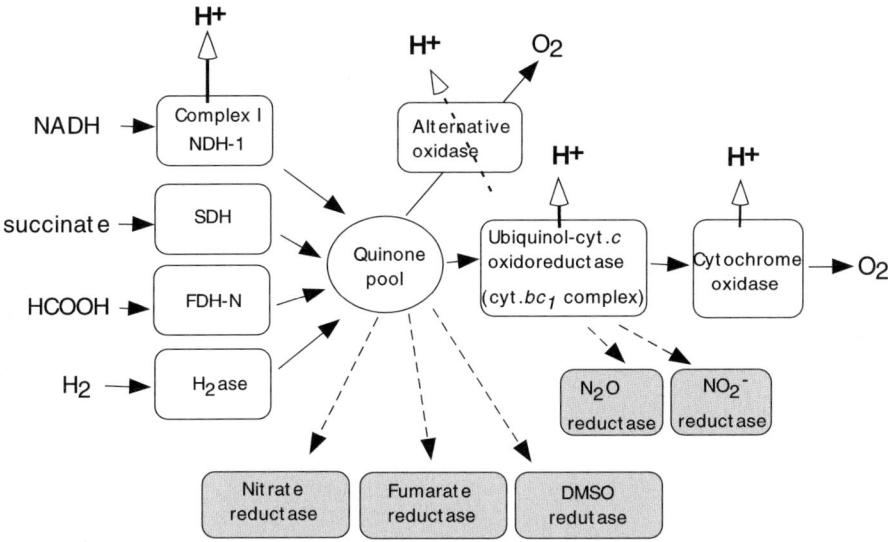

Fig. 6 Simplified and general scheme illustrating electron pathways in respiratory chains. *White boxes* indicate electron-input units and *black arrows* the influx of reducing equivalents in the membrane. The respiratory hydrogenase represented here is usually trimeric, the third subunit being a di-heme cytochrome *b*, which anchors the hydrogenase to the membrane, binds the quinone, and is the link for the transfer of H_2 electrons to the quinone (Cauvin et al. 1991; Bernhard et al. 1997; Gross et al. 2004). *Dashed arrows* indicate electron flux to output modules involved in anaerobic respiration (*shaded boxes*). Energy coupling sites are indicated by *arrows* showing vectorial proton ejection. Not shown is the Δp created across the membrane when fumarate reductase is reduced via the quinone (menaquinone, MK) with H_2 or formate (Kröger et al. 2002). The proton pumping activity of alternative oxidase (cyt *bo*-type) (Kömen et al. 1996) is indicated by a *dashed arrow* since it is not a common case among the quinol oxidases

idation is recovered by vectorial proton transfer at the level of the quinol oxidase (Kömen et al. 1996), cytochrome bc_1 complex and cytochrome oxidase (Paul et al. 1979; Porte and Vignais 1980). *W. succinogenes* performs oxidative phosphorylation with fumarate as terminal electron acceptor and H_2 (or formate) as electron donor. This fumarate respiration, catalyzed by an electron transport chain consisting of hydrogenase, menaquinone, and fumarate reductase (Fig. 2a), is coupled to the generation of an electrochemical proton potential (Δp) across the bacterial membrane generated by MK reduction with H_2 (Kröger et al. 2002; Lancaster et al. 2005). In the methanogenic archaeon *Methanosarcina mazei* Göl, the VhoGA uptake hydrogenase transfers electrons from H_2 to a cytochrome *b* (VhoC); the electrons are then channeled through methanophenazine to heterodisulfide reductase, which reduces the CoM–S–S–CoB heterodisulfide to produce CoB–SH, the reductant for the formation of methane from methyl–S–CoM (Ide et al. 1999) (Fig. 2c, Fig. 7).

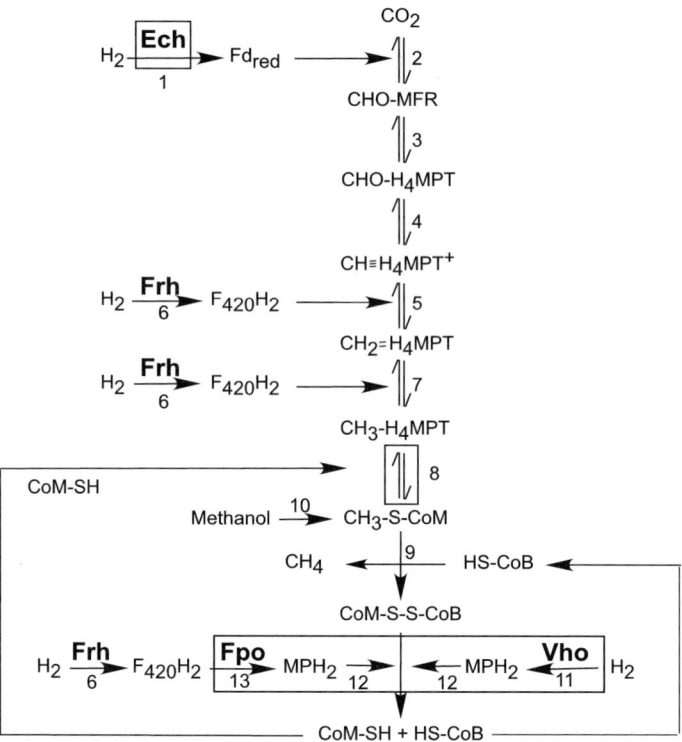

Fig. 7 Pathway of methanogenesis from $CO_2 + H_2$ and from methanol. Reactions catalyzed by membrane-bound energy-transducing enzyme complexes are *boxed*.
Abbreviations: $F_{420}H_2$ reduced form of coenzyme F_{420}; Fd_{red} reduced form of ferredoxin; MFR methanofuran; H_4MPT tetrahydromethanopterin; HS-CoM coenzyme M; HS-CoB coenzyme B; MPH_2 reduced form of methanophenazine.
Enzymes: *1* Ech, Ech hydrogenase; *2* formylmethanofuran dehydrogenase; *3* formyl-MFR:H_4MPT formyl transferase; *4* methenyl-H_4MPT cyclohydrolase; *5* methylene-H_4MPT dehydrogenase; *6* Frh, F_{420}-reducing hydrogenase; *7* methylene-H_4MPT reductase; *8* methyl-H_4MPT:HS-CoM methyltransferase; *9* methyl-CoM reductase; *10* soluble methyltransferases; *11* Vho, F_{420}-nonreducing hydrogenase; *12* heterodisulfide reductase; *13* Fpo, $F_{420}H_2$ dehydrogenase. Each type of hydrogenase *highlighted in bold* is shown with its specific electron acceptor (ferredoxin for Ech, cofactor F_{420} for Frh, and methanophenazine for Vho) (adapted from Deppenmeier 2004)

3.2
Energy-Conservation by Proton/Na^+ Translocation by Membrane-Bound, H_2 Evolving [NiFe]hydrogenases (Group 4)

The mechanism of energy conservation by Na^+/H^+ translocation has been best studied with [NiFe]hydrogenases from methanogens, which obtain most or all their energy for growth from the process of methane biosynthesis (methanogenesis), considered to be an anaerobic respiration (see Müller V in this volume and reviews by Deppenmeier 2002, 2004; Hedderich 2004; Hedderich and Forzi 2005).

Strictly anaerobic archaea of the genus *Methanosarcina* derive their metabolic energy from the conversion to methane of a restricted number of C_1 compounds and acetate (Deppenmeier 2002). Figure 7 shows how CH_4 is formed from $CO_2 + H_2$ via the CO_2-reducing pathway, or from methanol. Three types of [NiFe]hydrogenases, identified recently (see reviews by Hedderich 2004; Deppenmeier 2004), are involved in these two systems in which either H_2 or $F_{420}H_2$ are used as electron donor and the heterodisulfide CoM–S–S–CoB as electron acceptor (hence the term "disulfide respiration" used by Hedderich and Whitman (2005)). H_2 reduction of low-potential ferredoxin by Ech, thermodynamically unfavorable, requires the consumption of a membrane ion gradient and thus occurs by so-called reverse electron transport. Redox-driven proton translocation catalyzed by intrinsic membrane subunits of the Ech hydrogenase and Fpo dehydrogenase generates a protonmotive force and hence energy recovery during methanogenesis (Fig. 7). In acetoclastic methanogenesis, Ech couples the oxidation of reduced ferredoxin (arising from the oxidation of the carbonyl group of acetate) to the production of H_2.

Methanophenazine (MP), which acts in the membrane of the methanogen as the quinone in respiratory chains of bacteria and mitochondria, can be reduced either with H_2, by the F_{420}-non reducing hydrogenase VhoAG via its third subunit, VhoC, which interacts with MP (Fig. 2c), or with $F_{420}H_2$ by the $F_{420}H_2$ dehydrogenase (FpoDH), a multimeric complex encoded by the *fpo* genes, with subunits homologous to subunits of complex I (Table 1) (Fig. 4). The heterodisulfide reductase (HdrED) receives electrons from the reduced form of methanophenazine, MPH_2 (Fig. 2c). Each partial reaction, the reduction of MP by H_2 or $F_{420}H_2$ and the reduction of CoM–S–S–CoB by MPH_2 is coupled to the translocation of $2H^+/2e^-$. H^+-translocation in both reactions can occur via a redox-loop mechanism, while $F_{420}H_2$ dehydrogenase is thought to function as a proton pump (Ide et al. 1999; Bäumer et al. 2000).

Another member of group 4, the Mbh hydrogenase from *P. furiosus*, has been shown to couple electron transfer from reduced ferredoxin to both proton reduction and proton translocation. Oxidation of reduced ferredoxin by inverted membrane vesicles of *P. furiosus* generated both a $\Delta\psi$ and a ΔpH, which could be coupled to ATP synthesis (Sapra et al. 2003)

3.3
Disposal of Excess Reducing Equivalents

Growth of bacteria depends on dissimilatory and assimilatory processes. Oxidation of inorganic or organic substrates results in the formation of reducing power (NADH) and ATP, which is used to drive assimilatory processes leading to the synthesis of cell materials. Growth rates depend on ATP content and the (photo)phosphorylation rate is regulated by redox balance. "Over-reduction" or "over-oxidation" of the redox components of the electron transport chain (including the quinone pool) leads to inhibition of phosphorylation (Bose and Gest 1963). The requirement for a membrane redox poise close to the oxidation–reduction potential of the ubiquinone pool (Candela et al. 2001) can be explained by the involvement of a semiquinone intermediate in the Q cycle (Nicholls and Ferguson 1992; Dutton et al. 1998; Brandt 1999). To dissipate excess reducing equivalents from the photosynthetic membrane, bacteria such as *Rb. capsulatus* can use an alternative quinol oxidase, which allows the cell to control the redox state of the Q pool and the rate of photophosphorylation activity (Zannoni and Marrs 1981). In *Rb. capsulatus*, under anaerobic conditions in the light, excess reducing equivalents are transferred to NAD^+ by reverse electron flow through complex I (Klemme 1969; Dupuis et al. 1997). Reducing equivalents stored in NADH can be dissipated by metabolic systems such as CO_2 fixation (Calvin cycle), nitrogen fixation, or reduction of auxiliary oxidants (Hillmer and Gest 1977; Tichi et al. 2001). In the case of nitrogen fixation, which is catalyzed by nitrogenase, H_2 is produced as an intrinsic part of the enzymatic reaction and, in the absence of N_2, nitrogenase functions as a hydrogenase, reducing protons to H_2 (Vignais et al.1985; Willison 1993). Since nitrogenase is an ATP-dependent enzyme, this reaction dissipates energy as well as offering another means for disposal of excess reducing equivalents.

[NiFe]hydrogenases of group 3, which bind reduced coenzymes such as NADH, NADPH, and $F_{420}H_2$, can directly regenerate the oxidized coenzymes by using the protons of water as electron acceptors and then evolving H_2. H_2 production catalyzed by the cytoplasmic NAD(P)-dependent bidirectional [NiFe]hydrogenase (Hox) has indeed been observed with the cyanobacterium *Synechocystis* PCC6803. Significant H_2 production was observed when cells achieved anaerobiosis, the rate of H_2 production being higher in the presence of fermentative substrates such as glucose. The transient H_2 burst observed upon re-illumination probably reflected the increase in NAD(P)H concentration in response to photosystem I activity (Cournac et al. 2004). Appel et al. (2000) have proposed that the bidirectional hydrogenase functions as an electron valve for the disposal of low-potential electrons generated at the onset of illumination. Similarly, H_2 production has been observed when dark-adapted *Chlamydomonas reinhardtii* cells are illuminated (Cournac 2002). In that case, it is a [FeFe]hydrogenase which transfers electrons from a [2Fe − 2S] ferredoxin reduced by photosystem I. The [FeFe]hydrogenase is an electron "valve" that enables the algae to survive under anaerobic conditions (Happe et al. 2002).

4
Conclusions and Perspectives

Hydrogenases are a structurally and functionally diverse group of enzymes, and phylogenetic analysis has led to the identification of several phylogenetically distinct groups and subgroups that form the basis of a coherent system of classification. Their modular structure, their additional domains and subunits that have counterparts in other redox proteins and complexes, has long been a matter of speculation. Their relationships with NADH-ubiquinone oxidoreductase (complex I) of respiratory chains has gained renewed interest with the recently identified multisubunit, membrane-bound, energy-conserving [NiFe]hydrogenases of methanogenic archaea.

Whole genome sequencing is not only increasing significantly the number of available hydrogenase sequences but is also revealing the presence of multiple hydrogenases in *Bacteria* and *Archaea*. Postgenomic analysis (transcriptome, proteome, metabolome) has and will be essential for elucidating the metabolic roles of these enzymes and the regulation of their biosynthesis and activity. The chief role of [NiFe]hydrogenases is clearly the oxidation of H_2 or the reduction of protons, coupled to energy-conserving electron transfer chain reactions, which allow energy to be obtained either from H_2 or from the oxidation of substrates of lower potential. In the last decade, additional roles have been revealed. Thus, the so-called H_2 sensors hydrogenases are involved in regulating the biosynthesis of uptake [NiFe]hydrogenases in response to their substrate, H_2. Other, bidirectional hydrogenases able to bind directly reduced coenzymes and re-oxidize them using protons from water as electron acceptors can act as electron "valves" to control the redox poise of the respiratory chain at the level of the quinone pool. This is essential to ensure the correct functioning of the respiratory chain in the presence of excess reducing equivalents, in particular in photosynthetic microorganisms. Finally, hydrogenases from group 4, those originally thought to play a purely fermentative role and the newly discovered ones in methanoarchaea, are now known to be involved in membrane-linked energy conservation through the generation of a transmembrane protonmotive force.

The broad distribution of hydrogenases among existing microorganisms attests to the importance of H_2 metabolism in a wide range of environments, and suggests that hydrogenases may have appeared very early in evolution. Two newly formulated hypotheses propose that H_2 metabolism may have been the driving force that led to cellular symbiosis and fusion events involved in the formation of the first eukaryotic cells. The present day [FeFe]hydrogenases that are found in the organelles of unicellular eukaryotes (hydrogenosomes, chloroplasts) may be relics of these evolutionary events or the results of more recent lateral gene transfers. Their evolutionary origins is still unresolved and are the subject of current studies and debates. The discovery of hydrogenase-like sequences in the genomes of aerobic eukaryotes, including mammals, opens a new field of research. The encoded proteins, related to [FeFe]hydrogenases, appear to be involved in the maturation of Fe − S clusters, for insertion into the Fe − S proteins that are crucial for all cellular life. The cysteine desulfurase, a key enzyme of this pathway, located in the mitochondrion, appears to have originated from the mitochondrial endosymbiont. Comparative biochemical and genetic studies and determination of the localization of these proteins in hydrogenosomes and mitochondria will help to find out the reason why the host cell kept the endosymbiont: was it because of its ability to make Fe − S clusters for its host?

References

Adams MWW (1990) The structure and mechanism of iron-hydrogenases. Biochim Biophys Acta 1020:115
Adams MWW, Stiefel EI (2000) Organometallic iron: the key to biological hydrogen metabolism. Curr Opin Chem Biol 4:214
Albracht SPJ, Hedderich R (2000) Learning from hydrogenases: location of a proton pump and of a second FMN in bovine NADH–ubiquinone oxidoreductase (complex I). FEBS Lett 485:1
Andrews SC, Berks BC, McClay J, Ambler A, Quail MA, Golby P, Guest JR (1997) A 12-cistron *Escherichia coli* operon (*hyf*) encoding a putative proton-translocating formate hydrogenlyase system. Microbiology 143(11):3633–3647
Appel J, Schulz R (1998) Hydrogen metabolism in organisms with oxygenic photosynthesis: hydrogenases as important regulatory devices for a proper redox poising? J Photochem Photobiol B–Biol 47:1
Appel J, Phunpruch S, Steinmuller K, Schulz R (2000) The bidirectional hydrogenase of *Synechocystis* sp. PCC 6803 works as an electron valve during photosynthesis. Arch Microbiol 173:333
Atta M, Meyer J (2000) Characterization of the gene encoding the [Fe]-hydrogenase from *Megasphaera elsdenii*. Biochim Biophys Acta 1476:368
Barquera B, Nilges MJ, Morgan JE, Ramirez-Silva L, Zhou W, Gennis RB (2004) Mutagenesis study of the 2Fe − 2S center and the FAD binding site of the Na^+-translocating NADH:ubiquinone oxidoreductase from Vibrio cholerae. Biochemistry 43:12322
Bäumer S, Ide T, Jacobi C, Johann A, Gottschalk G, Deppenmeier U (2000) The F420H2 dehydrogenase from *Methanosarcina mazei* is a redox-driven proton pump closely related to NADH dehydrogenases. J Biol Chem 275:17968
Berks BC, Palmer T, Sargent F (2003) The Tat protein translocation pathway and its role in microbial physiology. Adv Microb Physiol 47:187
Berks BC, Palmer T, Sargent F (2005) Protein targeting by the bacterial twin-arginine translocation (Tat) pathway. Curr Opin Microbiol 8:174
Bernhard M, Benelli B, Hochkoeppler A, Zannoni D, Friedrich B (1997) Functional and structural role of the cytochrome *b* subunit of the membrane-bound hydrogenase complex of *Alcaligenes eutrophus* H16. Eur J Biochem 248:179
Bernhard M, Friedrich B, Siddiqui RA (2000) *Ralstonia eutropha* TF93 is blocked in tat-mediated protein export. J Bacteriol 182:581
Blattner FR, Plunkett G III, Bloch CA, Perna NT, Burland V, Riley M, Collado-Vides J, Glasner JD, Rode CK, Mayhew GF, Gregor J, Davis NW, Kirkpatrick HA, Goeden MA,

Rose DJ, Mau B, Shao Y (1997) The complete genome sequence of *Escherichia coli* K-12. Science 277:1453

Böhm R, Sauter M, Böck A (1990) Nucleotide sequence and expression of an operon in *Escherichia coli* coding for formate hydrogenlyase components. Mol Microbiol 4:231

Bose SK, Gest H (1963) Bacterial photophosphorylation: regulation by redox balance. Proc Natl Acad Sci USA 49:337

Brandt U (1999) Proton translocation in the respiratory chain involving ubiquinone – a hypothetical semiquinone switch mechanism for complex I. Biofactors 9:95

Brandt U (2006) Energy converting NADH:quinone oxidoreductase (complex I). Annu Rev Biochem 75:69-92

Brugna-Guiral M, Tron P, Nitschke W, Stetter KO, Burlat B, Guigliarelli B, Bruschi M, Giudici-Orticoni MT (2003) [NiFe] hydrogenases from the hyperthermophilic bacterium *Aquifex aeolicus*: properties, function, and phylogenetics. Extremophiles 7:145

Bui ETN, Bradley PJ, Johnson PJ (1996) A common evolutionary origin for mitochondria and hydrogenosomes. Proc Natl Acad Sci USA 93:9651

Calteau A, Gouy M, Perrière G (2005) Horizontal transfer of two operons coding for hydrogenases between bacteria and archaea. J Mol Evol 60:557

Candela M, Zaccherini E, Zannoni D (2001) Respiratory electron transport and light-induced energy transduction in membranes from the aerobic photosynthetic bacterium *Roseobacter denitrificans*. Arch Microbiol 175:168

Carroll J, Fearnley IM, Shannon RJ, Hirst J, Walker JE (2003) Analysis of the subunit composition of complex I from bovine heart mitochondria. Mol Cell Proteomics 2:117

Cauvin B, Colbeau A, Vignais PM (1991) The hydrogenase structural operon in *Rhodobacter capsulatus* contains a third gene, *hupM*, necessary for the formation of a physiologically competent hydrogenase. Mol Microbiol 5:2519

Cournac L, Mus F, Bernard L, Guedeney G, Vignais P, Peltier G (2002) Limiting steps of hydrogen production in *Chlamydomonas reinhardtii* and *Synechocystis* PCC 6803 as analysed by light-induced gas exchange transients. Int J Hydr Energ 27:1229

Cournac L, Guedeney G, Peltier G, Vignais PM (2004) Sustained photoevolution of molecular hydrogen in a mutant of *Synechocystis* sp. strain PCC 6803 deficient in the type I NADPH-dehydrogenase complex. J Bacteriol 186:1737

Deppenmeier U (2002) The unique biochemistry of methanogenesis. Prog Nucleic Acid Res Mol Biol 71:223

Deppenmeier U (2004) The membrane-bound electron transport system of *Methanosarcina* species. J Bioenerg Biomembr 36:55

Dimroth P, Cook GM (2004) Bacterial Na^+- or H^+-coupled ATP synthases operating at low electrochemical potential. Adv Microb Physiol 49:175

Dolla A, Pohorelic BK, Voordouw JK, Voordouw G (2000) Deletion of the *hmc* operon of *Desulfovibrio vulgaris* subsp. vulgaris Hildenborough hampers hydrogen metabolism and low-redox-potential niche establishment. Arch Microbiol 174:143

Dross F, Geisler V, Lenger R, Theis F, Krafft T, Fahrenholz F, Kojro E, Duchene A, Tripier D, Juvenal K et al. (1992) The quinone-reactive Ni/Fe-hydrogenase of *Wolinella succinogenes*. Eur J Biochem 206:93

Dubini APRL, Jack RL, Palmer T, Sargent F (2002) How bacteria get energy from hydrogen: a genetic analysis of periplasmic hydrogen oxidation in *Escherichia coli*. Int J Hydr Energ 27:1413

Dupuis A, Chevallet M, Darrouzet E, Duborjal H, Lunardi J, Issartel JP (1998) The complex I from *Rhodobacter capsulatus*. Biochim Biophys Acta 1364:147

Dupuis A, Prieur I, Lunardi J (2001) Toward a characterization of the connecting module of complex I. J Bioenerg Biomembr 33:159

Dutton PL, Moser CC, Sled VD, Daldal F, Ohnishi T (1998) A reductant-induced oxidation mechanism for complex I. Biochim Biophys Acta 1364:245

Embley TM, van der Giezen M, Horner DS, Dyal PL, Bell S, Foster PG (2003) Hydrogenosomes, mitochondria and early eukaryotic evolution. IUBMB Life 55:387

Fauque G, Peck HD Jr, Moura JJ, Huynh BH, Berlier Y, DerVartanian DV, Teixeira M, Przybyla AE, Lespinat PA, Moura I et al. (1988) The three classes of hydrogenases from sulfate-reducing bacteria of the genus *Desulfovibrio*. FEMS Microbiol Rev 4:299

Fearnley IM, Walker JE (1992) Conservation of sequences of subunits of mitochondrial complex I and their relationships with other proteins. Biochim Biophys Acta 1140:105

Flemming D, Stolpe S, Schneider D, Hellwig P, Friedrich T (2005) A possible role for ironsulfur cluster N2 in proton translocation by the NADH:ubiquinone oxidoreductase (complex I). J Mol Microbiol Biotechnol 10:208

Fontecilla-Camps JC, Frey M, Garcin E, Hatchikian C, Montet Y, Piras C, Vernede X, Volbeda A (1997) Hydrogenase: a hydrogen-metabolizing enzyme. What do the crystal structures tell us about its mode of action? Biochimie 79:661

Fox JD, Kerby RL, Roberts GP, Ludden PW (1996a) Characterization of the CO-induced, CO-tolerant hydrogenase from *Rhodospirillum rubrum* and the gene encoding the large subunit of the enzyme. J Bacteriol 178:1515

Fox JD, He Y, Shelver D, Roberts GP, Ludden PW (1996b) Characterization of the region encoding the CO-induced hydrogenase of *Rhodospirillum rubrum*. J Bacteriol 178: 6200

Friedrich B, Buhrke T, Burgdorf T, Lenz O (2005) A hydrogen-sensing multiprotein complex controls aerobic hydrogen metabolism in *Ralstonia eutropha*. Biochem Soc Trans 33:97

Friedrich T (2001) Complex I: a chimaera of a redox and conformation-driven proton pump? J Bioenerg Biomembr 33:169

Friedrich T, Weiss H (1997) Modular evolution of the respiratory NADH:ubiquinone oxidoreductase and the origin of its modules. J Theor Biol 187:529

Friedrich T, Scheide D (2000) The respiratory complex I of bacteria, archaea and eukarya and its module common with membrane-bound multisubunit hydrogenases. FEBS Lett 479:1

Friedrich T, Abelmann A, Brors B, Guenebaut V, Kintscher L, Leonard K, Rasmussen T, Scheide D, Schlitt A, Schulte U, Weiss H (1998) Redox components and structure of the respiratory NADH:ubiquinone oxidoreductase (complex I). Biochim Biophys Acta 1365:215

Garcin E, Vernede X, Hatchikian EC, Volbeda A, Frey M, Fontecilla-Camps JC (1999) The crystal structure of a reduced [NiFeSe] hydrogenase provides an image of the activated catalytic center. Structure 7:557

Gross R, Simon J, Kröger A (1999) The role of the twin-arginine motif in the signal peptide encoded by the *hydA* gene of the hydrogenase from *Wolinella succinogenes*. Arch Microbiol 172:227

Gross R, Pisa R, Sanger M, Lancaster CR, Simon J (2004) Characterization of the menaquinone reduction site in the diheme cytochrome *b* membrane anchor of *Wolinella succinogenes* NiFe-hydrogenase. J Biol Chem 279:274

Hackstein JH (2005) Eukaryotic Fe-hydrogenases – old eukaryotic heritage or adaptive acquisitions? Biochem Soc Trans 33:47

Hackstein JH, Akhmanova A, Voncken F, van Hoek A, van Alen T, Boxma B, Moonvan der Staay SY, van der Staay G, Leunissen J, Huynen M, Rosenberg J, Veenhuis M (2001) Hydrogenosomes: convergent adaptations of mitochondria to anaerobic environments. Zoology (Jena) 104:290

Hamamoto T, Hashimoto M, Hino M, Kitada M, Seto Y, Kudo T, Horikoshi K (1994) Characterization of a gene responsible for the Na^+/H^+ antiporter system of alkalophilic *Bacillus* species strain C-125. Mol Microbiol 14:939

Happe RP, Roseboom W, Pierik AJ, Albracht SP, Bagley KA (1997) Biological activation of hydrogen. Nature 385:126

Happe T, Hemschemeier A, Winkler M, Kaminski A (2002) Hydrogenases in green algae: do they save the algae's life and solve our energy problems? Trends Plant Sci 7:246

Hedderich R (2004) Energy-converting [NiFe] hydrogenases from archaea and extremophiles: ancestors of complex I. J Bioenerg Biomembr 36:65

Hedderich R, Forzi L (2005) Energy-converting [NiFe] hydrogenases: more than just H_2 activation. J Mol Microbiol Biotechnol 10:92

Hedderich R, Whitman WB (2006) Physiology and biochemistry of the methane-producing Archaea. In: Dworkin M, Falkow S, Rosenberg E, Schleifer KH, Stackebrandt E (eds) The prokaryotes, vol 2. Ecophysiology and biochemistry. Springer, Heidelberg Berlin New York, pp 1050–1079

Henry M-F, Vignais PM (1983) Electron pathways from H_2 to nitrate in *Paracoccus denitrificans*: Effects of inhibitors of the UQ-cytochrome *b* region. Arch Microbiol 169:98

Higuchi Y, Yagi T, Yasuoka N (1997) Unusual ligand structure in Ni–Fe active center and an additional Mg site in hydrogenase revealed by high resolution X-ray structure analysis. Structure 5:1671

Hillmer P, Gest H (1977) H_2 metabolism in the photosynthetic bacterium *Rhodopseudomonas capsulata*: H_2 production by growing cultures. J Bacteriol 129:724

Hinchliffe P, Carroll J, Sazanov LA (2006) Identification of a novel subunit of respiratory complex I from *Thermus thermophilus*. Biochemistry 45:4413

Hinkle PC (2005) P/O ratios of mitochondrial oxidative phosphorylation. Biochim Biophys Acta 1706:1

Hirst J, Carroll J, Fearnley IM, Shannon RJ, Walker JE (2003) The nuclear encoded subunits of complex I from bovine heart mitochondria. Biochim Biophys Acta 1604:135

Holt PJ, Morgan DJ, Sazanov LA (2003) The location of NuoL and NuoM subunits in the membrane domain of the *Escherichia coli* complex I: implications for the mechanism of proton pumping. J Biol Chem 278:43114

Horner DS, Foster PG, Embley TM (2000) Iron hydrogenases and the evolution of anaerobic eukaryotes. Mol Biol Evol 17:1695

Horner DS, Heil B, Happe T, Embley TM (2002) Iron hydrogenases – ancient enzymes in modern eukaryotes. Trends Biochem Sci 27:148

Ide T, Bäumer S, Deppenmeier U (1999) Energy conservation by the H_2:heterodisulfide oxidoreductase from *Methanosarcina mazei* Göl: identification of two proton-translocating segments. J Bacteriol 181:4076

Ito M, Guffanti AA, Oudega B, Krulwich TA (1999) *mrp*, a multigene, multifunctional locus in *Bacillus subtilis* with roles in resistance to cholate and to Na^+ and in pH homeostasis. J Bacteriol 181:2394

Kaneko T, Sato S, Kotani H, Tanaka A, Asamizu E, Nakamura Y, Miyajima N, Hirosawa M, Sugiura M, Sasamoto S, Kimura T, Hosouchi T, Matsuno A, Muraki A, Nakazaki N, Naruo K, Okumura S, Shimpo S, Takeuchi C, Wada T, Watanabe A, Yamada M, Yasuda M, Tabata S (1996) Sequence analysis of the genome of the unicellular cyanobacterium *Synechocystis* sp. strain PCC 6803. II. Sequence determination of the entire genome and assignment of potential protein-coding regions (supplement). DNA Res 3:185

Kashani-Poor N, Zwicker K, Kerscher S, Brandt U (2001) A central functional role for the 49-kDa subunit within the catalytic core of mitochondrial complex I. J Biol Chem 276:24082

Klemme JH (1969) Studies on the mechanism of NAD-photoreduction by chromatophores of the facultative phototroph, Rhodopseudomonas capsulata. Z Naturforsch B 24:67

Kömen R, Schmidt K, Zannoni D (1996) Hydrogen oxidation by membranes from autotrophically grown Alcaligenes eutrophus H16: role of the cyanide-resistant pathway in energy transduction. Arch Microbiol 165:418

Kosono S, Kajiyama Y, Kawasaki S, Yoshinaka T, Haga K, Kudo T (2006) Functional involvement of membrane-embedded and conserved acidic residues in the ShaA subunit of the multigene-encoded Na^+/H^+ antiporter in Bacillus subtilis. Biochim Biophys Acta 1758:627

Krebs W, Steuber J, Gemperli AC, Dimroth P (1999) Na^+ translocation by the NADH:ubiquinone oxidoreductase (complex I) from Klebsiella pneumoniae. Mol Microbiol 33:590

Kröger A, Biel S, Simon J, Gross R, Unden G, Lancaster CR (2002) Fumarate respiration of Wolinella succinogenes: enzymology, energetics and coupling mechanism. Biochim Biophys Acta 1553:23

Künkel A, Vorholt JA, Thauer RK, Hedderich R (1998) An Escherichia coli hydrogenase-3-type hydrogenase in methanogenic archaea. Eur J Biochem 252:467

Lancaster CR, Sauer US, Gross R, Haas AH, Graf J, Schwalbe H, Mäntele W, Simon J, Madej MG (2005) Experimental support for the "E pathway hypothesis" of coupled transmembrane e^- and H^+ transfer in dihemic quinol:fumarate reductase. Proc Natl Acad Sci USA 102:18860

Lill R, Mühlenhoff U (2005) Iron-sulfur-protein biogenesis in eukaryotes. Trends Biochem Sci 30:133

López-Garcia P, Moreira D (1999) Metabolic symbiosis at the origin of eukaryotes. Trends Biochem Sci 24:88

Lyon EJ, Shima S, Buurman G, Chowdhuri S, Batschauer A, Steinbach K, Thauer RK (2004) UV-A/blue-light inactivation of the "metal-free" hydrogenase (Hmd) from methanogenic archaea. Eur J Biochem 271:195

Martin W, Müller M (1998) The hydrogen hypothesis for the first eukaryote. Nature 392:37

Mathiesen C, Hägerhäll C (2002) Transmembrane topology of the NuoL, M and N subunits of NADH:quinone oxidoreductase and their homologues among membrane-bound hydrogenases and bona fide antiporters. Biochim Biophys Acta 1556:121

Mathiesen C, Hägerhäll C (2003) The "antiporter module" of respiratory chain complex I includes the MrpC/NuoK subunit – a revision of the modular evolution scheme. FEBS Lett 549:7

Matias PM, Soares CM, Saraiva LM, Coelho R, Morais J, Le Gall J, Carrondo MA (2001) [NiFe] hydrogenase from Desulfovibrio desulfuricans ATCC 27774: gene sequencing, three-dimensional structure determination and refinement at 1.8 A and modelling studies of its interaction with the tetrahaem cytochrome c_3. J Biol Inorg Chem 6:63

Meek L, Arp DJ (2000) The hydrogenase cytochrome b heme ligands of Azotobacter vinelandii are required for full H_2 oxidation capability. J Bacteriol 182:3429

Meuer J, Bartoschek S, Koch J, Künkel A, Hedderich R (1999) Purification and catalytic properties of Ech hydrogenase from Methanosarcina barkeri. Eur J Biochem 265:325

Montet Y, Amara P, Volbeda A, Vernede X, Hatchikian EC, Field MJ, Frey M, Fontecilla-Camps JC (1997) Gas access to the active site of Ni – Fe hydrogenases probed by X-ray crystallography and molecular dynamics. Nat Struct Biol 4:523

Moreira D, López-Garcia P (1998) Symbiosis between methanogenic archaea and delta-proteobacteria as the origin of eukaryotes: the syntrophic hypothesis. J Mol Evol 47:517

Mori H, Cline K (2001) Post-translational protein translocation into thylakoids by the Sec and DeltapH-dependent pathways. Biochim Biophys Acta 1541:80

Mori H, Cline K (2002) A twin arginine signal peptide and the pH gradient trigger reversible assembly of the thylakoid ΔpH/Tat translocase. J Cell Biol 157:205

Mühlenhoff U, Lill R (2000) Biogenesis of iron-sulfur proteins in eukaryotes: a novel task of mitochondria that is inherited from bacteria. Biochim Biophys Acta 1459:370

Müller M (1993) The hydrogenosome. J Gen Microbiol 139:2879

Nealson KH, Inagaki F, Takai K (2005) Hydrogen-driven subsurface lithoautotrophic microbial ecosystems (SLIMEs): do they exist and why should we care? Trends Microbiol 13:405

Nicholls DG, Ferguson SJ (1992) Bioenergetics 2. Academic Press, London

Nicolet Y, Cavazza C, Fontecilla-Camps JC (2002) Fe-only hydrogenases: structure, function and evolution. J Inorg Biochem 91:1

Nicolet Y, Lemon BJ, Fontecilla-Camps JC, Peters JW (2000) A novel FeS cluster in Fe-only hydrogenases. Trends Biochem Sci 25:138

Nicolet Y, Piras C, Legrand P, Hatchikian CE, Fontecilla-Camps JC (1999) *Desulfovibrio desulfuricans* iron hydrogenase: the structure shows unusual coordination to an active site Fe binuclear center. Structure 7:13

Ohnishi T, Sled VD, Yano T, Yagi T, Burbaev DS, Vinogradov AD (1998) Structure-function studies of iron-sulfur clusters and semiquinones in the NADH-Q oxidoreductase segment of the respiratory chain. Biochim Biophys Acta 1365:301

Palmer T, Sargent F, Berks BC (2005) Export of complex cofactor-containing proteins by the bacterial Tat pathway. Trends Microbiol 13:175

Paul F, Colbeau A, Vignais PM (1979) Phosphorylation coupled to H_2 oxidation by chromatophores from *Rhodopseudomonas capsulata*. FEBS Lett 106:29

Peters JW (1999) Structure and mechanism of iron-only hydrogenases. Curr Opin Struct Biol 9:670

Peters JW, Lanzilotta WN, Lemon BJ, Seefeldt LC (1998) X-ray crystal structure of the Fe-only hydrogenase (CpI) from *Clostridium pasteurianum* to 1.8 Å resolution. Science 282:1853

Pilak O, Mamat B, Vogt S, Hagemeier CH, Thauer RK, Shima S, Vonrhein C, Warkentin E, Ermler U (2006) The crystal structure of the apoenzyme of the iron-sulphur cluster-free hydrogenase. J Mol Biol 358:798

Pilkington SJ, Skehel JM, Gennis RB, Walker JE (1991) Relationship between mitochondrial NADH-ubiquinone reductase and a bacterial NAD-reducing hydrogenase. Biochemistry 30:2166

Porte F, Vignais PM (1980) Electron transport chain and energy transduction in *Paracoccus denitrificans* under autotrophic growth conditions. Arch Microbiol 127:1

Prieur I, Lunardi J, Dupuis A (2001) Evidence for a quinone binding site close to the interface between NUOD and NUOB subunits of complex I. Biochim Biophys Acta 1504:173

Rákhely G, Colbeau A, Garin J, Vignais PM, Kovács KL (1998) Unusual organization of the genes coding for HydSL, the stable [NiFe]hydrogenase in the photosynthetic bacterium *Thiocapsa roseopersicina* BBS. J Bacteriol 180:1460

Rákhely G, Kovács AT, Maróti G, Fodor BD, Csanádi G, Latinovics D, Kovács KL (2004) Cyanobacterial-type, heteropentameric, NAD^+-reducing NiFe hydrogenase in the purple sulfur photosynthetic bacterium *Thiocapsa roseopersicina*. Appl Environ Microbiol 70:722

Richardson DJ (2000) Bacterial respiration: a flexible process for a changing environment. Microbiology 146(Pt3):551

Rodrigue A, Chanal A, Beck K, Müller M, Wu LF (1999) Co-translocation of a periplasmic enzyme complex by a hitchhiker mechanism through the bacterial tat pathway. J Biol Chem 274:13223

Rodrigues R, Valente FM, Pereira IA, Oliveira S, Rodrigues-Pousada C (2003) A novel membrane-bound Ech [NiFe] hydrogenase in *Desulfovibrio gigas*. Biochem Biophys Res Commun 306:366

Rossi M, Pollock WB, Reij MW, Keon RG, Fu R, Voordouw G (1993) The *hmc* operon of *Desulfovibrio vulgaris* subsp. vulgaris Hildenborough encodes a potential transmembrane redox protein complex. J Bacteriol 175:4699

Sapra R, Verhagen MF, Adams MW (2000) Purification and characterization of a membrane-bound hydrogenase from the hyperthermophilic archaeon *Pyrococcus furiosus*. J Bacteriol 182:3423

Sapra R, Bagramyan K, Adams MW (2003) A simple energy-conserving system: proton reduction coupled to proton translocation. Proc Natl Acad Sci USA 100:7545

Sargent F, Berks BC, Palmer T (2002) Assembly of membrane-bound respiratory complexes by the Tat protein-transport system. Arch Microbiol 178:77

Sargent F, Bogsch EG, Stanley NR, Wexler M, Robinson C, Berks BC, Palmer T (1998) Overlapping functions of components of a bacterial Sec-independent protein export pathway. EMBO J 17:3640

Sauter M, Böhm R, Böck A (1992) Mutational analysis of the operon (*hyc*) determining hydrogenase 3 formation in *Escherichia coli*. Mol Microbiol 6:1523

Sawers RG, Blokesch M, Böck A (2004) Anaerobic formate and hydrogen metabolism. In: Curstiss IR (ed) EcoSal-*Escherichia coli* and *Salmonella*: Cellular and molecular biology. Online http://www.ecosal.org, Chapter 3.5.4. ASM, Washington, DC

Sayavedra-Soto LA, Arp DJ (1992) The *hoxZ* gene of the *Azotobacter vinelandii* hydrogenase operon is required for activation of hydrogenase. J Bacteriol 174:5295

Sazanov LA, Hinchliffe P (2006) Structure of the hydrophilic domain of respiratory complex I from *Thermus thermophilus*. Science 311(5766):1430–1436

Sazanov LA, Peak-Chew SY, Fearnley IM, Walker JE (2000) Resolution of the membrane domain of bovine complex I into subcomplexes: implications for the structural organization of the enzyme. Biochemistry 39:7229

Schmitz O, Boison G, Salzmann H, Bothe H, Schutz K, Wang SH, Happe T (2002) HoxE- a subunit specific for the pentameric bidirectional hydrogenase complex (HoxEFUYH) of cyanobacteria. Biochim Biophys Acta 1554:66

Schneider K, Schlegel HG (1976) Purification and properties of soluble hydrogenase from *Alcaligenes eutrophus* H 16. Biochim Biophys Acta 452:66

Schultz BE, Chan SI (2001) Structures and proton-pumping strategies of mitochondrial respiratory enzymes. Annu Rev Biophys Biomol Struct 30:23

Schwartz E, Henne A, Cramm R, Eitinger T, Friedrich B, Gottschalk G (2003) Complete nucleotide sequence of pHG1: a *Ralstonia eutropha* H 16 megaplasmid encoding key enzymes of H_2-based lithoautotrophy and anaerobiosis. J Mol Biol 332:369

Silva PJ, van den Ban EC, Wassink H, Haaker H, de Castro B, Robb FT, Hagen WR (2000) Enzymes of hydrogen metabolism in *Pyrococcus furiosus*. Eur J Biochem 267:6541

Skibinski DAG, Golby P, Chang Y-S, Sargent F, Hoffman R, Harper R, Guest JR, Attwood MM, Berks BC, Andrews SC (2002) Regulation of the hydrogenase-4 operon of *Escherichia coli* by the σ^{54}-dependent transcriptional activators FhlA and HyfR. J Bacteriol 184:6642

Soboh B, Linder D, Hedderich R (2002) Purification and catalytic properties of a CO-oxidizing:H_2-evolving enzyme complex from *Carboxydothermus hydrogenoformans*. Eur J Biochem 269:5712

Steuber J (2001) The Na+-translocating NADH:quinone oxidoreductase (NDH I) from *Klebsiella pneumoniae* and *Escherichia coli*: implications for the mechanism of redox-driven cation translocation by complex I. J Bioenerg Biomembr 33:179

Stolpe S, Friedrich T (2004) The *Escherichia coli* NADH:ubiquinone oxidoreductase (complex I) is a primary proton pump but may be capable of secondary sodium antiport. J Biol Chem 279:18377

Tamagnini P, Axelsson R, Lindberg P, Oxelfelt F, Wunschiers R, Lindblad P (2002) Hydrogenases and hydrogen metabolism of cyanobacteria. Microbiol Mol Biol Rev 66:1

Tersteegen A, Hedderich R (1999) *Methanobacterium thermoautotrophicum* encodes two multisubunit membrane-bound [NiFe] hydrogenases. Transcription of the operons and sequence analysis of the deduced proteins. Eur J Biochem 264:930

Tian F, Toon OB, Pavlov AA, De Sterck H (2005) A hydrogen-rich early Earth atmosphere. Science 308:1014

Tichi MA, Meijer WG, Tabita FR (2001) Complex I and its involvement in redox homeostasis and carbon and nitrogen metabolism in *Rhodobacter capsulatus*. J Bacteriol 183:7285

Tjaden J, Haferkamp I, Boxma B, Tielens AG, Huynen M, Hackstein JH (2004) A divergent ADP/ATP carrier in the hydrogenosomes of Trichomonas gallinae argues for an independent origin of these organelles. Mol Microbiol 51:1439

Vignais PM, Colbeau A (2004) Molecular biology of microbial hydrogenases. Curr Issues Mol Biol 6:159

Vignais PM, Colbeau A, Willison JC, Jouanneau Y (1985) Hydrogenase, nitrogenase, and hydrogen metabolism in the photosynthetic bacteria. Adv Microb Physiol 26:155

Vignais PM, Billoud B, Meyer J (2001) Classification and phylogeny of hydrogenases. FEMS Microbiol Rev 25:455

Vignais PM, Willison JC, Colbeau A (2004) H_2 respiration. In: Zannoni D (ed) Respiration in Archaea and Bacteria: Diversity of prokaryotic respiratory systems. Advances in photosynthesis and respiration, vol 2. Springer, Berlin Heidelberg New York, pp 233–260

Vignais PM, Elsen S, Colbeau A (2005) Transcriptional regulation of the uptake [NiFe]hydrogenase genes in *Rhodobacter capsulatus*. Biochem Soc Trans 33:28

Volbeda A, Charon MH, Piras C, Hatchikian EC, Frey M, Fontecilla-Camps JC (1995) Crystal structure of the nickel-iron hydrogenase from *Desulfovibrio gigas*. Nature 373:580

Volbeda A, Garcin E, Piras C, De Lacey AL, Fernandez VM, Hatchikian EC, Frey M, Fontecilla-Camps JC (1996) Structure of the [NiFe] hydrogenase active site: evidence for biologically uncommon Fe ligands. J Am Chem Soc 118:12989

Weidner U, Geier S, Ptock A, Friedrich T, Leif H, Weiss H (1993) The gene locus of the proton-translocating NADH: ubiquinone oxidoreductase in *Escherichia coli*. Organization of the 14 genes and relationship between the derived proteins and subunits of mitochondrial complex I. J Mol Biol 233:109

Weiner JH, Bilous PT, Shaw GM, Lubitz SP, Frost L, Thomas GH, Cole JA, Turner RJ (1998) A novel and ubiquitous system for membrane targeting and secretion of cofactor-containing proteins. Cell 93:93

Willison JC (1993) Biochemical genetics revisited: the use of mutants to study carbon and nitrogen metabolism in the photosynthetic bacteria. FEMS Microbiol Rev 10:1

Wu LF, Chanal A, Rodrigue A (2000) Membrane targeting and translocation of bacterial hydrogenases. Arch Microbiol 173:319

Yagi T (1993) The bacterial energy-transducing NADH-quinone oxidoreductases. Biochim Biophys Acta 1141:1

Yagi T, Yano T, Di Bernardo S, Matsuno-Yagi A (1998) Procaryotic complex I (NDH-1), an overview. Biochim Biophys Acta 1364:125

Yahr TL, Wickner WT (2001) Functional reconstitution of bacterial Tat translocation in vitro. Embo J 20:2472

Yano T, Ohnishi T (2001) The origin of cluster N2 of the energy-transducing NADH-quinone oxidoreductase: comparisons of phylogenetically related enzymes. J Bioenerg Biomembr 33:213

Yen MR, Tseng YH, Nguyen EH, Wu LF, Saier MH Jr (2002) Sequence and phylogenetic analyses of the twin-arginine targeting (Tat) protein export system. Arch Microbiol 177:441

Zannoni D, Marrs B (1981) Redox chain and energy transduction in chromatophores from *Rhodopseudomonas capsulata* cells grown anaerobically in the dark on glucose and dimethylsulfoxide. Biochim Biophys Acta 637:96

Zirngibl C, Van Dongen W, Schworer B, Von Bunau R, Richter M, Klein A, Thauer RK (1992) H2-forming methylenetetrahydromethanopterin dehydrogenase, a novel type of hydrogenase without iron-sulfur clusters in methanogenic archaea. Eur J Biochem 208:511

A Structural Perspective on Mechanism and Function of the Cytochrome bc_1 Complex

Carola Hunte (✉) · Sozanne Solmaz · Hildur Palsdóttir · Tina Wenz

Dept. Molecular Membrane Biology, Max Planck Institute of Biophysics,
60438 Frankfurt, Germany
carola.hunte@mpibp-frankfurt.mpg.de

Abstract The cytochrome bc_1 complex is a fundamental component of the energy conversion machinery of respiratory and photosynthetic electron transfer chains. The multisubunit membrane protein complex couples electron transfer from hydroquinone to cytochrome c to the translocation of protons across the membrane, thereby substantially contributing to the proton motive force that is used for ATP synthesis. Considerable progress has been made with structural and functional studies towards complete elucidation of the Q cycle mechanism, which was originally proposed by Mitchell 30 years ago. Yet, open questions regarding key steps of the mechanism still remain. The role of the complex as a major source of reactive oxygen species and its implication in pathophysiological conditions has recently gained interest.

Abbreviations

cyt	Cytochrome
EESEM	Electron spin echo envelope modulation
E_m	Redox midpoint potential
ENDOR	Electron nuclear double resonance
EPR	Electron paramagnetic resonance
HHDBT	5-n-heptyl-6-hydroxy-4,7-dioxobenzothiazole
HYSCORE	Hyperfine sublevel correlation spectroscopy
ROS	Reactive oxygen species
UHDBT	5-n-undecyl-6-hydroxy-4,7-dioxobenzothiazole

1
Introduction

The enzyme ubihydroquinone: cytochrome c oxidoreductase (E.C.1.10.2.2) or cytochrome bc_1 complex (cyt bc_1) belongs to the superfamily of cytochrome bc complexes, which are multi-subunit membrane protein complexes that serve as central components of the energy conversion machinery of photosynthesis and respiration in chloroplasts, mitochondria, and bacteria (Berry et al. 2000; Crofts 2004; Darrouzet et al. 2001; Schutz et al. 2000). Mitochondrial cyt bc_1 is a target for fungicidal agrochemicals. Furthermore, the complex is a source of reactive oxygen species that are implicated in neurodegenerative diseases and aging processes. Several mutations of cyt bc_1 are correlated with mitochondrial myopathies. Common to all cyt bc complexes is the electron transfer from a lipophilic ubihydroquinone-type two-electron/two-proton carrier such as ubihydroquinone, menahydroquinone or plastohydroquinone to a one-electron carrier protein, a c-type cytochrome or plastocyanin. The integral membrane protein complexes couple electron transfer reactions to proton translocation across a membrane and thereby conserve the free energy of the redox reactions in an electrochemical proton gradient. The latter is utilized to drive energy-dependent processes like ATP synthesis or secondary transport.

In mitochondria, cyt bc_1 catalyzes electron transfer from ubihydroquinone to cytochrome c (cyt c). It is also named "complex III" of the mitochondrial respiratory chain. Cyt bc_1 has a similar function in aerobic respiration of bacteria. In anoxygenic photosynthetic bacteria, cyt bc_1 is an integral component of light-driven electron transfer and dark respiration. The homologous enzyme in chloroplasts from plants and algae and in cyanobacteria is the cytochrome b_6f complex (cyt b_6f), also called plastohydroquinone: plastocyanin oxidoreductase (Cramer et al. 2006). It is a component of oxygenic photosynthesis and mediates electron transfer between photosystem II and photosystem I by oxidizing plastohydroquinone and reducing soluble plastocyanin.

The catalytic core of all cyt bc_1 complexes comprises three redox-active subunits, cytochrome b (cyt b) with two b-type hemes, cytochrome c_1 (cyt c_1) with a c-type heme, and the iron-sulfur protein with a 2Fe-2S cluster (Rieske protein). The enzyme mechanism was first described by Mitchell as proton-motive Q cycle (Mitchell 1975, 1976) and further developed to the now generally accepted scheme depicted in Fig. 1 (Brandt and Trumpower 1994; Crofts 1985). The general mechanism is well understood, but important mechanistic details are still in question. There are two spatially separated catalytic sites for ubiquinone and ubihydroquinone. Ubihydroquinone oxidation occurs at center P (also called Q_o site) located at the electropositive side of the membrane, so that the protons of the reaction are released to the intermembrane space or the periplasm. The two electrons are transferred in a bifurcated manner, a highly debated step of the reaction. One electron is passed via the high-potential electron-transfer chain, from the 2Fe–2S cluster to heme c_1. The latter acts as donor for reduction of the substrate cyt c. Surprisingly, the second electron follows the low-potential electron-transfer chain, a thermodynamically less favorable route, and is recycled within the enzyme. It is sequentially transferred via heme b_L and heme b_H to center N (also called Q_i site), at which ubiquinone is reduced. The resulting stable semiquinone is fully reduced after oxidation of a second ubihydroquinone molecule at center P. Protons are taken up from the electronegative side upon ubiquinone reduction, i.e., from the matrix or the cytoplasm. Taken together, proton translocation is enabled by transmembrane electron transfer from the positive to the negative side and the free diffusion of lipophilic

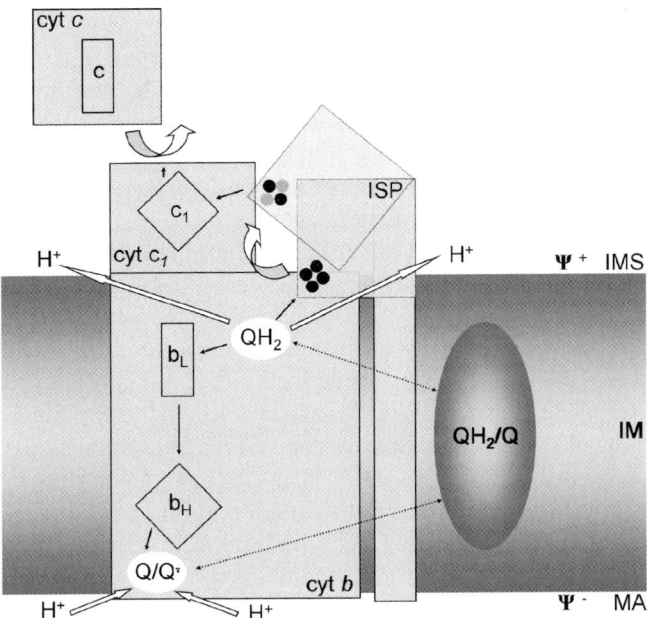

Fig. 1 The Q cycle. Ubihydroquinone (QH$_2$) oxidation at center P results in bifurcated electron transfer and release of two protons. One electron passes through the high potential chain, which requires the domain movement of the cluster bearing iron-sulfur protein (ISP) as electron shuttle to cyt c_1 and ultimately reduces cyt c. The second electron passes through heme b_L to heme b_H, the low-potential chain, and reduces ubiquinone (Q) at center N to semiquinone. The cycle is completed with oxidation of a second QH$_2$ molecule in the same manner, resulting in QH$_2$ formation at center N and uptake of two protons from the electronegative side. Electron and proton transfer are indicated with *black* and *white arrows*, respectively. Matrix, inner membrane and intermembrane space are indicated with MA, IM, IMS, respectively

ubiquinone/ubihydroquinone, whereby the transfer across the membrane occurs via the reduced form of the substrate.

The net reaction can be summarized as follows:

$$QH_2 + 2\,cyt\ c(Fe^{3+}) + 2H^+_N \rightarrow Q + 2\,cyt\ c(Fe^{2+}) + 4H^+_P\,.$$

2
Structural Characterization of cyt bc_1

The development of purification protocols for stable and pure protein preparations of the large multi-subunit cyt bc_1 was a prerequisite for its structure determination. The first methods to purify cyt bc_1 were based on bile salt detergent solubilization and differential ammonium sulfate precipitation (Rieske 1976; Siedow et al. 1978; Yu et al. 1974). In the following, chromatographic approaches, such as hydroxyapatite chromatography with Triton X-100 (Engel et al. 1980; Riccio et al. 1976) and anion-exchange chromatography of dodecyl-maltopyranoside solubilized protein (Berry and Trumpower 1985) were successfully applied. The latter procedure has been modified and optimized for preparations from diverse sources, including animals (Ljungdahl et al. 1989), plants (Berry et al. 1991), bacteria (Montoya et al. 1999), and yeast (Hunte et al. 2000; Palsdottir and Hunte 2003).

The multi-subunit integral membrane protein complexes are homodimeric (Fig. 2A,B). The first electron micrographs of the mitochondrial complex from *Neurospora crassa* already indicated the dimeric nature of the enzyme (Leonard et al. 1981). In the late nineties, the first X-ray structures of mitochondrial complexes from bovine heart (Iwata et al. 1998; Xia et al. 1997), chicken (Zhang et al. 1998) and the yeast *Saccharomyces cerevisiae* (Hunte et al. 2000) were determined. More recently, structures of prokaryotic cyt bc_1 from *Rhodobacter capsulatus* and (Berry et al. 2004) and *Rh. sphaeroides* (Esser et al. 2006) became available. The redox-active subunits cyt b, cyt c_1 and the Rieske-

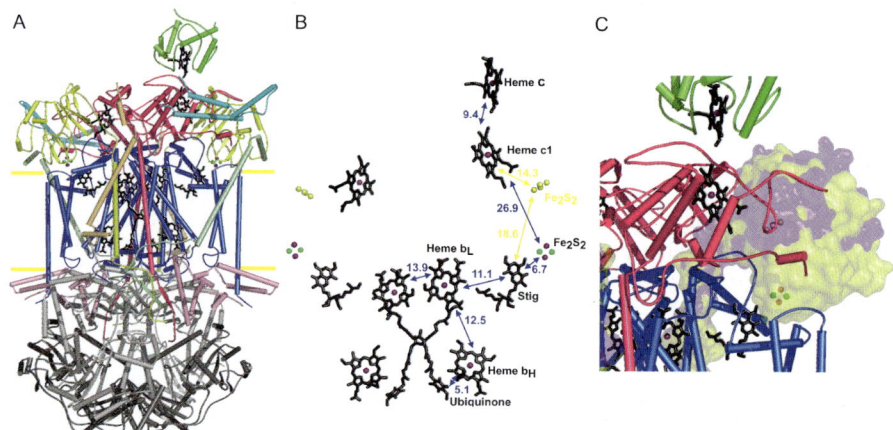

Fig. 2 Structure of cyt bc_1 from *Saccharomyces cerevisiae* with cyt c bound. **A** The homodimer consist per monomer of nine subunits: cyt b (*blue*), cyt c_1 (*pink*) and Rieske protein (*yellow*) and their respective cofactors, the core proteins 1 and 2 (*grey*), and the small subunits QCR6 (*cyan*), QCR7 (*faint purple*), QCR8 (*light green*), and QCR9 (*orange*). Stigmatellin bound at center P fixes the Rieske protein in b position. Only one molecule cyt c (*green*) is bound. Coordinates are taken from PDB-ID 1KYO (Lange and Hunte 2002). Ubiquinone at center N was included by superimposition with 1KB9 (Lange et al. 2001). **B** Close-up view of cofactors, substrate and inhibitor molecules as in (**A**). The 2Fe–2S cluster in c_1 position (*yellow*) is taken from superimposition with a structure of avian cyt bc_1 (PDB-ID 1BCC, Zhang et al. 1998). Edge-to-edge distances between the cofactors are given in Å. (**C**), Domain movement of the Rieske head group facilitates electron transfer between center P and heme c_1. Close-up view of superimposed structures (as in Fig. 2B) with Rieske protein in b and c_1 position shown in *yellow* and *purple*, respectively

protein form the core of the complex (Fig. 2A). The cofactor-carrying helical bundles of cyt b are positioned at the center of the dimeric assembly. Cyt c_1 and the Rieske protein are peripherally attached with a single transmembrane helix each; their respective catalytic domains are located in the intermembrane space. Two or three additional peripheral single-transmembrane helices are present from small supernumerary subunits in the yeast and bovine complex, respectively (Hunte et al. 2000; Iwata et al. 1998). The so-called hinge protein, a small acidic subunit probably involved in binding of cyt c, is part of the intermembrane portion of the complex (Kim and King 1983). Core proteins 1 and 2, and one additional subunit, named Qcr7p in yeast, comprise the large domain extruding in the matrix. The 7–8 supernumerary subunits that lack redox centers have proposed roles in regulation, assembly, and stability (Zara et al. 2004). Assembly of cyt bc_1 is believed to occur in regulated steps of precomplex formation guided by chaperones (Cruciat et al. 1999). Cyt b is the only mitochondrially encoded subunit.

The prokaryotic complexes are considerably smaller in size. Homologous to the eukaryotic ones, they are dimers. A monomer either comprises the three redox-active subunits or the catalytic core with one supernumerary subunit (Berry et al. 2004). The additional subunit IV of the *Rh. sphaeroides* complex is thought to be important for ubiquinone binding and structural integrity of the complex (Yu et al. 1999). Both the preliminary structure of *Rh. capsulatus*, which was determined at 3.8 Å resolution (Berry et al. 2004) and the 3.2 Å resolution structure of the *Rh. sphaeroides* complex contain the redox-active subunits only (Esser et al. 2006). For the latter, a highly active and stable cyt bc_1 complex variant has been used, in which a double mutation stabilizes the interface between cyt b and Rieske protein (Elberry et al. 2006).

The biological unit of the complex is an obligatory intertwined dimer with the symmetry axis along the membrane normal. Each functional unit is composed of cyt b, cyt c_1 and the 2Fe–2S cluster-bearing extrinsic domain of the Rieske protein. The transmembrane anchor of the latter is associated with the other monomer, i.e., the Rieske protein provides a cross link between the functional units. In principle, both functional units are fully active, but a regulatory interplay is discussed (see below).

The structure of chicken cyt bc_1 revealed a large-scale movement of the 2Fe–2S domain of the Rieske protein, which is required to shuttle electrons between the site of ubihydroquinone oxidation at center P to cytochrome c_1 (Zhang et al. 1998). Alternate positions of the Rieske domain were observed in different structures depending on crystal form and center P occupancy (Fig. 2B,C, Fig. 4): docked on cyt b (b position), intermediate positions, and close to heme c_1 (c_1 position) (Iwata et al. 1998; Kim et al. 1998; Zhang et al. 1998). Inhibitors such as stigmatellin and alkyl-hydroxydioxobenzothiazol (HDBT) lock the Rieske domain in b position (Hunte et al. 2000; Kim et al. 1998; Palsdottir et al. 2003; Zhang et al. 1998). It is generally accepted that the mobility of the Rieske domain is crucial for cyt bc_1 activity (Darrouzet et al. 2001). Mutations that restrict the flexibility of the linker region, which connects extrinsic domain and transmembrane anchor, limit enzyme activity (Darrouzet et al. 2000; Nett et al. 2000; Tian et al. 1998). Formation of intersubunit disulfide bonds between cyt *b* and Rieske protein fully blocks the enzyme (Xiao et al. 2000). Molecular dynamics simulation predicted a steered but stochastic movement (Izrailev et al. 1999). In contrast, EPR studies of oriented membranes suggest a redox-dependent movement of the domain (Brugna et al. 2000). The actual mechanism underlying this functionally important domain movement is still unknown.

2.1
Redox-Active Subunits

The two *b* hemes are non-covalently bound in a four-helical bundle of transmembrane helices A–D of cyt *b*. The low-potential heme b_L (b_{566}) is closer to the intermembrane space and center P, whereas the high-potential heme b_H (b_{562}) is close to center N (Fig. 2A,B). They can be distinguished based on their redox-specific absorbance maxima in the visible spectra. The conserved iron-coordinating ligands are His82 (helix B) and His183 (helix D) for heme b_L and His96 (helix B) and His197 (helix D) for heme b_H[1]. Heme c_1 is a prosthetic group covalently linked to cyt c_1 with thioether bonds from heme c_1 vinyl groups to the conserved residues Cys101 and Cys104. Conserved axial ligands of the iron atom are His105 and Met225. Midpoint potentials of + 116 mV, – 4 mV, and + 270 mV have been determined for heme b_H, heme b_L and heme c_1, respectively, from yeast cyt bc_1 near neutral pH (T'sai and Palmer 1983).

The extrinsic catalytic domain of the Rieske protein is formed by three layers of antiparallel β sheets, which fold into a compact structure held together by a conserved disulfide bond between residues Cys164 and Cys180 at the tip where the metal cluster is bound (Iwata et al. 1996). Several strong and conserved bonds interact with the cluster and contribute to its high midpoint potential. One of the iron atoms of the 2Fe–2S cluster is coordinated by the Nδ atoms of two histidine ligands, His161 and His181, whereas the other is coordinated by the sulfur atoms of two cysteine residues, Cys159 and Cys178. The one electron reduction of the $[2Fe-2S]^{2+}$ cluster is of localized nature, so that the histidine coordinated iron atom is reduced and the second iron remains ferric. The midpoint potential of the Rieske cluster (∼ 285 mV) decreases with increasing pH. The effect is attributed to the protonation state of the two histidine cluster ligands. Experimental evidence and theoretical calculations agree that there are two pK_a values of ∼ 7.5 and ∼ 9 in the oxidized state, whereas both values are > 11 in the reduced state (Link et al. 1992; Ugulava and Crofts 1998; Ullmann et al. 2002; Zu et al. 2003). Studies showed that an altered midpoint

---

[1] The numbering of residues is given for *S. cerevisiae* throughout the text.

potential of the Rieske protein can influence the rate and direction of electron transfer in the high-potential electron transfer chain (Snyder et al. 1999). Variants with lowered Rieske midpoint potential exhibit decreased catalytic activity, consistent with ubihydroquinone oxidation being the rate-limiting step in catalysis (Denke et al. 1998; Guergova-Kuras et al. 2000).

Within the low- and high-potential chain, all but one edge-to-edge distances between the respective cofactors are below 14 Å (Fig. 2C), and therefore close enough to ensure electron tunnelling at physiological rates (Moser et al. 1992; Page et al. 1999). The Rieske protein in the b position has the 2Fe–2S cluster at a distance exceeding efficient electron tunnelling to heme c_1, whereas the cluster is close enough for rapid tunnelling with the Rieske protein in c_1 position (Hunte et al. 2000; Iwata et al. 1998; Kim et al. 1998; Zhang et al. 1998, 2000). This explains why the domain movement is required for electron flow in the high-potential chain. In all orientations of the Rieske protein, the distance between heme b_L and 2Fe–2S cluster is large enough to prevent unproductive direct electron transfer and to accommodate the substrate ubihydroquinone.

Substrate and Inhibitor Binding

Site-specific inhibitors have been crucial to elucidate the mechanism of cyt bc_1 with applications in functional and structural studies. Three types of center P inhibitors can be distinguished: ligands binding to the proximal domain and therefore perturbing the spectroscopic properties of heme b_L (Q_o-I e.g., myxothiazol, MOA-stilbene), those binding to the distal domain and affecting the Rieske [2Fe–2S] EPR lineshape (Q_o-II e.g. UHDBT), and compounds exhibiting both effects (Q_o-III e.g., stigmatellin) (von Jagow 1986). Kinetic and binding studies indicated that binding of many of these inhibitors is mutually exclusive, suggesting overlapping binding sites (Brandt et al. 1991; von Jagow and Onishi 1985; von Jagow et al. 1986). This was confirmed in crystal structures where the head group of type I and type II/III inhibitors is stabilized in different positions in the bilobal Q_o site, binding proximal and distal to heme b_L, respectively, whereas their side chains overlap and tail into a common tunnel, which gradually opens up into the cavity at the dimer interface (Crofts et al. 1999a; Esser et al. 2004; Hunte et al. 2000; Kim et al. 1998; Lancaster et al. 2007; Palsdottir et al. 2003). Position and dimension of the ubiquinone binding pocket have been deduced from cyt bc_1 structures with bound center N specific inhibitors such as antimycin A (Huang et al. 2005) and also with bound substrate (Gao et al. 2003; Huang et al. 2005; Hunte et al. 2000).

Center P and N are mainly formed by cyt b. The ubihydroquinone binding pocket is flanked by transmembrane helices C and F, helices $cd1$ and ef, and by the PEWY loop ($271^{cyt\ b}$–$274^{cyt\ b}$). Residues 248–270 of the ef loop form a cap and separate the binding site from the intermembrane space. The pocket extends toward helix D, opening to the substrate exchange cavity, a hydrophobic cleft present at each side of the dimer interface. Center P is occluded from the aqueous environment, when the Rieske domain is docked on in b position. In that case, the 2Fe–2S cluster ligand His181 contributes to the binding pocket. Center N is flanked by helix A, D, E, and a of cyt b and heme b_H. It is occluded from the matrix by the de loop and residues of the N-terminus (Hunte et al. 2003) and it is open to the substrate exchange cavity. Each of the two exchange cavities is shared between the monomers, i.e., center N of one monomer and center P of the other are open toward the same exchange cavity. This architectural feature could enhance substrate exchange between the two sites in case of physical constraints imposed by supramolecular organization (see below).

The substrate ubiquinone occurs with different length of the isoprenoid chain. In *S. cerevisiae*, it contains six isoprene units and is also called coenzyme Q_6, whereas a four units longer isoprenoid chain is present in bovine and human mitochondria (coenzyme Q_{10}). The length and nature of the side chain affect the catalytic efficiency. Yet, the isoprenoid chain is not essential and can be replaced with an alkyl group (Yu 1985). Rates of ubihydroquinone oxidation are found to increase proportionally with the length of the alkyl side chain until a critical length is reached and the rates decrease again. The partition coefficient of 2,3-dimethoxy quinones is similar for oxidized and reduced state (Rich and Harper 1990) possibly balanced by intramolecular hydrogen bonding between the hydroxy and methoxy groups. Interestingly, a naphthoquinol oxidizing cyt bc_1 was recently discovered and purified from hyperthermophilic *Aquifex aeolicus* (Schutz et al. 2003).

2.2
Tight Binding of Phospholipids to the cyt bc_1

Several studies demonstrate the importance of phospholipids for full functionality of cyt bc_1. Increased delipidation of the protein leads to a gradual decrease in enzyme activity up to complete inactivation and destabilization (Schägger et al. 1990; Yu and Yu 1980). Furthermore, specific removal of cardiolipin was found to reversibly inactivate the enzyme (Gomez and Robinson 1999). The X-ray structure of yeast cyt bc_1 revealed tightly bound phospholipid molecules including cardiolipin (Lange et al. 2001; Palsdottir and Hunte 2004; Palsdottir et al. 2003). These endogenous lipids have been co-purified and are consistently found in all structures of the yeast complex. Their individual binding sites suggest specific roles in facilitating structural and functional integrity of the enzyme. The position of the complex in respect to the membrane can be exactly deduced from the location of annular lipids on both bilayer sides. Site-directed mutagenesis substitution of positively charged amino acid residues that ligate the phosphodiester moiety of the interhelical lipid phosphatidyl-inositol and of cardiolipin destabilize the complex and confirm the importance of the specific interactions (Lange et al. 2001). Cardiolipin appears to stabilize the architecture of the proton-conducting environment at center N and may be involved in proton uptake (Klingen et al. 2007; Lange et al. 2001).

3
Mechanistic Considerations

Ubiquinone Reduction at Center N

The mechanism at center N links the single electron path of the b-hemes to the two-electron/two-proton chemistry of ubiquinone reduction featuring a semiquinone as reaction intermediate. A thermodynamically stable radical can be generated. The yield depends on redox poise and pH, but was always substoichiometric in experiments. It was suggested that a fraction of the radical signals is "silenced" by spin coupling with oxidized heme b_H when monitoring the radical by EPR (de la Rosa and Palmer 1983; Rich et al. 1990). Details of semiquinone and heme b_H interaction, equilibrium and rate constants, and ubiquinone/ubihydroquinone binding affinities have been previously summarized (Crofts 2004). Not all of it is entirely known, so that the exact description of center N catalysis has yet to be completed.

Important information has been provided by structural analysis. Ubiquinone binding at center N was characterized in X-ray structures of chicken (Zhang et al. 1998), yeast (Hunte et al. 2000) and bovine cyt bc_1 (Gao et al. 2003; Huang et al. 2005). In all cases, the natural ubiquinone substrate has been co-purified and co-crystallized. Higher average B factors of the ubiquinone molecules indicate substoichiometric occupancy of the binding site and/or heterogeneity in the binding mode. Yet, the orientation of the quinone ring plane is well defined and the structures agree that Asp229, His202 and Ser206 of cyt b provide either direct or water-molecule bridged hydrogen-bonds to the ubiquinone carbonyl oxygen atoms (Asp, His) and a methoxy group (Ser). The difference in involvement of water molecules may appear as minor detail, but is mechanistically important for proton delivery and semiquinone binding while avoiding formation of deleterious reactive oxygen species (ROS). In the structure of the yeast complex, both carbonyl oxygen atoms have water molecules as direct ligands. On one side, the water molecule is stabilized by His202. On the other side, the Asp229 side chain is close to the ubiquinone carbonyl and the water molecule (Fig. 3). Therefore, water molecules have been suggested as primary proton donors for ubiquinone reduction (Hunte 2001; Hunte et al. 2000, 2003). Water molecules also exist as direct ligands of the carbonyl oxygen atoms in one structure of bovine cyt bc_1 (Gao et al. 2003), whereas direct hydrogen bonds with His202 and Asp229 side chains were described for other structures from chicken and bovine complexes (Huang et al. 2005; Zhang et al. 1998). X-ray structures with

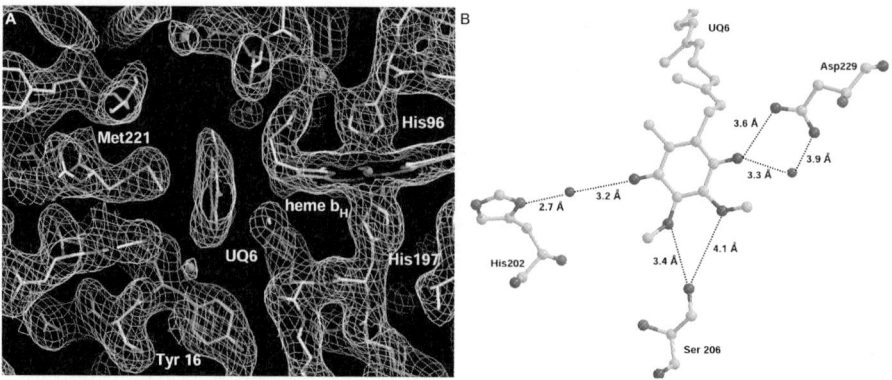

Fig. 3 Ubiquinone bound at center N. **A** Top-view from the matrix side on the binding pocket of the yeast complex with the natural substrate coenzyme Q6 (UQ6) bound. Structure and electron density map (2Fo–Fc) are shown (PDB ID 1EZV, Hunte et al. 2000). **B** In this structure, water molecules and His202, Asp229 of cyt *b* provide contacts to the carbonyl oxygen atoms, and Ser206 is close to the methoxy groups

higher occupancy and defined binding mode of substrate or reaction intermediates could provide further insight.

Precise information about the environment of the paramagnetic ubisemiquinone can be obtained with EPR techniques. ENDOR analysis with proton/deuterium exchange showed strongly hyperfine-coupled exchangeable protons indicating hydrogen-bond interactions with the radical oxygen atoms (Salerno et al. 1990). Pulsed EPR spectra of *Rh. sphaeroides* cyt bc_1 indicated a histidine nitrogen atom in close distance to the ubisemiquinone carbonyl (Kolling et al. 2003). ESEEM and HYSCORE analysis showed the presence of three exchangeable protons in interaction with ubisemiquinone (Dikanov et al. 2004). The spectra indicate a direct nitrogen-atom ligand and they were interpreted as direct hydrogen bonds between carbonyl oxygen atoms and His/Asp side chains and also with a methoxy group.

The structurally observed differences in ligation led to mechanistic speculations in which side-chain and water interactions change to stabilize different reaction intermediates (Gao et al. 2003; Kolling et al. 2003). However, there is no experimental proof that the ligation of different ubiquinone species varies through the catalytic cycle and further experiments are required to resolve mechanistic details and role of the radical.

3.1
Ubihydroquinone Oxidation at Center P

The key step of the Q cycle is the bifurcated electron transfer into thermodynamically different pathways at center P. The thermodynamically favorable but energetically wasteful reduction of the Rieske protein by both electrons of ubihydroquinone is circumvented by the bifurcated electron flow at center P, in which the second electron enters the low-potential electron transfer chain. The divergent electron flow was demonstrated with the oxidant-induced reduction of heme b_L in the presence of center N inhibitor antimycin A, which could be terminated by center P specific inhibitors (Wikstrom and Berden 1972). No structural data are available up to now for substrate binding at center P, but characterization with substrate analogs, inhibitors and mutants provided valuable information for the development of mechanistic models.

The enzyme substrate complex is formed between cyt b, ubihydroquinone and Rieske protein, however, its exact molecular structure is not known. E272 of the highly conserved cyt b PEWY motif (Hauska et al. 1988) and H181 of the Rieske protein have been suggested as direct ligands of ubihydroquinone (Fig. 4A) and as primary acceptors for the protons released upon substrate oxidation (Crofts et al. 1999a,c, 2003; Ding et al. 1992; Hunte et al. 2000; Palsdottir et al. 2003; Rich 2004). The rotational displacement of E272 observed in cyt bc_1 structures with different inhibitors and with non-occupied center P led to the proposal that the E272 side chain upon protonation rotates towards heme b_L facilitating a thermodynamically favorable coupled proton–electron transfer in that direction, the initial step for proton release (Crofts et al. 1999a,c; Hunte et al. 2000; Palsdottir et al. 2003) (Fig. 4 B). The

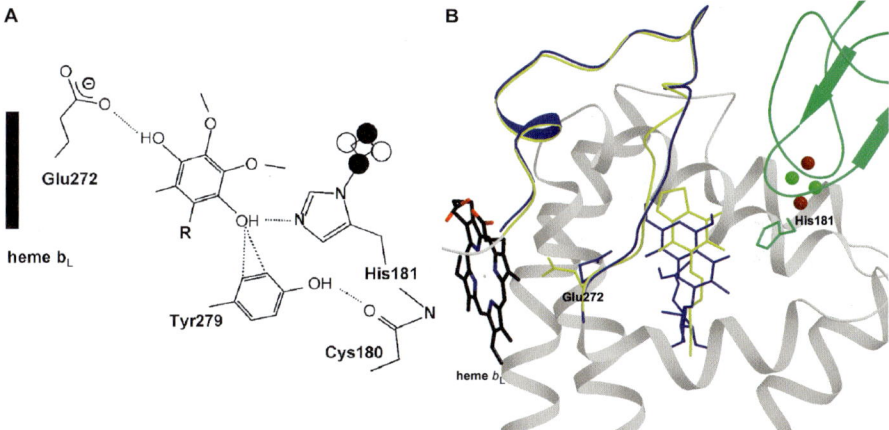

Fig. 4 Model of the enzyme substrate complex at center P. **A** Ubihydroquinone is stabilized by hydrogen bonds with side chains of E272 of cyt b and His181 of the Rieske protein. 2Fe-2S cluster and heme b_L are in oxidized state. Tyr279 of cyt b provides a hydrogen bond to the backbone carbonyl oxygen atom of Cys180 of the Rieske protein, thereby stabilizing the closed conformation. Tyr279 could provide weak hydrogen bonds to stabilize deprotonated substrate (Palsdottir et al. 2003). **B** Orientation of the Glu272 side chain depends on center P occupancy and might facilitate parallel proton electron transfer toward heme b_L. Structures of yeast cyt bc_1 with bound stigmatellin (*blue*, PDB 1EZV, Hunte et al. 2000) and HHDBT (*yellow*, PDB 1P84, Palsdottir et al. 2003) have been superimposed (from Wenz et al. 2006)

"conformational gating" model integrates E272 as a switch, which directs the stabilization of substrate and reaction intermediates by appropriate side chain conformation (Mulkidjanian 2005).

The proposed mechanistic role of E272 was scrutinized by mutagenesis studies (Crofts et al. 1999c; Osyczka et al. 2006; Wenz et al. 2006). Detailed characterization of yeast variants E272D and E272Q, which were obtained by site-directed mutagenesis of the mitochondrially encoded cyt b gene, showed that the residue governs efficient ubihydroquinone oxidation (Wenz et al. 2006). The substitutions do affect binding and oxidation of ubihydroquinone, alter the pH dependence of the enzyme activity, promote energy-wasting electron short-circuit reactions resulting in superoxide production (Wenz et al. 2006), and affect pre-steady-state kinetics of the electron transfer (Wenz et al. 2007). The altered pH dependent activity profile fits the loss of a protonable group with a pK_a of ~ 6, supporting the suggested role of E272 in proton exchange. E272 is fully conserved in mitochondrial cyt b, but substituted by valine or proline in β- and γ-proteobacteria. It was suggested that in those bacteria a glutamate at homologous position to yeast H253 is conserved, which could take over the proton transfer function (Wenz et al. 2006). Multiple substitutions of E295 (corresponds to yeast E272) in purple photosynthetic *Rh. capsulatus* resulted also in substantially decreased hydroquinone oxidation rates, but mutants showed little or no detectable effects on substrate exchange and binding at center P (Osyczka et al. 2006). It has been argued that the lack of effects of glutamate removal on midpoint potential and reduced pK values of heme b_L rules out any mechanism that relies on redox and proton-coupled charge interactions between heme b_L and the PEWY glutamate. The authors point out the considerable resilience of center P activity to mutational changes of the binding pocket in general and suggest that residues at center P are a product of natural selection to develop high specificity by raising barriers to prevent semiquinone loss or short-circuit reactions. The latter was shown for the yeast E272Q and E272D variants, in which the substitutions cause substantial superoxide anion production of $\sim 50\%$ and $\sim 30\%$, respectively (Wenz et al. 2006). E272 clearly is mechanistically important but not essential. The experimental evidence that E272 is a ligand of ubihydroquinone is still lacking.

Based on structural, functional, and mutagenesis data combined with thermodynamic and kinetic information, numerous models were created that show considerable disagreement concerning the presence of one or two substrate molecules in the binding site, a sequential or concerted electron transfer, the nature of the rate-limiting step or possible gating reactions (Crofts 2004; Osyczka et al. 2005; Rich 2004).

The kinetically controlled reaction model employs the formation of a highly unstable semiquinone, which represents the activation barrier for ubihydroquinone oxidation. This transition state is suggested as the rate-limiting step (Crofts and Wang 1989). In agreement with this model, kinetic studies indicate that the movement of the extrinsic Rieske domain is not the source of the activation barrier and the rate-limiting step is earlier in center P catalysis, not restricted by the domain movement (Crofts et al. 1999b). With the structural data at hand, the movement of semiquinone in the bilobal binding pocket was included in the model (Crofts et al. 2000; Hong et al. 1999). Another single occupancy model is the "three-rate-constant" model, which is based on the lack of a detectable semiquinone and the notion that semiquinone is thermodynamically unstable, so that the rate constants of the second electron transfer would drive the reaction (Junemann et al. 1998). This model was advanced to the "logic-gated" mechanism, in which correct hydrogen bonding restricts the formation of an active enzyme substrate complex, so that reversibility of the reaction and prevention of by-pass reactions is accounted for (Rich 2004). Ubihydroquinone would only bind productively, when heme b_L and the 2Fe–2S cluster are oxidized. The latter is also indicated by the binding mode of substrate analogs in the active site (Palsdottir et al. 2003). The "concerted"-mechanism assumes that neither electron is transferred independently, but rather the semiquinone is so unstable that ubiquinol can not reduce the Rieske center unless the semiquinone reduces heme b_L. In such a mechanism the concentration of ubisemiquinone is so low as to be almost non-existent (Snyder et al. 2000).

Based on evaluation of EPR line shapes of the Rieske protein, a double occupancy model was put forward (Ding et al. 1992, 1995). The "proton-gated mechanism" of Brandt took up the simultaneous presence of two quinones, it assumes a quinhydrone-type intermediate and assigns the rate-limiting step to the deprotonation of ubihydrochinone (Brandt 1996). NMR-based quantification of ubiquinone displaced with inhibitors supports the presence of two substrate molecules in the center P pocket (Bartoschek et al. 2001), though direct evidence for an enzyme substrate complex involving both molecules is still lacking. In the "proton-gated affinity change" model (Link 1997), binding of semiquinone anion at center P results in an elevated E_m of the Rieske cluster, similar to the effect of distal center P inhibitors such as stigmatellin. The E_m up-shift is considered as gating mechanism, as the reduced Rieske protein stabilizes the radical after the initial coupled proton–electron transfer to the Rieske protein. The electron on the 2Fe–2S cluster will only be transferred to cyt c_1, once the second electron is passed from semiquinone to the low-potential chain. In agreement with the Brandt model, the first ubihydroquinone deprotonation is considered as the rate-controlling step as it limits the availability of single deprotonated ubihydroquinone. This model is supported by the observation that the position of the Rieske domain and interactions with center P occupants affect the midpoint potential of the 2Fe–2S cluster (Cooley et al. 2004; Darrouzet et al. 2002). A redox-regulated movement of the Rieske domain is an attractive candidate for gating quinol oxidation reactions. Kinetic studies indicate that the oxidized form of the Rieske protein has higher mobility than the reduced one (Zhang et al. 2000). Functional studies of variants suggest that surface exposed loops at the active site construct physical barriers for fine-tuning the large-scale movement. Supporting evidence is provided by a number of compensatory variants where mutations in the Rieske domain are found to suppress functionally impaired cytochrome *b* mutants and vice versa (Brasseur et al. 2004; Darrouzet and Daldal 2003). One should note that the movement of the Rieske domain is essential for multiple turnover and steady-state enzymatic activity but it is not required for the single quinol oxidation event (Darrouzet and Daldal 2002).

Several of the above-mentioned models did not take into account that the reactions for ubiquinol oxidation are fully reversible in a millisecond timescale and that by-pass reactions are controlled under physiological conditions. These two fundamental principles were elegantly demonstrated by a study with cofactor knockouts (Osyczka et al. 2004). The authors suggest a "double-gated" model in which either a kinetically concerted two-electron process—i.e., both electron transfers within femtoseconds—avoids the semiquinone intermediate, or a conformational gating of semiquinone exists for both forward and reverse electron transfer. The latter sequential electron transfer would only be possible if the 2Fe–2S cluster and heme b_L have the same redox state, and could be controlled by hydroquinone and quinone binding or modulated semiquinone properties. In line with the reversibility of the reaction is the identification of a reverse operating cyt bc_1 in the acidophilic chemolithotrophic organism *Acidothiobacillus ferrooxidans* (Elbehti et al. 1999). For the mechanism that involves an intermediate semiquinone that moves in the binding pocket, a gating mechanism has been put forward that allows for reversible ubihydroquinone oxidation but prevents bypass reactions; Coulombic interaction prevents the semiquinone anion from close approach to heme b_L, when the latter is reduced (Crofts et al. 2006).

Obviously, the formation and presence of a semiquinone as intermediate of ubihydroquinone oxidation, which was an inherent feature of the initial Q cycle model (Mitchell 1976), is an important issue for the detailed mechanistic models. Ubiquinone/ubihydroquinone has an E_m of 110 mV. Splitting the two-electron transfer steps in two redox couples, the E_m of the first one (semiquinone/hydroquinone) is considerably more positive close to that of the high-potential chain, whereas the second (quinone/semiquinone) is negative and close to the low-potential chain, so that the stability constant is considered small (Cape et al. 2005; Osyczka et al. 2005). Whereas the semiquinone at center N is easy to detect as transient EPR radical and can be specifically assigned by its cancellation with the center N specific inhibitor antimycin, the semiquinone at center P remained elusive despite consistent attempts from various groups. A single study on beef heart submitochondrial particles reported a semiquinone-EPR signal, which was attributed to center P (de Vries et al. 1981). This signal was later shown to be insensitive to center P inhibitors and it derives most likely from other respiratory complexes (Junemann et al. 1998). The authors showed that under conditions of oxidant-induced reduction, in which both b-hemes are reduced and cyt c_1 is oxidized, center P is primarily occupied by ubihydroquinone with the [2Fe-2S] center oxidized or possibly by an antiferromagnetically coupled semiquinone/reduced [2Fe-2S] center pair, which is EPR silent. A recent EPR study of freeze-quenched probes obtained with an ultra-fast mixer indicate simultaneous reduction of Rieske protein and heme b_L, and did not detect any center P specific semiquinone (Zhu et al. 2007). However, a center P generated ubisemiquinone anion was recently detected by continuous wave and pulsed EPR under anaerobic conditions and elevated pH (Cape et al. 2007). The radical is trapped with a yield of 1–10%. It is sensitive to center P specific inhibitors and mutations as well as to oxygen. The latter suggests that it may be the intermediate responsible for superoxide production and would explain why the other group failed to detect the radical working under aerobic conditions. A previous study showed that the enthalpies of activation were almost identical between Q cycle and superoxide production and suggested a common rate limiting step (Forquer et al. 2006). A likely candidate for this reaction intermediate common between Q cycle and superoxide production is a semiquinone anion (Cape et al. 2007). The authors point out that their results contradict all concerted mechanisms and those including a highly stabilized semiquinone, but they are consistent with a model that maintains a highly unstable semiquinone to limit superoxide production (Cape et al. 2007). A simultaneous study made use of a genetic heme b_H knockout of *Rhodobacter* cyt bc_1 to arrest the transient semiquinone, trap it by rapid freeze-quenching and analyze it by EPR. The radical is light induced in photosynthetic membranes of the mutant under anaerobic and mildly alkaline conditions, i.e., by flashing the photosynthetic reaction center providing oxidized cyt c_2 and hydroquinone. The signal is redox-poise dependent, stigmatellin-sensitive and neither appears in wild type chromatophores or those with a double mutation that also knocks heme c_1, clearly indicating that it is a center P generated semiquinone. The spectral properties indicate some spin interaction with the nearby Rieske cluster. The signal can be generated in 1% of the complex. The authors argue that the effective stability constant range is sufficient to support sub-millisecond electron tunnelling and catalysis given the likely electron transfer distances at center P. The redox properties of the semiquinone would thereby enable fast productive electron transfer while keeping the radical concentration as low as possible to suppress superoxide generation (Zhang et al. 2007). For both studies, it remains to be shown that the observed semiquinone is the source of reduction of molecular oxygen and that semiquinone is a genuine intermediate in center P catalysis. In addition, the issue of spin coupling needs to be addressed.

Short-circuit reactions under steady-state conditions are extremely rare (Darrouzet et al. 2001; Osyczka et al. 2004). However, by-pass reactions occur when the bifurcation reaction is perturbed, for instance by addition of center N inhibitor Antimycin A (Muller et al. 2002; Sun and Trumpower 2003) or by mutation of the proposed ubihydroquinone ligand E272 (Wenz et al. 2006). It has been discussed that the Q cycle serves a dual role in both energy transduction and the avoidance of deleterious side reactions, as reactive oxygen species (ROS) can damage proteins and mitochondrial DNA (Cape et al. 2006). In man, ROS are implicated in several diseases like Parkinson's, Alzheimer's and cancer. Furthermore, it was proposed that in mitochondria, the electron-transfer chain acts as an oxygen sensor by releasing (ROS) in response to hypoxia; the primary site of ROS production seems to be cyt bc_1 with center P dominating center N in radical contribution (Guzy and Schumacker 2006).

The dimeric nature of cyt bc_1 raises the question whether both monomers operate as independent or coupled functional units. There is growing experimental evidence for long-range interactions between center P and N (Cooley et al. 2005; Covian and Trumpower 2006; Wenz et al. 2007). Mechanisms are discussed that involve a dimeric Q cycle or half-of-the-sites activity of the dimeric complex.

4
Interaction of cyt c with cyt bc_1

Cyt c is the final electron acceptor of cyt bc_1. It shuttles the electrons to cyt c oxidase by reversibly docking to donor and acceptor. The interaction of cyt bc_1 and cyt c is transient and efficient, enabling catalytic turnover higher than 100 per second. Cyt c reduction is not the rate-limiting step of the reaction, as cyt c is present in high concentration, diffusion rates are high, and association as well as dissociation processes are fast. As expected for a transient complex, the affinity of cyt bc_1 for cyt c is low. A K_M of 3.5 μM was determined for the yeast complex (Schägger and Pfeiffer 2000). Furthermore, the multi-functional electron carrier cyt c reacts with multiple reaction partners and the interaction requires sufficient specificity for any of them.

The 3-Å resolution X-ray structure of yeast cyt bc_1 with cyt c and an F_V-fragment bound provided the first description of the binding interface between cyt c and subunit cyt c_1 (Lange and Hunte 2002) (Fig. 2A,C). The interface is especially small with an area of 886 Å^2 and it surrounds the heme clefts. The tight and specific interactions are mainly mediated by nonpolar forces. A central cation–π interaction between Arg19 of cyt c and Phe230 of cyt c_1 appears to be an important and conserved feature of cyt c binding. Complementary charged residues are located peripheral to the contact sites, but no salt bridges or hydrogen bonds are observed. The weak electrostatic interactions are probably important for the preorientation of the complex. An electrostatic contribution to the binding interaction was expected, as cyt c reduction by cyt bc_1 strongly depends on ionic strength (Hunte et al. 2002; Speck et al. 1979). Crystallization conditions of the complex are within the physiological range and the structure is considered as the native bound state of the electron-transfer complex. A small interface of hydrophobic interactions surrounded by charged residue pairs appear to be the dominant features of transient electron-transfer complexes, and are also observed for the cyt c peroxidase: cyt c and bacterial reaction center: cyt c_2 complexes (Axelrod et al. 2002; Pelletier and Kraut 1992). A close spatial arrangement of the c-type hemes is observed for the cyt bc_1: cyt c complex, with a distance as close as 4.5 Å between the heme CBC atoms and an edge-to-edge distance of 9.3 Å. This suggests a rapid and direct heme-to-heme electron transfer at a calculated rate of up to 8.6×10^6 s^{-1} (Lange and Hunte 2002; Page et al. 1999). The redox potentials of cyt c_1 and cyt c are equal (Cutler et al. 1987; T'sai and Palmer 1983) and electron transfer can occur in both directions. However, it is not known whether the binding interface is affected by the redox state of the partners as to favor formation of the productive complex.

Unexpectedly, only one molecule of cyt c is bound to one of the two identical binding sites in the structure of the cyt bc_1: cyt c complex (Lange and Hunte 2002). The half-of-the-sites binding mode appears to be genuine and not affected by crystal contacts or sterical hindrance. Furthermore, cyt c binding coincides with ubiquinone occupancy at center N, so that a coordinated binding of both electron acceptors was suggested. The binding mode fits into the context of the alternating half-of-the-sites hypothesis, which implies that only one half of the dimeric enzyme is active at a time (Trumpower 2002).

5
Respiratory Supercomplexes

The supramolecular organization of the oxidative phosphorylation (OXPHOS) system has been controversially discussed (Boekema and Braun 2007; Lenaz and Genova 2007). In the random collision model, the respiratory chain complexes are freely distributed in the lipid bilayer and functionally connected by lateral diffusion of the small redox carriers coenzyme Q and cyt c (Hackenbrock et al. 1986). The model is supported by kinetic data and by the fact that the components of the respiratory chain are active as isolated complexes. However, there is growing evidence for a stable supramolecular organization of the respiratory complexes in the membrane, the 'respirasomes' (Schägger and Pfeiffer 2000). Early isolation protocols provided first evidence that the respiratory chain assembles into supercomplexes. Complex III (cyt bc_1) was co-purified with complex II (succinate : ubiquinone oxidoreductase) (Yu et al. 1974) and also with complex IV (cyt c oxidase) (Cruciat et al. 2000; Schägger and Pfeiffer 2000). The analysis of digitonin extracts from membranes with blue native polyacrylamide gel electrophoresis (BN-PAGE) suggested specific assemblies of mammalian respiratory chain complexes, so that Schägger introduced the model of the "respirasomes" (Schägger and Pfeiffer 2000). Supramolecular assemblies have also been shown for complexes from bacterial sources (Schägger 2002). More recently, single-particle electron microscopy studies of mitochondrial supercomplexes provided first structural insights for supercomplex I (NADH : ubiquinone oxidoreductase) + III from *Arabidopsis thaliana* (Dudkina et al. 2005), supercomplexes I + III and I + III + IV from bovine heart (Schäfer et al. 2006) and supercomplex III + IV from yeast (Heinemeyer et al. 2007). The characteristic structural features of the individual complexes allowed their relative positioning within the assembly. However, the exact positioning is hampered by the limited resolution. For these studies, isolated supercomplexes were used, solubilized with non-ionic detergents and purified either with sucrose-gradient centrifugation or BN-PAGE. Based on electron microscopic analysis of whole mitochondria, Schägger and colleagues suggested a model for an even higher supramolecular association of respirasomes into "respiratory strings" (Wittig et al. 2006).

Experimental evidence for the in vivo organization in the membrane is still scarce (Boumans et al. 1998; Genova et al. 2003). Yet, the structural information clearly supports the view of stable supramolecular associations. Substrate channeling (Boumans et al. 1998), stability of individual complexes for instance of complex I (Schägger et al. 2004), regulation of respiratory activity, mitochondrial ultrastructure, and optimized high protein content of the membrane are discussed as potential roles of the supercomplexes (Boekema and Braun 2007; Lenaz and Genova 2007). Mitochondrial supercomplexes are also implicated in pathological conditions, e.g., they are destabilized in Barth syndrome patients with defects in the cardiolipin metabolism (McKenzie et al. 2006). Cardiolipin has been shown to stabilize respiratory supercomplexes (Pfeiffer et al. 2003; Zhang et al. 2002). The dynamic collision and the static model of the OXPHOS system seem to be contradictory, but they may only describe two possible extreme conditions. Supercomplexes may be dynamic rather than static assemblies, and the relative stability may vary between organisms, tissues or with physiological conditions.

Acknowledgements This work was supported by the German Research Foundation (DFG, SFB 472). We thank U. Brandt and A.R. Crofts for critical comments on the manuscript.

References

Axelrod HL, Abresch EC, Okamura MY, Yeh AP, Rees DC, Feher G (2002) X-ray structure determination of the cytochrome c(2): reaction center electron-transfer complex from *Rhodobacter sphaeroides*. J Mol Biol 319:501–515

Bartoschek S, Johansson M, Geierstanger BH, Okun JG, Lancaster CRD, Humpfer E, Yu L, Yu CA, Griesinger C, Brandt U (2001) Three molecules of ubiquinone bind specifically to mitochondrial cytochrome bc(1) complex. J Biol Chem 276:35231–35234

Berry EA, Guergova-Kuras M, Huang LS, Crofts AR (2000) Structure and function of cytochrome bc complexes (Review). Ann Rev Biochem 69:1005–1075

Berry EA, Huang LS, DeRose VJ (1991) Ubiquinol-cytochrome c oxidoreductase of higher plants. Isolation and characterization of the $bc1$ complex from potato tuber mitochondria. J Biol Chem 266:9064–9077

Berry EA, Huang LS, Saechao LK, Pon NG, Valkova-Valchanova M, Daldal F (2004) X-ray structure of Rhodobacter capsulatus cytochrome $bc(1)$: comparison with its mitochondrial and chloroplast counterparts. Photosynthesis Res 81:251–275

Berry EA, Trumpower BL (1985) Isolation of ubiquinol oxidase from Paracoccus denitrificans and resolution into cytochrome $bc1$ and cytochrome c-aa3 complexes. J Biol Chem 260:2458–2467

Boekema EJ, Braun HP (2007) Supramolecular structure of the mitochondrial oxidative phosphorylation system [Review]. J Biol Chem 282:1–4

Boumans H, Grivell LA, Berden JA (1998) The respiratory chain in yeast behaves as a single functional unit. J Biol Chem 273:4872–4877

Brandt U (1996) Energy conservation by bifurcated electron-transfer in the cytochrome-$bc1$ complex. Biochim Biophys Acta 1275:41–46

Brandt U, Haase U, Schägger H, von Jagow G (1991) Significance of the 'Rieske' iron-sulfur protein for formation and function of the ubiquinol oxidation pocket of mitochondrial cytochrome c reductase (bc_1 complex). J Biol Chem 266:19958–19964

Brandt U, Trumpower BL (1994) The protonmotive Q cycle in mitochondria and bacteria. Crit Rev Biochem 29:165–197

Brasseur G, Lemesle-Meunier D, Reinaud F, Meunier B (2004) Q(O) site deficiency can be compensated by extragenic mutations in the hinge region of the iron-sulfur protein in the $bc(1)$ complex of Saccharomyces cerevisiae. J Biol Chem 279:24203–24211

Brugna M, Rodgers S, Schricker A, Montoya G, Kazmeier M, Nitschke W, Sinning I (2000) A spectroscopic method for observing the domain movement of the Rieske iron-sulfur protein. Proc Natl Acad Sci USA 97:2069–2074

Cape JL, Bowman MK, Kramer DM (2006) Understanding the cytochrome bc complexes by what they don't do. The Q-cycle at 30 [Review]. Trends Plant Sci 11:46–55

Cape JL, Bowman MK, Kramer DM (2007) A semiquinone intermediate generated at the Q_o site of the cytochrome $bc1$ complex: importance for the Q-cycle and superoxide production. Proc Natl Acad Sci USA 104:7887–7892

Cape JL, Strahan JR, Lenaeus MJ, Yuknis BA, Le TT, Shepherd JN, Bowman MK, Kramer DM (2005) The respiratory substrate rhodoquinol induces Q-cycle bypass reactions in the yeast cytochrome $bc(1)$ complex—mechanistic and physiological implications. J Biol Chem 280:34654–34660

Cooley JW, Ohnishi T, Daldal F (2005) Binding dynamics at the quinone reduction (Q_i) site influence the equilibrium interactions of the iron-sulfur protein and hydroquinone oxidation (Q(o)) site of the cytochrome $bc(1)$ complex. Biochemistry 44:10520–10532

Cooley JW, Roberts AG, Bowman MK, Kramer DM, Daldal F (2004) The raised midpoint potential of the [2Fe–2S] cluster of cytochrome $bc1$ is mediated by both the Q(o) site occupants and the head domain position of the Fe-S protein subunit. Biochemistry 43:2217–2227

Covian R, Trumpower BL (2006) Regulatory interactions between ubiquinol oxidation and ubiquinone reduction sites in the dimeric cytochrome $bc(1)$ complex. J Biol Chem 281:30925–30932

Cramer WA, Zhang HM, Yan JS, Kurisu G, Smith JL (2006) Transmembrane traffic in the cytochrome b(6)f complex (Review). Ann Rev Biochem 75:769–790

Crofts AR (1985) The mechanism of ubiquinol:cytochrome c oxidoreductases of mitochondria and of *Rhodopseudomonas sphaeroides*. In: Martonosi AN (ed) The Enzymes of Biological Membranes. Plenum Publishing Corporation, New York, pp 347–382

Crofts AR (2004) The cytochrome $bc(1)$ complex: function in the context of structure [Review]. Ann Rev Physiol 66:689–733

Crofts AR, Barquera B, Gennis RB, Kuras R, Guergova-Kuras M, Berry EA (1999a) Mechanism of ubiquinol oxidation by the $bc(1)$ complex: different domains of the quinol binding pocket and their role in the mechanism and binding of inhibitors. Biochemistry 38:15807–15826

Crofts AR, Guergova-Kuras M, Huang LS, Kuras R, Zhang ZL, Berry EA (1999b) Mechanism of ubiquinol oxidation by the $bc(1)$ complex: role of the iron-sulfur protein and its mobility. Biochemistry 38:15791–15806

Crofts AR, Guergova-Kuras M, Kuras R, Ugulava N, Li JY, Hong SJ (2000) Proton-coupled electron transfer at the Q(o) site: what type of mechanism can account for the high activation barrier? Biochim Biophys Acta Bioenerg 1459:456–466

Crofts AR, Hong SJ, Ugulava N, Barquera B, Gennis R, Guergova-Kuras M, Berry EA (1999c) Pathways for proton release during ubihydroquinone oxidation by the $bc(1)$ complex. Proc Natl Acad Sci USA 96:10021–10026

Crofts AR, Lhee S, Crofts SB, Cheng J, Rose S (2006) Proton pumping in the $bc(1)$ complex: a new gating mechanism that prevents shirt circuits. Biochim Biophys Acta 1757:1019–1034

Crofts AR, Shinkarev VP, Kolling DRJ, Hong S (2003) The modified Q-cycle explains the apparent mismatch between the kinetics of reduction of cytochromes c1 and bH in the $bc(1)$ complex. J Biol Chem 278:36191–36201

Crofts AR, Wang Z (1989) How rapid are the internal reactions of the ubiquinol:cytochrome c2 oxidoreductase? Photosynthesis Res 22:69–87

Cruciat CM, Brunner S, Baumann F, Neupert W, Stuart RA (2000) The cytochrome $bc(1)$ and cytochrome c oxidase complexes associate to form a single supracomplex in yeast mitochondria. J Biol Chem 275:18093–18098

Cruciat CM, Hell K, Folsch H, Neupert W, Stuart RA (1999) Bcs1p, an AAA-family member, is a chaperone for the assembly of the cytochrome $bc(1)$ complex. EMBO J 18:5226–5233

Cutler RL, Pielack GJ, Mauk AG, Smith M (1987) Replacement of cystein-107 of *Saccharomyces cerevisiae* iso-1-cytochrome c with threonine: improved stability of the mutant protein. Protein Eng 1:95–99

Darrouzet E, Daldal F (2002) Movement of the iron-sulfur subunit beyond the EF loop of cytochrome b is required for multiple turnovers of the bc1 complex but not for single turnover Q(o) site catalysis. J Biol Chem 277:3471–3476

Darrouzet E, Daldal F (2003) Protein–protein interactions between cytochrome b and the Fe–S protein subunits during QH(2) oxidation and large-scale domain movement in the $bc(1)$ complex. Biochemistry 42:1499–1507

Darrouzet E, Moser CC, Dutton PL, Daldal F (2001) Large-scale domain movement in cytochrome $bc(1)$: a new device for electron transfer in proteins [Review]. Trends Biochem Sci 26:445–451

Darrouzet E, Valkova-Valchanova M, Daldal F (2002) The [2Fe–2S] cluster E-m as an indicator of the iron-sulfur subunit position in the ubihydroquinone oxidation site of the cytochrome $bc(1)$ complex. J Biol Chem 277:3464–3470

Darrouzet E, Valkova-Valchanova M, Moser CC, Dutton PL, Daldal F (2000) Uncovering the [2Fe2S] domain movement in cytochrome $bc(1)$ and its implications for energy conversion. Proc Natl Acad Sci USA 97:4567–4572 de la Rosa FF, Palmer G (1983) Reductive titration of CoQ-depleted complex III from Baker's yeast. FEBS Lett 163:140–143

de Vries S, Albracht SPJ, Berden JA, Slater EC (1981) A new species of bound ubisemiquinone anion in QH2 – cytochrome-c oxidoreductase. J Biol Chem 256:1996–1998

Denke E, Merbitzzahradnik T, Hatzfeld OM, Snyder CH, Link TA, Trumpower BL (1998) Alteration of the midpoint potential and catalytic activity of the Rieske iron-sulfur protein by changes of amino acids forming hydrogen bonds to the iron-sulfur cluster. J Biol Chem 273:9085–9093

Dikanov SA, Samoilova RI, Kolling DRJ, Holland JT, Crofts AR (2004) Hydrogen bonds involved in binding the Q(i)-site semiquinone in the $bc(1)$ complex, identified through deuterium exchange using pulsed EPR. J Biol Chem 279:15814–15823

Ding H, Robertson DE, Daldal F, Dutton PL (1992) Cytochrome $bc - 1$ complex cluster and its interaction with ubiquinone and ubihydroquinone at the q-o site a double-occupancy q-o site model. Biochemistry 31:3144–3158

Ding H, Robertson DE, Daldal F, Dutton PL (1995) Cytochrome b1 complex [2Fe–2S] cluster and its interaction with ubiquinone and ubihydroquinone at the Qo site: a double occupancy Qo site model. Biochemistry 31:3144–3158

Dudkina NV, Eubel H, Keegstra W, Boekema EJ, Braun HP (2005) Structure of a mitochondrial supercomplex formed by respiratory-chain complexes I and III. Proc Natl Acad Sci USA 102:3225–3229

Elbehti A, Nitschke W, Tron P, Michel C, Lemesle-Meunier D (1999) Redox components of cytochrome bc-type enzymes in acidophilic prokaryotes I. Characterization of the cytochrome $bc(1)$-type complex of the acidophilic ferrous ion-oxidizing bacterium *Thiobacillus ferrooxidans*. J Biol Chem 274:16760–16765

Elberry M, Xiao KH, Esser L, Xia D, Yu L, Yu CA (2006) Generation, characterization and crystallization of a highly active and stable cytochrome $bc(1)$ complex mutant from *Rhodobacter sphaeroides*. Biochim Biophys Acta Bioenerg 1757:835–840

Engel WD, Schägger H, Jagow G (1980) Ubiquinol-cytochrome c reductase (E.C. 1.10.2.2). Isolation in triton X-100 by hydroxyapatite and gel chromatography. Structural and functional properties. Biochim Biophys Acta 592:211–222

Esser L, Gong X, Yang SQ, Yu L, Yu CA, Xia D (2006) Surface-modulated motion switch: capture and release of iron-sulfur protein in the cytochrome $bc(1)$ complex. Proc Natl Acad Sci USA 103:13045–13050

Esser L, Quinn B, Li YF, Zhang MQ, Elberry M, Yu L, Yu CA, Xia D (2004) Crystallographic studies of quinol oxidation site inhibitors: a modified classification of inhibitors for the cytochrome $bc(1)$ complex. J Mol Biol 341:281–302

Forquer I, Covian R, Bowman MK, Trumpower BL, Kramer DM (2006) Similar transition states mediate the Q cycle and superoxide production by the cytochrome $bc1$ complex. J Biol Chem 281:38459–38465

Gao XG, Wen XL, Esser L, Quinn B, Yu L, Yu CA, Xia D (2003) Structural basis for the quinone reduction in the $bc(1)$ complex: a comparative analysis of crystal structures of mitochondrial cytochrome $bc(1)$ with bound substrate and inhibitors at the Qi site. Biochemistry 42:9067–9080

Genova MW, Bianchi C, Lenaz G (2003) Structural organization of the mitochondrial respiratory chain. Ital J Biochem 52:58–61

Gomez B, Robinson NC (1999) Phospholipase digestion of bound cardiolipin reversibly inactivates bovine cytochrome $bc(1)$. Biochemistry 38:9031–9038

Guergova-Kuras M, Kuras R, Ugulava N, Hadad I, Crofts AR (2000) Specific mutagenesis of the Rieske iron-sulfur protein in *Rhodobacter sphaeroides* shows that both the thermodynamic gradient and the pK of the oxidized form determine the rate of quinol oxidation by the $bc(1)$ complex. Biochemistry 39:7436–7444

Guzy RD, Schumacker PT (2006) Oxygen sensing by mitochondria at complex III: the paradox of increased reactive oxygen species during hypoxia. Exp Physiol 91:807–819

Hackenbrock CR, Chazotte B, Gupte SS (1986) The random collision model and a critical assessment of diffusion and collision in mitochondrial electron transport. J Bioenerg Biomembranes 18:331–368

Hauska G, Nitschke W, Hermann RG (1988) Amino acid identities in the three redox center-carrying polypeptides of cytochrome bc_1/b_6f complexes. J Bioenerg Biomembranes 20:211–228

Heinemeyer J, Braun HP, Boekema EJ, Kouril R (2007) A structural model of the cytochrome c reductase/oxidase supercomplex from yeast mitochondria. J Biol Chem 282:12240–12248

Hong SJ, Ugulava N, Guergova-Kuras M, Crofts AR (1999) The energy landscape for ubihydroquinone oxidation at the Q(o) site of the $bc(1)$ complex in *Rhodobacter sphaeroides*. J Biol Chem 274:33931–33944

Huang LS, Cobessi D, Tung EY, Berry EA (2005) Binding of the respiratory chain inhibitor antimycin to the mitochondrial $bc(1)$ complex: a new crystal structure reveals an altered intramolecular hydrogen-bonding pattern. J Mol Biol 351:573–597

Hunte C (2001) Insights from the structure of the yeast cytochrome $bc(1)$ complex: crystallization of membrane proteins with antibody fragments. FEBS Lett 504(3 Special Issue):126–132

Hunte C, Koepke J, Lange C, Rossmanith T, Michel H (2000) Structure at 2.3 angstrom resolution of the cytochrome $bc(1)$ complex from the yeast *Saccharomyces cerevisiae* co-crystallized with an antibody Fv fragment. Struct Fold Des 8:669–684

Hunte C, Palsdottir H, Trumpower BL (2003) Protonmotive pathways and mechanisms in the cytochrome $bc(1)$ complex [Review]. FEBS Lett 545:39–46

Hunte C, Solmaz S, Lange C (2002) Electron transfer between yeast cytochrome $bc(1)$ complex and cytochrome c: a structural analysis. Biochim Biophys Bioenerg 1555(1–3 Special Issue):21–28

Iwata S, Lee JW, Okada K, Lee JK, Iwata M, Rasmussen B, Link TA, Ramaswamy S, Jap BK (1998) Complete structure of the 11-subunit bovine mitochondrial cytochrome $bc1$ Complex. Science 281:64–71

Iwata S, Saynovits M, Link TA, Michel H (1996) Structure of a water soluble fragment of the Rieske iron-sulfur protein of the bovine heart mitochondrial cytochrome Bc(1) complex determined by mad phasing at 1.5 Angstrom resolution. Structure 4:567–579

Izrailev S, Crofts AR, Berry EA, Schulten K (1999) Steered molecular dynamics simulation of the Rieske subunit motion in the cytochrome $bc(1)$ complex. Biophys J 77:1753–1768

Junemann S, Heathcote P, Rich PR (1998) On the mechanism of quinol oxidation in the Bc(1) Complex. J Biol Chem 273:21603–21607

Kim CH, King TE (1983) A mitochondrial protein essential for the formation of the cytochrome c1-c complex. Isolation, purification, and properties. J Biol Chem 258:13543–13551

Kim H, Xia D, Yu CA, Xia JZ, Kachurin AM, Zhang L, Yu L, Deisenhofer J (1998) Inhibitor binding changes domain mobility in the iron-sulfur protein of the mitochondrial Bc1 complex from bovine heart. Proc Natl Acad Sci USA 95:8026–8033

Klingen AR, Palsdottir H, Hunte C, Ullmann GM (2007) Redox-linked protonation state changes in cytochrome $bc(1)$ identified by Poisson-Boltzmann electrostatics calculations. Biochim Biophys Acta Bioenerg 1767:204–221

Kolling DRJ, Samoilova RI, Holland JT, Berry EA, Dikanov SA, Crofts AR (2003) Exploration of ligands to the Q(i) site semiquinone in the $bc(1)$ complex using high-resolution EPR. J Biol Chem 278:39747–39754

Lancaster CR, Hunte C, Kelley J 3rd, Trumpower BL, Ditchfield R (2007) A comparison of stigmatellin conformations, free and bound to the photosynthetic reaction center and the cytochrome $bc1$ complex. J Mol Biol 368:197–208

Lange C, Hunte C (2002) Crystal structure of the yeast cytochrome $bc(1)$ complex with its bound substrate cytochrome c. Proc Natl Acad Sci USA 99:2800–2805

Lange C, Nett JH, Trumpower BL, Hunte C (2001) Specific roles of protein-phospholipid interactions in the yeast cytochrome $bc(1)$ complex structure. EMBO J 20:6591–6600

Lenaz G, Genova ML (2007) Kinetics of integrated electron transfer in the mitochondrial respiratory chain: random collisions vs. solid state electron channeling. Am J Physiol 292:C1221–C1239

Leonard K, Wingfield P, Arad T, Weiss H (1981) Three-dimensional structure of ubiquinol: cytochrome c reductase from Neurospora mitochondria determined by electron microscopy of membrane crystals. J Molec Biol 149:259–274

Link TA (1997) The role of the Rieske iron-sulfur protein in the hydroquinone oxidation (Qp) site of the cytochrome $bc(1)$ complex – the proton-gated affinity change mechanism [Review]. FEBS Lett 412:257–264

Link TA, Hagen WR, Pierik AJ, Assmann C, von Jagow G (1992) Determination of the redox properties of the Rieske [2Fe-2S] cluster of bovine heart $bc1$ complex by direct electrochemistry of a water-soluble fragment. Eur J Biochem 208:685–691

Ljungdahl PO, Pennoyer JD, Robertson DE, Trumpower BL (1989) Mutational analysis of the mitochondrial Rieske iron-sulfur protein of *Saccharomyces cerevisiae*. II. Biochemical Characterization of temperature-sensitive RIP1-mutations. J Biol Chem 264:3723–3731

McKenzie M, Lazarou M, Thorburn DR, Ryan MT (2006) Mitochondrial respiratory chain supercomplexes are destabilized in Barth Syndrome patients. J Mol Biol 361:462–469

Mitchell P (1975) The proton motive Q cycle: a general formulation. FEBS Lett 59:137–139

Mitchell P (1976) Possible molecular mechanisms of the protonmotive function of cytochrome c systems. J Theor Biol 62:327–367

Montoya G, te Kaat K, Rodgers S, Nitschke W, Sinning I (1999) The cytochrome $bc(1)$ complex from *Rhodovulum sulfidophilum* is a dimer with six quinones per monomer and an additional 6-kDa component. Eur J Biochem 259:709–718

Moser CC, Keske JM, Warncke K, Farid RS, Dutton PL (1992) Nature of biological electron transfer. Nature (London) 355:796–802

Mulkidjanian AY (2005) Ubiquinol oxidation in the cytochrome $bc(1)$ complex: reaction mechanism and prevention of short-circuiting [Review]. Biochim Biophys Acta Bioenerg 1709:5–34

Muller F, Crofts AR, Kramer DM (2002) Multiple Q-cycle bypass reactions at the Q(o) site of the cytochiome $bc(1)$ complex. Biochemistry 41:7866–7874

Nett JH, Hunte C, Trumpower BL (2000) Changes to the length of the flexible linker region of the Rieske protein impair the interaction of ubiquinol with the cytochrome $bc(1)$ complex. Eur J Biochem 267:7266

Osyczka A, Moser CC, Daldal F, Dutton PL (2004) Reversible redox energy coupling in electron transfer chains. Nature 427:607–612

Osyczka A, Moser CC, Dutton PL (2005) Fixing the Q cycle. Trends Biochem Sci 30:176–182

Osyczka A, Zhang HB, Mathe C, Rich PR, Moser CC, Dutton PL (2006) Role of the PEWY glutamate in hydroquinone-quinone oxidation-reduction catalysis in the Q(o) site of cytochrome $bc(1)$. Biochemistry 45:10492–10503

Page CC, Moser CC, Chen XX, Dutton PL (1999) Natural engineering principles of electron tunnelling in biological oxidation-reduction. Nature 402:47–52

Palsdottir H, Hunte C (2003) Purification of the cytochrome $bc1$ complex from yeast. In: Hunte C, von Jagow G, Schägger H (eds) Membrane protein purification and crystallization: a practical guide. Academic Press, New York, USA, pp 191–203

Palsdottir H, Hunte C (2004) Lipids in membrane protein structures [Review]. Biochim Biophys Acta Biomembranes 1666:2–18

Palsdottir H, Lojero CG, Trumpower BL, Hunte C (2003) Structure of the yeast cytochrome $bc(1)$ complex with a hydroxyquinone anion Q(o) site inhibitor bound. J Biol Chem 278:31303–31311

Pelletier H, Kraut J (1992) Crystal structure of a complex between electron transfer partners, cytochrome c peroxidase and cytochrome c. Science 258:1748–1755

Pfeiffer K, Gohil V, Stuart RA, Hunte C, Brandt U, Greenberg ML, Schägger H (2003) Cardiolipin stabilizes respiratory chain supercomplexes. J Biol Chem 278:52873–52880

Riccio P, Schägger H, Engel WD, von Jagow G (1976) $bc1$-Complex from beef heart. One-step purification by hydroxyapatite chromatography in Triton X-100, polypeptide pattern and respiratory chain characteristics. Biochim Biophys Acta 459:250–262

Rich PR (2004) The quinone chemistry of bc complexes. Biochim Biophys Acta Bioenerg 1658(1–2 Special Issue):165–171

Rich PR, Harper R (1990) Partition coefficients of quinones and hydroquinones and their relation to biochemical reactivity. FEBS Lett 269:139–144

Rich PR, Jeal AE, Madgwick SA, Moody J (1990) Inhibitor effects on redox-linked protonations of the b haemes of the mitochondrial bc_1 complex. Biochim Biophys Acta 1018:29–40

Rieske JS (1976) Composition, structure, and function of complex III of the respiratory chain. Biochim Biophys Acta 456:195–247

Salerno JC, Osgood M, Liu Y, Taylor H, Scholes CP (1990) Electron nuclear double resonance (ENDOR) of the Qc.-ubiqsemiquinone radical in the mitochondrial electron transport chain. Biochemistry 29:6987

Schäfer E, Seelert H, Reifschneider NH, Krause F, Dencher NA, Vonck J (2006) Architecture of active mammalian respiratory chain supercomplexes. J Biol Chem 281:15370–15375

Schägger H (2002) Respiratory chain supercomplexes of mitochondria and bacteria. Biochim Biophys Acta Bioenerg 1555(1–3 Special Issue):154–159

Schägger H, de Coo R, Bauer MF, Hofmann S, Godinot C, Brandt U (2004) Significance of respirasomes for the assembly/stability of human respiratory chain complex I. J Biol Chem 279:36349–36353

Schägger H, Hagen T, Roth B, Brandt U, Link TA, von Jagow G (1990) Phospholipid specificity of bovine heart $bc1$ complex. Eur J Biochem 190:123–130

Schägger H, Pfeiffer K (2000) Supercomplexes in the respiratory chains of yeast and mammalian mitochondria. EMBO J 19:1777–1783

Schutz M, Brugna M, Lebrun E, Baymann F, Huber R, Stetter KO, Hauska G, Toci R, Lemesle-Meunier D, Tron P, Schmidt C, Nitschke W (2000) Early evolution of cytochrome bc complexes [Review]. J Mol Biol 300:663–675

Schutz M, Schoepp-Cothenet B, Lojou E, Woodstra M, Lexa D, Tron P, Dolla A, Durand MC, Stetter KO, Baymann F (2003) The naphthoquinol oxidizing cytochrome $bc(1)$ complex of the hyperthermophilic knallgasbacterium *Aquifex aeolicus*: properties and phylogenetic relationships. Biochemistry 42:10800–10808

Siedow JN, Power S, de la Rosa FF, Palmer G (1978) The preparation and characterization of highly purified, enzymatically active complex III from baker's yeast. J Biol Chem 253:2392–2399

Snyder CH, Gutierrez-Cirlos EB, Trumpower BL (2000) Evidence for a concerted mechanism of ubiquinol oxidation by the cytochrome $bc(1)$ complex. J Biol Chem 275:13535–13541

Snyder CH, Merbitz-Zahradnik T, Link TA, Trumpower BL (1999) Role of the Rieske iron-sulfur protein midpoint potential in the protonmotive Q-cycle mechanism of the cytochrome $bc(1)$ complex. J Bioenerg Biomembranes 31:235–242

Speck SH, Ferguson-Miller S, Osheroff N, Margoliash E (1979) Definition of cytochrome c binding domains by chemical modification: kinetics of reaction with beef mitochondrial reductase and functional organization of the respiratory chain. Proc Natl Acad Sci USA 76:155–159

Sun H, Trumpower BL (2003) Superoxide anion generation by the cytochrome $bc(1)$ complex. Arch Biochem Biophys 419:198–206

T'sai A, Palmer G (1983) Potentiometric studies on yeast complex III. Biochim Biophys Acta 722:349–363

Tian H, Yu L, Mather MW, Yu CA (1998) Flexibility of the neck region of the Rieske iron-sulfur protein is functionally important in the cytochrome Bc(1) complex. J Biol Chem 273:27953–27959

Trumpower BL (2002) A concerted, alternating sites mechanism of ubiquinol oxidation by the dimeric cytochrome $bc(1)$ complex. Biochim Biophys Acta Bioenerg 1555(1–3 Special Issue):166–173

Ugulava NB, Crofts AR (1998) CD-monitored redox titration of the Rieske Fe-S protein of *Rhodobacter sphaeroides*: pH dependence of the midpoint potential in isolated $bc1$ complex and in membranes. FEBS Lett 440:409–413

Ullmann GM, Noodleman L, Case DA (2002) Density functional calculation of pK(a) values and redox potentials in the bovine Rieske iron-sulfur protein. J Biol Inorg Chem 7:632–639

von Jagow G, Gribble GW, Trumpower BL (1986) Mucidin and strobilurin A are identical and inhibit electron transfer in the cytochrome $bc1$ complex of the mitochondrial respiratory chain at the same site as myxothiazol. Biochemistry 25:775–780

von Jagow G, Link TA (1986) Use of specific inhibitors on the mitochondrial $bc1$ complex. Method Enzymol 126:253–271

von Jagow G, Ohnishi T (1985) The chromone inhibitor stigmatellin – binding to the ubiquinol oxidation center at the C-side of the mitochondrial membrane. FEBS Lett 185:311–315

Wenz T, Covian R, Hellwig P, MacMillan F, Meunier B, Trumpower BL, Hunte C (2007) Mutational analysis of cytochrome b at the ubiquinol oxidation site of yeast complex III. J Biol Chem 282:3977–3988

Wenz T, Hellwig P, MacMillan F, Meunier B, Hunte C (2006) Probing the role of E272 in quinol oxidation of mitochondrial complex III. Biochemistry 45:9042–9052

Wikstrom MK, Berden JA (1972) Oxidoreduction of cytochrome b in the presence of antimycin. Biochim Biophys Acta 283:403–420

Wittig I, Carrozzo R, Santorelli FM, Schägger H (2006) Supercomplexes and subcomplexes of mitochondrial oxidative phosphorylation. Biochim Biophys Acta Bioenerg 1757:1066–1072

Xia D, Yu CA, Kim H, Xian JZ, Kachurin AM, Zhang L, Yu L, Deisenhofer J (1997) Crystal structure of the cytochrome Bc(1) complex from bovine heart mitochondria. Science 277:60–66

Xiao KH, Yu L, Yu CA (2000) Confirmation of the involvement of protein domain movement during the catalytic cycle of the cytochrome $bc1$ complex by the formation of an intersubunit disulfide bond between cytochrome b and the iron-sulfur protein. J Biol Chem 275:38597–38604

Yu CA, Gu LQ, Lin YZ, Yu L (1985) Effect of alkyl side chain variation on the electron-transfer activity of ubiquinone derivatives. Biochemistry 24:3897–3902

Yu CA, Yu L (1980) Structural role of phospholipids in ubiquinol-cytochrome c reductase. Biochemistry 19:5715–5720

Yu CA, Yu L, King TE (1974) Soluble cytochrome bc_1 complex and the reconstitution of succinate-cytochrome-c reductase. J Biol Chem 249:4905–4910

Yu L, Tso SC, Shenoy SK, Quinn BN, Xia D (1999) The role of the supernumerary subunit of *Rhodobacter sphaeroides* cytochrome $bc(1)$ complex. J Bioenerg Biomembranes 31:251–257

Zara V, Palmisano I, Conte L, Trumpower BL (2004) Further insights into the assembly of the yeast cytochrome $bc(1)$ complex based on analysis of single and double deletion mutants lacking supernumerary subunits and cytochrome b. Eur J Biochem 271:1209–1218

Zhang H, Osyczka A, Dutton PL, Moser CC (2007) Exposing the complex III Qo semiquinone radical. Biochim Biophys Acta 1767:883–887

Zhang L, Tai CH, Yu L, Yu CA (2000) pH-induced intramolecular electron transfer between the iron-sulfur protein and cytochrome $c(1)$ in bovine cytochrome $bc(1)$ complex. J Biol Chem 275:7656–7661

Zhang M, Mileykovskaya E, Dowhan W (2002) Gluing the respiratory chain together: cardiolipin is required for supercomplex formation in the inner mitochondrial membrane. J Biol Chem 277:43553–43556

Zhang ZL, Huang LS, Shulmeister VM, Chi YI, Kim KK, Hung LW, Crofts AR, Berry EA, Kim SH (1998) Electron transfer by domain movement in cytochrome bc_1. Nature 392:677–684

Zhu H, Egawa T, Yeh SR, Yu LD, Yu CA (2007) Simultaneous reduction of iron-sulfur protein and cytochrome b(L) during ubiquinol oxidation in cytochrome $bc(1)$ complex. Proc Natl Acad Sci USA 104:4864–4869

Zu YB, Couture MMJ, Kolling DRJ, Crofts AR, Eltis LD, Fee JA, Hirst J (2003) Reduction potentials of Rieske clusters: importance of the coupling between oxidation state and histidine protonation state. Biochemistry 42:12400–12408

Results Probl Cell Differ (45)
G. Schäfer, H. S. Penefsky: Bioenergetics
DOI 10.1007/400_2007_043/Published online: 17 November 2007
© Springer-Verlag Berlin Heidelberg 2007

Regulatory Mechanisms of Proton-Translocating F_OF_1-ATP Synthase

Boris A. Feniouk[1,2] (✉) · Masasuke Yoshida[2,3] (✉)

[1]ATP System Project, Exploratory Research for Advanced Technology, Japan Science and Technology Corporation (JST), 5800-3 Nagatsuta, Midori-ku, 226-0026 Yokohama, Japan

[2]Chemical Resources Laboratory, Tokyo Institute of Technology, 4259 Nagatsuta, Midori-ku, 226-8503 Yokohama, Japan

[3]ICORP ATP-Synthesis Regulation Project (Japanese Science and Technology Agency), National Museum of Emerging Science and Innovation, 2-41 Aomi, Koto-ku, 135-0064 Tokyo, Japan
feniouk@atp.miraikan.jst.go.jp, myoshida@res.titech.ac.jp

Abstract H^+-F_OF_1-ATP synthase catalyzes synthesis of ATP from ADP and inorganic phosphate using the energy of transmembrane electrochemical potential difference of proton ($\Delta\tilde{\mu}_{H^+}$). The enzyme can also generate this potential difference by working as an ATP-driven proton pump. Several regulatory mechanisms are known to suppress the ATPase activity of F_OF_1:

1. Non-competitive inhibition by MgADP, a feature shared by F_OF_1 from bacteria, chloroplasts and mitochondria.
2. Inhibition by subunit ε in chloroplast and bacterial enzyme
3. Inhibition upon oxidation of two cysteines in subunit γ in chloroplast F_OF_1
4. Inhibition by an additional regulatory protein (IF_1) in mitochondrial enzyme

In this review we summarize the information available on these regulatory mechanisms and discuss possible interplay between them.

1
Introduction

H^+-F_OF_1-ATP synthase (also known as F-type H^+-ATPase, or simply F_OF_1) is a multisubunit membrane enzyme. It synthesizes ATP from ADP and inorganic phosphate (P_i) using the energy of transmembrane electrochemical potential difference of proton ($\Delta\tilde{\mu}_{H^+}$). In Eukaryota the enzyme is found in mitochondrial inner membrane and in chloroplast thylakoid membrane; in bacteria F_OF_1 is located in the cytoplasmatic membrane.

The conditions under which the enzyme operates vary significantly between different organisms. In mitochondria the $\Delta\tilde{\mu}_{H^+}$ is constantly generated by respiratory chain enzymes and the chemical composition of the milieu on both sides of the coupling membrane is controlled by the cell, so the enzyme environment is more or less stable. In chloroplasts the $\Delta\tilde{\mu}_{H^+}$ is high during daytime, but during the night the membrane is de-energized so that no ATP synthesis is possible. The pH on both sides of the thylakoid membrane also varies during the day–night cycle (see (Kramer et al. 1999) and references therein). In bacteria, the conditions are most variable; the cell has a very limited control over the chemical composition of the milieu on the periplasmatic side of the membrane, and the magnitude of $\Delta\tilde{\mu}_{H^+}$ may vary significantly in response to such factors as concentrations of oxygen, nutrients, ions (pH), temperature, etc.

The need to regulate the activity of ATP synthase, primarily to avoid ATPase activity upon decrease in $\Delta\tilde{\mu}_{H^+}$ that may result in wasteful ATP hydrolysis, is evident. Indeed, there are several regulatory features present in $F_O F_1$. This review summarizes the experimental data on these regulatory features and describes how a common catalytic core of the enzyme was tuned to the specific needs of different organisms.

2
Structure and Rotary Catalysis: a Brief Summary

2.1
Structure

Before proceeding to the regulation of $F_O F_1$, it is necessary to briefly outline the main structural and functional features of the enzyme.

The enzyme is composed of two distinct portions: membrane-embedded F_O and hydrophilic F_1 that protrudes by more than 100 Å from the membrane plane. Both portions are multisubunit complexes. The F_1 portion is involved in nucleotide and P_i binding/release, while the F_O portion is responsible for trans-membrane proton transport. The two portions are connected by two "stalks", one of which is located approximately in the center, and the other is on the periphery of the enzyme (Fig. 2). The two portions can be separated (e.g. by sonication in the absence of Mg^{2+}) and reconstituted back. Isolated F_1 portion can hydrolyze ATP at high rate, and therefore is often named "F_1-ATPase"; isolated F_O portion performs passive proton transport downhill $\Delta\tilde{\mu}_{H^+}$.

The catalytic core of F_1 is capable of high rate ATP hydrolysis and is composed of three kinds of subunits in stoichiometry $\alpha_3\beta_3\gamma$. The structure was solved in 1994 for bovine enzyme by X-ray crystallography (Abrahams et al. 1994). Studies revealed that three $\alpha\beta$ pairs form a spherical hexamer with a cavity in the middle. The cavity is filled by part of the elongated γ subunit; the rest of subunit γ protrudes towards the membrane and composes the central stalk in $F_O F_1$. The primary structure of subunits α, β and γ is highly conserved in ATP synthases from various organisms. Biochemical data strongly indicate that the catalytic mechanism is also highly conserved.

There are six nucleotide-binding sites located in the clefts between subunits α and β (Abrahams et al. 1994). Only three of them are directly involved in catalysis (Cross and Nalin 1982; Yoshida and Allison 1986) and reside mostly on β subunits; the other three are located mostly on α subunits and are probably involved in regulation of the enzyme.

Besides the $\alpha_3\beta_3\gamma$, there are other smaller subunits in F_1. One of them (named ε in bacterial and chloroplast enzyme, but δ in the mitochondrial F_OF_1) is part of the central stalk connecting F_O and F_1, and is indispensable for coupling between proton transport and ATP synthesis/hydrolysis. In bacterial and chloroplast enzyme this subunit also has regulatory functions, which are discussed in detail below (for a recent review see Feniouk et al. 2006).

The functional core of the F_O portion is composed of a ring-shaped oligomer of c-subunits, and of a-subunit located on the periphery of the c-ring. Subunit c is a small hairpin-like protein with two transmembrane helices and a short hydrophilic loop connecting them. Proton transport occurs on the interface of subunit a with the c-ring. The central stalk connecting F_O and F_1 is composed of subunits γ and ε that are bound to the c-ring. The second, peripheral stalk is composed of other subunits; their number, stoichiometry, and nomenclature differs among bacterial, chloroplast, and mitochondrial enzymes. However, the structure itself is quite similar – a complex with transmembrane helices bound to subunit a; a protruding long α-helical stretch reaching the very distant part of F_1 and attached to the latter in part directly, and in part through an additional small F_1 subunit (δ in bacteria/chloroplasts and oligomycin sensitivity-conferring protein, OSCP, in mitochondria).

2.2
Catalytic Mechanism

An enormous contribution to our understanding of the ATP synthase catalytic mechanism was made by Paul Boyer and colleagues. They have demonstrated that the energy-requiring step was not the chemical step of ATP synthesis, but the binding of P_i and the release of the tightly bound ATP from the enzyme (Boyer et al. 1973). Later they found that F_OF_1 showed a strong dependence of catalytic events and product(s) release at one site on the binding of substrate(s) at a second site (Kayalar et al. 1977). This general principle of highly cooperative multisite catalysis was later confirmed by lots of functional and structural evidence and is usually referred to as "binding change mechanism" (see Boyer 1997, 2002; Senior et al. 2002, and the references therein for details).

The molecular implementation of the binding change mechanism in F_OF_1 involves rotation of subunit γ inside the $\alpha_3\beta_3$ hexamer. Such a rotary mechanism was predicted from the structural data (Abrahams et al. 1994) and later got support from the biochemical (Duncan et al. 1995) and biophysical (Sabbert et al. 1996) studies. Finally, ATP-driven rotation of subunit γ was directly visualized in the $\alpha_3\beta_3\gamma$ complex from *Bacillus* PS3 in single-molecule experiments (Noji et al. 1997). More single molecule data followed, demonstrating

ATP-driven rotation in F_OF_1 that was sensitive to the F_O-inhibitor tributyltin (Ueno et al. 2005), and ATP synthesis driven by mechanical rotation of subunit γ in immobilized F_1 (Itoh et al. 2004; Rondelez et al. 2005). The results of single-molecule FRET experiments with *E. coli* F_OF_1 incorporated into liposomes suggested that rotation of subunit γ also occurs during ATP synthesis driven by artificially imposed $\Delta\tilde{\mu}_{H^+}$ (Diez et al. 2004; Zimmermann et al. 2005).

Combination of the data from single-molecule experiments with structural information from X-ray crystallographic studies allowed reconstruction of a rather detailed molecular mechanism of ATP hydrolysis in isolated F_1. Hydrolysis of one ATP molecule drives a 120°-unit rotation of subunit γ and, therefore, hydrolysis of three ATP molecules is required for the one complete 360° revolution (Yasuda et al. 1998). Analysis of rotation with a high speed camera (Yasuda et al. 1998; Shimabukuro et al. 2003), a slow-hydrolysis mutant F_1 (Shimabukuro et al. 2003; Nishizaka et al. 2004), and direct observation of binding/release of fluorescently labeled nucleotide during rotation (Nishizaka et al. 2004) suggest the following reaction sequence as a plausible model (Fig. 1; see Adachi et al. 2007; Ariga et al. 2007 for more details). Three β subunits are designated as β_I, β_{II}, and β_{III}. When F_1 is waiting for ATP, it is assumed that the catalytic sites of β_I, β_{II}, and β_{III} contain none, ATP, and ADP/P_i, respectively (states I and V in Fig. 1). The angular position of the subunit γ in this state is set to be 0°:

1. ATP binds to an empty catalytic site of β_I (Fig. 1, transition I → II).
2. Binding induces an 80° rotation of subunit γ. This rotation leads to simultaneous release of ADP from the catalytic site of β_{III} (Fig. 1, transition II→III).
3. Two catalytic events, each with a lifetime of ~1 ms, occur at the 80° position. One of these is hydrolytic cleavage of ATP into ADP and P_i at a catalytic site of β_{II} (state III in Fig. 1). The other event is not known but we assume it to be P_i release from β_{III} (state IV in Fig. 1). The order of the two events is not determined (in Fig. 1 ATP hydrolysis precedes P_i release, but the opposite event sequence is also probable).
4. A 40° rotation occurs to complete one 120° rotation (Fig. 1, transition IV → V). ATP binds to the newly emptied catalytic site of β_{III}, and the cycle repeats.

In this model, all three β subunits participate to drive a 120° rotation (active β subunits are marked as filled in the cartoon representation of Fig. 1), and catalytic turnover of one particular ATP molecule needs 360° rotation; the events on β_I are ATP-binding at 0°, ATP-cleavage at 200°, ADP-release at 240°–320°, and P_i release at 320°. Recent crystal structure of yeast F_1 with two catalytic sites occupied by AMPPNP and one occupied with P_i (Kabaleeswaran et al. 2006) may represent state III in Fig. 1, blocked on the level of ATP hydrolysis in β_{II}.

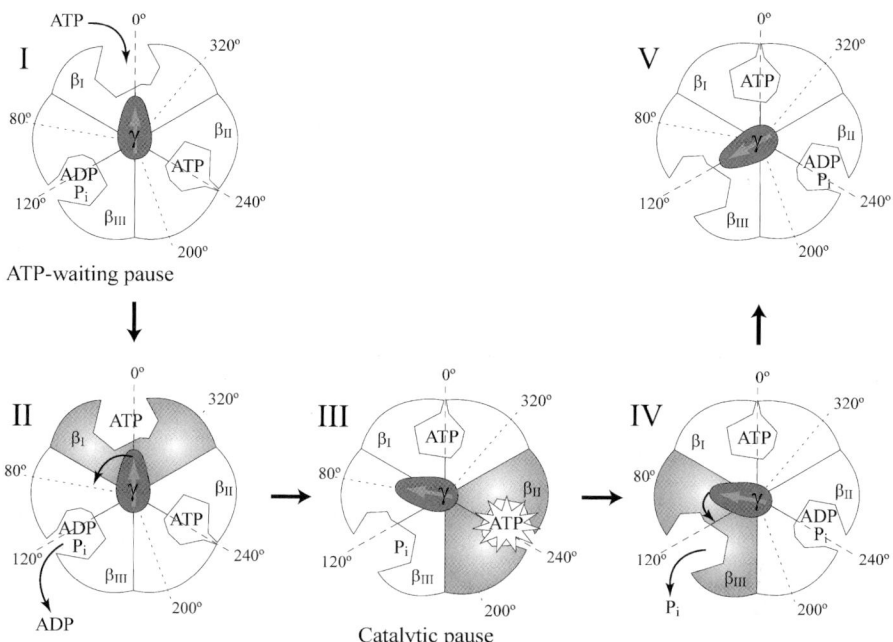

Fig. 1 Hypothetical catalytic mechanism of rotary ATP hydrolysis. F_1 is depicted as seen from the membrane; only the three catalytic nucleotide binding sites are shown. The *filled* αβ-pairs represent the power stroke step that presumably drives subunit γ rotation. Only 1/3 of full γ subunit revolution corresponding to hydrolysis of one particular ATP molecule is shown; state *V* is identical to state *I* (just rotated by 120°). See details in text (Sect. 2.2)

In the whole F_OF_1, subunit γ is bound to the ring-shaped oligomer of *c*-subunits. In the case of ATP synthesis the proton flow driven by $\Delta\tilde{\mu}_{H^+}$ powers the rotation of the *c*-ring with subunit γ (and with subunit ε in bacterial and chloroplast F_OF_1, or with δε complex in the mitochondrial F_OF_1) relative to other subunits. This rotation induces the cyclic conformational changes of the catalytic sites on F_1 that result in ATP synthesis. Although hypothetical mechanisms of proton translocation and torque generation by F_O were proposed (Junge et al. 1997; Vik et al. 1998), the experimental evidence supporting them is still insufficient. It is likely that F_O operates as an entropic machine, as proposed by Junge and collaborators (Junge et al. 1997). This model and its later modifications (Dimroth et al. 1998; Elston et al. 1998) correspond well to the experimental data. A detailed study on *Rhodobacter capsulatus* membranes confirmed that the rotary model can quantitatively describe the proton transport through isolated F_O (Feniouk et al. 2004).

The coupling between the F_O and F_1 is rather tight. For example, DCCD (*N,N*-dicyclohexylcarbodiimide), a specific inhibitor of F_O, blocks >75% ATPase activity of F_OF_1 from *E. coli* (Fillingame 1975) or *Bacillus* PS3 (Suzuki et al. 2002); an even higher degree of inhibition is observed in other organisms. No detectable proton leak was observed through Rb. capsulatus F$_O$F$_1$ in the presence of $\Delta\tilde{\mu}_{H^+}$ under conditions where the F$_1$ portion was blocked, e.g., by specific inhibitors (Feniouk et al. 2001) or in the absence of nucleotides in the medium (Feniouk et al. 2005).

Such tight coupling ensures that factors affecting the proton transport function of the enzyme also affect the ATP synthesis/hydrolysis and vice versa.

3
ADP-Inhibition: a Common Regulatory Mechanism

As mentioned above, ATP synthase is capable of both $\Delta\tilde{\mu}_{H^+}$-driven ATP synthesis and ATP-driven $\Delta\tilde{\mu}_{H^+}$ generation. In mitochondria, chloroplasts, and aerobic/photosynthetic bacteria the former activity is primary (but see Matsuyama et al. 1998; St Pierre et al. 2000; Lefebvre-Legendre et al. 2003, for

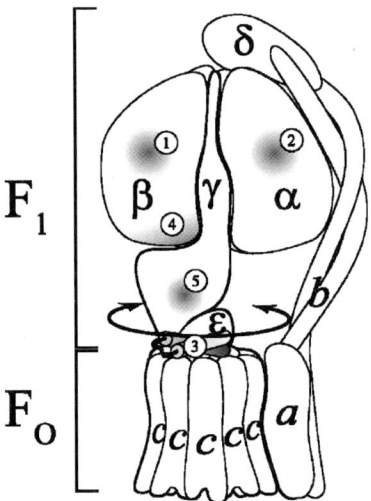

Fig. 2 Cartoon representation of bacterial/chloroplast F$_O$F$_1$. Zones involved in regulation are marked:
1. Catalytic sites occlude MgADP without P$_i$ and the enzyme lapses into ADP-inhibited state (Sect. 3)
2. Binding of ATP or pyrophosphate to non-catalytic sites counteracts ADP-inhibition (Sect. 3.2)
3. Subunit ε C-terminal α-helical domain is responsible for inhibition of ATPase activity (Sect. 4)
4. Acid residues of βDELSEED are involved in inhibition exerted by subunit ε C-terminal domain (Sect. 4)
5. In chloroplast F$_O$F$_1$ oxidation/reduction of a special cysteine pair modulates the enzyme activity (Sect. 5)

the importance of the reverse activity in mitochondria). The universal way to modulate the ATP synthesis activity is by changing the magnitude of $\Delta\tilde{\mu}_{H^+}$. It is well documented that $\Delta\tilde{\mu}_{H^+}$ above a certain thermodynamic threshold is necessary for ATP synthesis, and that further increase in $\Delta\tilde{\mu}_{H^+}$ results in acceleration of ATP production (Graber and Witt 1976; Slooten and Vandenbranden 1989; Junesch and Graber 1991; Turina et al. 1991; Pitard et al. 1996). Therefore, regulation of ATP synthesis activity can be achieved via regulation of $\Delta\tilde{\mu}_{H^+}$ magnitude either by modulation of respiratory/photosynthetic $\Delta\tilde{\mu}_{H^+}$-generating protein complexes or by changing the proton permeability of the membrane.

There are several regulatory mechanisms controlling ATP hydrolysis (Fig. 2); most of them are aimed at blocking the ATPase activity of F_OF_1 upon decrease in $\Delta\tilde{\mu}_{H^+}$, decrease in ATP concentration, or decrease in the ATP/ADP ratio. This is hardly surprising for aerobic/photosynthetic organisms, where such mechanisms are essential to protect the cellular ATP pool from wasting upon membrane de-energization. However, in many bacteria the primary function of F_OF_1 is ATP-driven proton pumping that provides $\Delta\tilde{\mu}_{H^+}$ necessary for ion transport, flagella rotation, and other vital processes. Nevertheless, certain regulatory features limiting the ATPase activity of F_OF_1 are present in these organisms as well.

3.1
Mechanism of ADP-Inhibition

One of the most well-known unidirectional regulatory factors influencing the activity of F_OF_1 is MgADP: it not only serves as a substrate for ATP synthesis, but also inhibits ATPase activity of the enzyme in a non-competitive manner. Such inhibition (denoted hereafter as "ADP-inhibition") is described for F_OF_1 from chloroplasts (Carmeli and Lifshitz 1972; Dunham and Selman 1981b; Feldman and Boyer 1985; Zhou et al. 1988; Creczynski-Pasa and Graber 1994), mitochondria (Minkov et al. 1979; Fitin et al. 1979; Roveri et al. 1980; Drobinskaya et al. 1985), and bacteria (Yoshida and Allison 1983; Hyndman et al. 1994), and is clearly distinct from simple product inhibition. It is observed not only in the whole enzyme or F_1-portion, but also in the $\alpha_3\beta_3\gamma$ complex (Jault et al. 1995; Hirono-Hara et al. 2001), indicating that this regulatory feature is embedded in the very catalytic core of F_1.

Numerous biochemical studies indicate that ADP-inhibition is caused by tight binding of MgADP without P_i at a high-affinity catalytic site (Minkov et al. 1979; Fitin et al. 1979; Smith et al. 1983; Drobinskaya et al. 1985; Milgrom and Boyer 1990; Hyndman et al. 1994). It is noteworthy that the presence of ADP without P_i in the tight binding catalytic site is not inhibitory by itself, but is a prerequisite for slow transition into the ADP-inhibited state, which probably includes an additional conformational change that is affected by Mg^{2+} (Bulygin and Vinogradov 1991).

Single-molecule experiments on $\alpha_3\beta_3\gamma$ complex from *Bacillus* PS3 revealed that ADP-inhibition results in long pauses in ATP-driven rotation of subunit γ (Hirono-Hara et al. 2001). These pauses occur with subunit γ blocked in the angular position of 80° relative to the "ATP-waiting" state. Spontaneous re-activation occurs in the tens of seconds time scale, but was completely abolished if the angular position of subunit γ was fixed at 80° by external forces. Therefore, it was proposed that spontaneous activation is due to stochastic rotational fluctuations of subunit γ. This proposal was strongly supported by the finding that forced rotation of subunit γ by > 40° in the hydrolysis direction relieved ADP-inhibition (Hirono-Hara et al. 2005). Numerous experimental studies on F_OF_1 from various organisms demonstrated that the tightly bound inhibitory ADP can be expelled by $\Delta\tilde{\mu}_{H^+}$ (Strotmann et al. 1976; Graber et al. 1977; Shoshan and Selman 1979; Sherman and Wimmer 1984; Creczynski-Pasa and Graber 1994; Feniouk et al. 2005). This phenomenon underlies the so-called "activation by $\Delta\tilde{\mu}_{H^+}$", or increase in the ATPase activity of the enzyme after brief membrane energization (Carmeli and Lifshitz 1972; Baltscheffsky and Lundin 1979; Turina et al. 1992; Galkin and Vinogradov 1999; Fischer et al. 2000; Zharova and Vinogradov 2004). In view of the single-molecule data, it is conceivable that such activation is caused by $\Delta\tilde{\mu}_{H^+}$-driven rotation of the γ subunit (see below for a detailed discussion).

ADP-inhibition is likely to be a common feature of all ATP synthases. However, there are many factors that influence ADP-inhibition. As a result, the ATPase activity of F_OF_1 is finely regulated to match the needs of different cells at various physiological conditions.

3.2
Factors Affecting ADP-Inhibition

Phosphate

The role of P_i in the regulation of F_OF_1, as well as the details of P_i binding/release during catalysis, has many unclear aspects. In a pioneer study on P_i binding it was revealed that the mitochondrial F_1 (with ADP bound at a catalytic site and two nucleotides in the non-catalytic sites) reversibly binds a single P_i anion with a high affinity (K_d of 80 µM) (Penefsky 1977). Many factors such as pH, Mg^{2+}, inorganic anions, and nucleotides affected the binding. It was also documented that nucleotide-free mitochondrial F_1 binds P_i poorly, and that binding of P_i requires the presence of tightly bound ADP in the same catalytic site (Kozlov and Vulfson 1985).

There are two points concerning the data above. First, during normal catalysis P_i is likely to be bound/released at an open, not high affinity catalytic site. Second, in a living cell the enzyme is always in the medium with a millimolar concentration of nucleotides. Therefore, the measurements of P_i binding in the presence of ADP and ATP are more physiologically relevant. In the case of mitochondrial F_1, 150 μM of each nucleotide inhibited the high affinity P_i binding by approximately 50% (Penefsky 1977). It is worthy of note that non-hydrolyzable ATP analog AMP-PNP was a markedly stronger inhibitor of P_i binding, confirming that P_i was bound in the position where the γ-phosphate of ATP resides.

However, a detailed study of mitochondrial F_1 revealed that there is a second binding site for P_i with K_d of ~5 mM (Kasahara and Penefsky 1978). Recently Penefsky confirmed that *E. coli* F_1 also has two P_i-binding sites with K_d in the range of 0.1 mM (Penefsky 2005). This result contradicts the earlier failure to observe P_i binding to *E. coli* enzyme (al Shawi and Senior 1992) and was supposedly due to a rapid dissociation of the bound P_i during the the centrifuge column separation procedure. Studies of chloroplast F_OF_1 incorporated into liposomes also provided evidence for existence of two P_i binding sites on the enzyme (Grotjohann and Graber 2002).

Until recently the high resolution structures of F_1 solved by X-ray crystallography have not revealed any bound P_i. However, a short time ago Walker's group solved the X-ray structure of yeast F_1 that has a phosphate (or sulfate) bound at an "empty" catalytic site. The location of the anion is close to the expected position of ATP γ-phosphate, indicating that P_i might be bound in the empty catalytic site (Kabaleeswaran et al. 2006).

As a substrate of ATP synthesis, P_i was demonstrated to have K_m in the range 0.2–10 mM in enzymes from various sources (Kayalar et al. 1976; Hatefi et al. 1982; McCarthy and Ferguson 1983; Junge 1987; Strotmann et al. 1990; Perez and Ferguson 1990a,b; Richard et al. 1995; al Shawi et al. 1997; Etzold et al. 1997; Grotjohann and Graber 2002; Tomashek et al. 2003). However, P_i in millimolar concentrations does not significantly inhibit the ATPase activity of the enzyme, suggesting that the affinity to P_i is different for ATP synthesis and for uncoupled ATP hydrolysis. Indeed, the affinity of F_OF_1 to P_i is strongly enhanced in the presence of $\Delta\bar{\mu}_{H^+}$ (Kayalar et al. 1976; Hatefi et al. 1982; McCarthy and Ferguson 1983; al Shawi et al. 1997), in line with the suggestion of Boyer et al. that binding of P_i is one of the main energy-requiring steps during ATP synthesis (Rosing et al. 1977; Rosen et al. 1979).

Interestingly, many experimental studies documented a higher ATPase activity of F_OF_1 in the presence of P_i (Carmeli and Lifshitz, 1972; Melandri et al. 1975; Moyle and Mitchell 1975; Dunham and Selman 1981a; Turina et al. 1992; Zharova and Vinogradov 2004). A pioneering study by Carmeli and Lifshitz on chloroplast F_OF_1 provided evidence that such an increase occurs because P_i counteracts ADP-inhibition (Carmeli and Lifshitz 1972). Later, it was found that P_i also relieves ADP-inhibition in isolated mitochondrial (Drobinskaya et al. 1985; Kalashnikova et al. 1988) and bacterial (Bald et al. 1999; Mitome et al. 2002) F_1, although the concentration of P_i necessary to relieve inhibition was rather high: > 20 mM for *Bacillus* PS3 (Mitome et al. 2002) and > 5 mM for the mitochondrial F_1 (Drobinskaya et al. 1985).

The mechanism of such inhibition relief is not completely clear. It is likely that the presence of P_i in the same site where ADP is bound prevents conformational transition to the ADP-inhibited state. Indeed, it has been demonstrated that in the high-affinity catalytic site ATP is in equilibrium with ADP+P_i, so if P_i can bind to the high-affinity site having ADP, it is expected to keep the enzyme in the active state.

It should be noted that the experimental evidence available is insufficient to determine if P_i can facilitate the re-activation of the enzyme once it has lapsed into ADP-inhibited form, or if P_i only prevents ADP-inhibition of the active enzyme. We find the latter possibility more likely, since in the case of mitochondrial F_1 the P_i concentration necessary to relieve ADP-inhibition (5 mM) matched the experimentally estimated affinity of the second P_i-binding site (Kasahara and Penefsky 1978), which is distinct from the high-affinity catalytic site.

Binding of Nucleotides or Pyrophosphate to Non-catalytic Sites

As mentioned above, there are six nucleotide-binding sites on F_1. Three of them can rapidly exchange nucleotides with the medium, while the other three exhibit slow nucleotide exchange rates, and were named "non-catalytic sites" (Cross and Nalin 1982). The details of nucleotide/pyrophosphate binding to the non-catalytic sites are not completely clear. Early studies have revealed that in mitochondrial F_1 all three non-catalytic sites can be occupied with ATP (Kironde and Cross 1987). The crystal structure confirmed this finding showing AMP-PNP (an ATP analog) in all non-catalytic sites (Abrahams et al. 1994). Experiments with chloroplast F_1 (activated by heat treatment at 60 °C, since the non-activated chloroplast F_1 has almost no ATPase activity) also indicated that all three sites can be filled with ATP, but that ADP is able to fill only two (Milgrom et al. 1991). Several other studies have pointed out that the three non-catalytic sites differ in their binding properties. Experiments with nucleotide-depleted *E. coli* enzyme indicated that F_1 binds a maximum of two ATP, ADP, or GTP molecules at non-catalytic sites, whereas all three sites can be occupied only by a mixture of nucleotide di- and triphosphates (Hyndman et al. 1994). However, a study by Weber and coworkers on the mutant *E. coli* F_1 yielded occupancy of 2.8 and 2.6 non-catalytic sites by MgATP and MgADP, respectively (Weber et al. 1994). In chloroplast F_1 that was not heat-treated, one non-catalytic site was found to tightly bind ADP, while the other two could bind both ADP and ATP, albeit with different affinities (Malyan and Allison 2002). The dissociation of ADP from the latter two sites was much faster than that of ATP.

In chloroplasts, binding of F_1 to F_O was demonstrated to significantly modify the nucleotide occupancy of the non-catalytic sites, decreasing the ATP/ADP ratio for bound nucleotides (Malyan, 2006). Magnesium ions were also found to influence the nucleotide binding to the non-catalytic sites (Weber et al. 1994; Malyan 2005).

Experimental studies revealed that the occupancy of the non-catalytic sites has a marked effect on the activity of F_OF_1. It was demonstrated on isolated F_1 from mitochondria, chloroplasts, and *Bacillus* PS3 that binding of ATP to these sites stimulates the ATPase activity of the enzyme (Milgrom et al. 1990; Jault and Allison 1993; Jault et al. 1995). This stimulation is due to attenuation of ADP-inhibition: binding of ATP to the non-catalytic sites facilitates the release of the inhibitory ADP from the high-affinity catalytic site (Murataliev and Boyer 1992; Milgrom and Cross 1993; Jault et al. 1995). Binding of pyrophosphate to the non-catalytic sites has a similar effect (Kalashnikova et al. 1988; Jault et al. 1994). In contrast to ATP and pyrophosphate, ADP was demonstrated to promote hysteretic inhibition of mitochondrial F_1 when bound to non-catalytic sites, presumably by blocking the binding of ATP to these sites and thereby preventing the activation mentioned above (Jault and Allison 1994).

$\Delta\tilde{\mu}_{H^+}$

Corresponding to thermodynamic considerations, in well-coupled membranes $\Delta\tilde{\mu}_{H^+}$ acts as a back-pressure that limits the rate of ATP hydrolysis catalyzed by F_OF_1. This effect is documented in many experimental studies demonstrating stimulation of ATPase activity by uncouplers. But, $\Delta\tilde{\mu}_{H^+}$ is also known to stimulate ATP hydrolysis by F_OF_1. This phenomenon was first documented in chloroplasts, where the enzyme has only traces of ATPase activity (albeit competent in ATP synthesis) (Jagendorf and Avron 1958; Avron and Jagendorf 1959), but can be activated by $\Delta\tilde{\mu}_{H^+}$ (Kaplan et al. 1967; Schwartz 1968; Carmeli and Avron 1972; Bakker-Grunwald and Van Dam 1974; Smith et al. 1976; Komatsu-Takaki 1986). A similar increase in the ATPase activity induced by $\Delta\tilde{\mu}_{H^+}$ was also documented for mitochondrial and bacterial enzymes (Turina et al. 1992; Galkin and Vinogradov 1999; Fischer et al. 2000; Pacheco-Moises et al. 2000; Zharova and Vinogradov 2004).

Stimulation of F_OF_1 ATPase activity by $\Delta\tilde{\mu}_{H^+}$ combines two distinct phenomena. First, $\Delta\tilde{\mu}_{H^+}$ promotes the release of the tightly bound ADP from the enzyme (Strotmann et al. 1976; Graber et al. 1977; Sherman and Wimmer 1984; Feniouk et al. 2005) and therefore relieves ADP-inhibition (Sherman and Wimmer 1984; Zharova and Vinogradov 2004). In view of the single-molecule experiments described in Sect. 3.1, it is highly conceivable that the enzyme is relieved from ADP-inhibition by $\Delta\tilde{\mu}_{H^+}$-powered rotation of subunit γ. Second, the steady-state ATPase activity is also stimulated by $\Delta\tilde{\mu}_{H^+}$ (Turina et al. 1992; Zharova and Vinogradov 2004; Feniouk et al. 2007), although this phenomenon is partially masked by suppression of ATP hydrolysis by $\Delta\tilde{\mu}_{H^+}$ back-pressure. Interestingly, the latter stimulation (unlike

$\Delta\tilde{\mu}_{H^+}$-driven release of inhibitory ADP) is observed only in the presence of P_i (Zharova and Vinogradov 2004; Feniouk et al. 2007). In a recent study we investigated this phenomenon and found that the P_i-dependent stimulation of the steady-state ATPase activity by $\Delta\tilde{\mu}_{H^+}$ in F_OF_1 from *Bacillus* PS3 is due to relief of ADP-inhibition (Feniouk et al. 2007). It is likely that such stimulation occurs because $\Delta\tilde{\mu}_{H^+}$ induces an increase in the affinity of F_OF_1 to P_i (Kayalar et al. 1976; Hatefi et al. 1982; McCarthy and Ferguson 1983; al Shawi et al. 1997). In turn, P_i binding protects the enzyme from ADP-inhibition, as described above. A scheme illustrating such regulatory interplay between ADP-inhibition, $\Delta\tilde{\mu}_{H^+}$, and P_i (and other factors discussed below) is presented in Fig. 3.

As already mentioned, a prerequisite for ADP-inhibition is ADP bound at a high-affinity catalytic site without P_i (D-state in Fig. 3). Because the order of ATP hydrolysis product release is unclear, we include both possible pathways for ADP and P_i liberation from a catalytic site: **DP** → **D** → **O** and **DP** → **P** → **O**. In the latter pathway ADP-inhibition requires binding of ADP to the opened site, since the D-state does not occur.

A high P_i concentration or increased affinity of the enzyme to P_i caused by $\Delta\tilde{\mu}_{H^+}$ can increase the rate of ATP hydrolysis by increasing the probability of the **DP** → **P** → **O** transition that excludes transition to the **DI** state. P_i binding to the **D** state (in the case of both high and low affinity catalytic sites) is expected to accelerate the **D** → **DP** transition and therefore also prevent the enzyme from lapsing into the **DI** state.

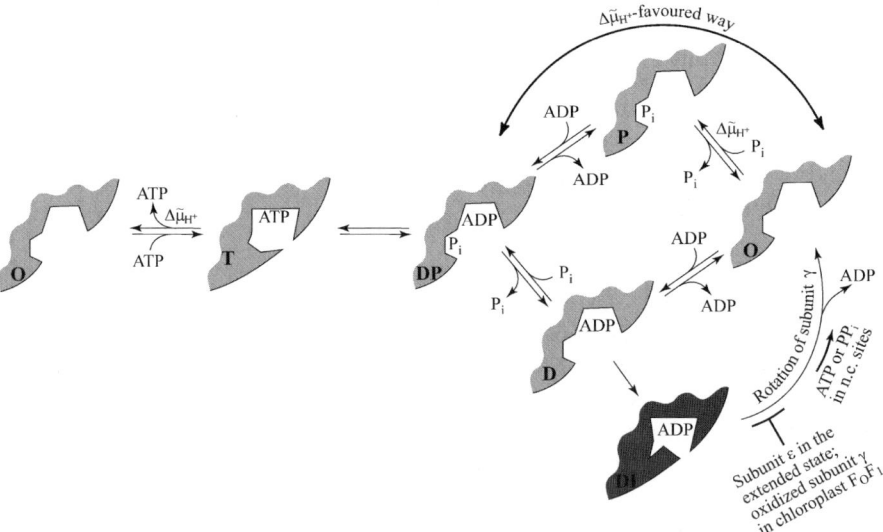

Fig. 3 Scheme of ATP hydrolysis regulation for bacterial/chloroplast F_OF_1 (extended from Feniouk et al. 2007) See text for details (Sect. 3.2)

If ATP is the nucleotide bound to the empty site after release of ADP and P_i, then ATP hydrolysis proceeds. However, binding of ADP (O → D transition) might lead to ADP inhibition. Again, a high concentration of P_i or an increased affinity to P_i diminishes the probability of the O → D-transition (and therefore, of ADP inhibition) by biasing the reaction towards the O → P-transition.

It is probable that ATP or pyrophosphate binding to the non-catalytic (n.c.) sites might destabilize the ADP-inhibited state. Structurally such destabilization might be achieved by facilitating the rotation of subunit γ inside the $\alpha_3\beta_3$ hexamer. Further studies are necessary to clarify this point.

Extending this rationale, one could presume that factors stabilizing the angular position of subunit γ corresponding to the ADP-inhibited state would enhance ADP-inhibition. Below we discuss such factors in detail.

Subunit ϵ (in Bacterial and Chloroplast F_0F_1)

It was proposed by Feniouk and Junge that in the bacterial and chloroplast F_0F_1 the ADP-inhibition might be enhanced by subunit ε (Feniouk and Junge 2005), which is part of the central stalk in F_0F_1 (see below for details). Single-molecule experiments on cyanobacterial F_1 confirmed that subunit ε blocks the rotation of subunit γ at the same angular position as ADP inhibition does (Konno et al. 2006). Biochemical studies on the F_0F_1 from *Bacillus* PS3 also confirmed that ADP-inhibition is enhanced by ε, presumably because the latter subunit stabilizes the ADP-inhibited state (Feniouk et al. 2007). However, subunit ε affects the ATPase activity of F_0F_1 also in the absence of ADP, so we have summarized the data on the inhibitory role of this subunit in Sect. 4.

4
Subunit ϵ in Bacterial and Chloroplast Enzyme

4.1
Structure of Subunit ϵ

Subunit ε (subunit δ in mitochondrial F_0F_1) is a small protein consisting of the N-terminal β-sandwich domain and the C-terminal domain composed of two α-helices. Structural NMR and X-ray studies revealed that in *E. coli* subunit ε the two C-terminal helices form a hairpin (Wilkens et al. 1995; Uhlin et al. 1997). The location of the subunit within F_1 was also determined in a high-resolution X-ray structure of bovine mitochondrial F_1 (Gibbons et al. 2000). The latter structure demonstrated a striking similarity in the fold of *E. coli* subunit ε and its homolog in bovine mitochondrial F_0F_1 (subunit δ).

Subunit ε plays a dual role in F_0F_1 from bacteria and chloroplasts (for reviews see Capaldi and Schulenberg 2000; Vik 2000; Feniouk et al.

2006). On one hand, subunit ε is indispensable for coupling between proton translocation though F_O and ATP synthesis/hydrolysis in F_1. On the other hand, subunit ε has a regulatory role inhibiting the ATPase activity of the enzyme. These two functions are structurally separated: the N-terminal β-sandwich domain is responsible for the coupling function, while the C-terminal α-helical domain is responsible for inhibition of ATP hydrolysis (but see Cipriano and Dunn 2006, for some evidence on the influence of the C-terminal domain on coupling efficiency in *E. coli* F_OF_1). In this review we discuss only the inhibitory function of subunit ε. We would also like to emphasize that there is no sound evidence for a similar regulatory role of mitochondrial F_OF_1 subunit δ (homologous to the bacterial/chloroplast ε). It is therefore likely that this regulatory feature is present exclusively in bacterial and chloroplast F_OF_1.

4.2
Inhibition of ATP Hydrolysis by Subunit ϵ

In 1972 Nelson et al. reported that subunit ε inhibits ATP hydrolysis in chloroplast F_1 (Nelson et al. 1972). Later, a similar inhibitory effect was documented (Smith et al. 1975) and studied in detail (Smith and Sternweis 1977; Laget and Smith 1979) on *E. coli* F_1. The possibility of performing mutagenesis makes bacteria a powerful experimental system for studies of protein function, and most of the data on subunit ε inhibitory role come from studies on *E. coli* or *Bacillus* PS3 F_OF_1.

It was revealed that the inhibitory effect of bacterial subunit ε is lost upon truncation of its C-terminal domain ($\varepsilon^{\Delta C}$-mutant) (Kuki et al. 1988; Keis et al. 2006; Cipriano and Dunn 2006). However, the details of the inhibitory effect vary among different species. In *E. coli* $F_OF_1\varepsilon^{\Delta C}$ mutation leads to 1.5-fold increase in the ATP hydrolysis rate, and the inhibitory effect is constant in the ATP concentration range from 50 μM to 5 mM (Cipriano and Dunn 2006). Markedly stronger stimulation was observed in $\varepsilon^{\Delta C}$-mutant enzyme from *Bacillus* PS3 (Kato-Yamada et al. 1999): at 50 μM ATP the activity is more than fourfold higher in the mutant. However, at 2 mM ATP the steady-state activity was the same in the $\varepsilon^{\Delta C}$-mutant and in the wild-type enzyme (but the initial lag in the onset of ATPase activity present in the wild type was lacking in the mutant). In F_OF_1 from thermoalkaliphilic *Bacillus* TA2.A1 the inhibition was also dependent on ATP concentration and decreased from a factor of seven at 50 μM ATP to ∼three at 2 mM ATP (Keis et al. 2006). These findings indicate that there is a pronounced difference between the inhibitory effects of subunit ε in different bacteria.

In chloroplast enzyme the inhibitory effect of subunit ε C-terminal domain is very strong: at 5 mM ATP the ATPase activity of $\varepsilon^{\Delta C}$-F_OF_1 in spinach thylakoids was more than sixfold higher than that of the wild-type enzyme (Nowak and McCarty 2004).

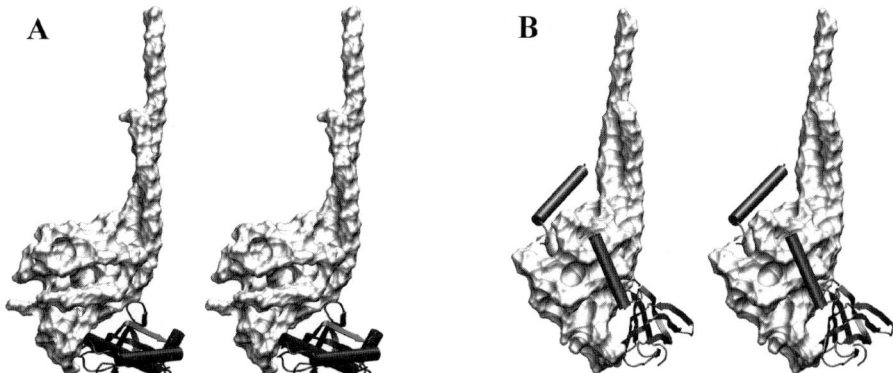

Fig. 4 Two conformations of bacterial F_OF_1 subunit ε C-terminal domain (stereopairs): *A* Contracted hairpin state (bovine mitochondrial F_1, coordinates from PDB entry 1E79). *B* Extended state (*E. coli* F_1, PDB entry 1JNV). The backbone of subunit γ is shown in surface representation (colored *light gray*); subunit ε (mitochondrial δ) is shown in cartoon representation (colored *dark gray*). The image was generated with VMD software package (Humphrey et al. 1996)

4.3
Conformational Transitions of Subunit ε C-Terminal Domain

An important advance in the understanding of the molecular mechanism of the inhibitory effect of subunit ε was initiated by a publication reporting the structure of the γε complex from *E. coli* F_OF_1 (Rodgers and Wilce 2000). In this structure the α-helices of subunit ε C-terminus were not folded in a hairpin (A in Fig. 4), but were stretched along subunit γ towards the $α_3β_3$ hexamer (B in Fig. 4). The existence of such conformation in the whole F_OF_1 was later confirmed by Tsunoda et al. in cross-linking experiments (Tsunoda et al. 2001).

Similar cross-linking experiments performed in our group demonstrated that in *Bacillus* PS3 F_OF_1 the C-terminus of subunit ε can be stretched even further, reaching the N-terminus of subunit γ (Suzuki et al. 2003). Moreover, it was revealed that in the mutant where both the extended and the contracted hairpin conformations of subunit ε C-terminus could be fixed by a cross-link, the extended conformation prevailed in the absence of ATP, while the contracted conformation was induced by ATP. Functional studies of the mutants with one of the ε conformations fixed by a cross-link revealed that subunit ε in the extended conformation inhibited the ATPase activity of F_OF_1 but had no significant effect on ATP synthesis (Suzuki et al. 2003), in agreement with the results obtained on *E. coli* F_OF_1 (Tsunoda et al. 2001). In the contracted hairpin conformation subunit ε had no effect on either activity (Suzuki et al. 2003). This result explained the earlier data indicating two distinct states of *Bacillus* PS3 subunit ε, of which only one was inhibiting ATP hydrolysis (Kato et al. 1997).

Experiments on fluorescence resonance energy transfer between labels introduced in *Bacillus* PS3 F_1 on the N-terminus of subunit γ and on the C-terminus of subunit ε confirmed that the transition from extended to contracted state is induced by ATP and correlates with the increase in the ATPase activity (Iino et al. 2005).

The findings described above indicate that in *Bacillus* PS3 F_OF_1 subunit ε might play a regulatory role, and that the molecular mechanism of the regulation involves large conformational transitions of the C-terminal α-helical domain triggered by ATP. Although no sound evidence on similar transitions in F_OF_1 from other organisms has been published, there are several studies reporting conformational changes of subunit ε in response to nucleotides, P_i, and $\Delta\tilde{\mu}_{H^+}$ in the *E. coli* enzyme (Mendel-Hartvig and Capaldi 1991; Wilkens and Capaldi 1994; Aggeler and Capaldi 1996). $\Delta\tilde{\mu}_{H^+}$-induced changes in subunit ε conformations are also reported for chloroplast F_OF_1 (Richter and McCarty 1987; Komatsu-Takaki 1989; Nowak and McCarty 2004).

4.4
The Role of βDESLEED Region in Inhibition Mediated by Subunit ε

The demonstration of conformational transitions of the ε C-terminus does not provide information on the interactions responsible for the inhibitory effect. The latter issue was partially clarified by a study in our group demonstrating that in *Bacillus* PS3 F_OF_1 the inhibitory effect of ε was dependent on the presence of basic, positively charged residues on the second C-terminal α-helix of subunit ε and of the negatively charged acid residues in the DELSDED[1] segment of subunit β (Hara et al. 2001). Alanine replacements of either basic residues in the ε C-terminus or acidic residues in the βDELSDED segment led to a dramatic decrease of the inhibitory effect. The same effect of alanine replacements in subunit ε was reported in a recent study on *Bacillus* TA2.A1 F_OF_1 (Keis et al. 2006). It should be noted that in *E. coli* F_OF_1 the replacement of the first glutamate in the βDELSEED to cysteine also led to a marked increase in the ATPase activity (Garcia and Capaldi 1998). It is tempting to speculate that interactions of the βDELSEED segment with the C-terminal domain of subunit ε is a common inhibitory mechanism in bacterial and probably chloroplast F_OF_1.

In support of the latter suggestion, a marked decrease in the inhibitory effect of ε was observed in chloroplast enzyme upon truncation of the tenth C-terminal residue (Arg, marked bold in the sequence motif below), while the truncation of the previous nine (non-basic) residues had a much weaker effect (Shi et al. 2001). It should be noted that the AXLAL(R/K)RAXXR motif in the second C-terminal helix of ε is present both in chloroplast F_OF_1 and in the enzyme from the bacteria of *Bacillus* genera (Feniouk et al. 2006). It is prob-

---

[1] DELSEED in most other organisms; corresponds to the *E. coli* ^{380}DELSEED^{386} of subunit β.

able that the mechanism of ATPase activity inhibition mediated by subunit ε in chloroplast and in *Bacillus* F_OF_1 is the same. This suggestion is further supported by experiments demonstrating that *Bacillus* PS3 F_OF_1 is effectively inhibited by chimeric ε with the C-terminus replaced by that from chloroplast enzyme (Konno et al. 2004).

Although the enzymes from chloroplasts and from *Bacillus* bacteria share a conservative motif in the subunit ε second C-terminal α-helix, the latter region is conserved neither in length nor in its amino acid composition among bacteria (Feniouk et al. 2006). Moreover, in subunit ε from *E. coli* ATP synthase the deletion of the second C-terminal α-helix alone does not have a detectable effect on the inhibition, and only the deletion of both helices leads to a pronounced decrease of inhibition (Kuki et al. 1988; Xiong et al. 1998; Cipriano and Dunn 2006). This implies that the role and the inhibitory power of subunit ε might differ substantially among bacteria. It is likely that the conservative C-terminal positive residues mentioned above are necessary for a strong inhibitory effect in photosynthetic/aerobic organisms, while a less "inhibitory" C-terminus is present in species that use F_OF_1 as an ATP-driven $\Delta\tilde{\mu}_{H^+}$ generator (Feniouk et al. 2006). In line with this hypothesis, the whole C-terminal domain is absent in subunit ε from some anaerobic bacteria (e.g. of *Bacteroides*, *Bifidobacterium*, or *Chlorobium* genera) (Feniouk et al. 2006).

It should be noted that isolated subunit ε from bacteria of *Bacillus* genera can directly bind ATP with $K_d \sim 1-2$ mM at optimal growth temperature, and that the C-terminal domain is critically important for the binding (Kato-Yamada and Yoshida 2003; Iino et al. 2005; Kato-Yamada 2005). Such binding was proposed to stabilize the contracted conformation of subunit ε and thereby prevent the inhibition of ATPase activity (Iino et al. 2005). Recent high resolution crystal structure of *Bacillus* PS3 subunit ε with bound ATP is also in line with this hypothesis (Yagi et al. 2007). It remains unclear if subunit ε has ATP-binding properties in the whole F_OF_1 and if these properties are also present in F_OF_1 from other organisms.

5
Thiol Regulation in Chloroplast Enzyme

Chloroplast F_OF_1 has a distinctive redox regulatory feature absent in bacterial and mitochondrial enzymes (for reviews see Evron et al. 2000; Hisabori et al. 2002, 2003; Richter 2004). Early studies revealed that latent ATPase activity of chloroplasts is markedly stimulated by reduction with thiol reagents (Petrack et al. 1965; Kaplan and Jagendorf 1968). Later study by Mills and Mitchell demonstrated that ATP synthesis was also stimulated by the reduction of the enzyme under conditions of limiting $\Delta\tilde{\mu}_{H^+}$, suggesting that the $\Delta\tilde{\mu}_{H^+}$ required for activation of the chloroplast F_OF_1 is larger than that required thermodynamically for ATP synthesis (Mills and Mitchell 1982). This suggestion was confirmed by experiments with flashing light excitation of thylakoid membranes showing that the $\Delta\tilde{\mu}_{H^+}$ threshold for release of the inhibitory ADP, for activation of ATP hydrolysis, and for initiation of ATP synthesis was higher than the phosphate potential of the medium, especially in the oxidized F_OF_1 (Hangarter et al. 1987). Increase of ATP concentration from 10 µM to 1.5 mM had no detectable effect on $\Delta\tilde{\mu}_{H^+}$-induced release of the inhibitory ADP from reduced thylakoid membranes, indicating that the phosphate potential has no effect on activation. Assessment of the activation $\Delta\tilde{\mu}_{H^+}$ value done in the same study yielded $\sim$42 kJ/mol and $\sim$51 kJ/mol for reduced and oxidized enzyme, respectively. Experiments with acid–base transitions on thylakoids indicated that the ΔpH necessary for half-maximal activation of reduced F_OF_1 was 2.2, but increased to 3.4 for the oxidized enzyme (Junesch and Graber 1987).

The stimulation of chloroplast F_OF_1 ATPase activity correlates with reduction of two cysteine residues in subunit γ (Arana and Vallejos 1982; Nalin and McCarty 1984). These two cysteines specific for chloroplast enzyme are located in a $\sim$30 residue long "regulatory region" in subunit γ that is not found in bacterial or mitochondrial enzymes (Hisabori et al. 2002; Hong and Pedersen 2003). It is probable that the formation of a disulfide bond between these two cysteines markedly elevates the $\Delta\tilde{\mu}_{H^+}$ threshold necessary for release of the inhibitory ADP from chloroplast F_OF_1, and stabilizes the ADP-inhibited state. However, this disulfide bond does not affect ATP synthesis rate at high $\Delta\tilde{\mu}_{H^+}$ (Junesch and Gräber 1985, 1987; Hangarter et al. 1987). Therefore, it is tempting to suggest that the thiol regulation of chloroplast F_OF_1 is also partially due to the modulation of the ADP-inhibition efficiency. It is likely that the formation of the disulfide bond impedes the rotation of subunit γ necessary to expel ADP from the high-affinity catalytic site.

Besides the modulation of ADP-inhibition strength, oxidation/reduction of subunit γ also influences the inhibitory effect of subunit ε on ATPase activity of the chloroplast F_OF_1. It has been demonstrated that reduction of the disulfide bond on subunit γ enhances the dissociation of subunit ε from F_1 (Duhe and Selman 1990; Soteropoulos et al. 1992). In turn, subunit ε protects the SS-bond from reduction when bound to F_1. Noteworthy, the truncated ε lacking the C-terminal domain does not protect subunit γ from reduction (Nowak and McCarty 2004). The influence of subunit ε C-terminal domain on redox regulation in chloroplast F_OF_1 is supported by experiments on the introduction of subunit γ regulatory region into *Bacillus* PS3 enzyme (Konno et al. 2004). It was found that the redox regulation emerged only when the regulatory region was introduced together with the C-terminal domain of chloroplast subunit ε. This finding indicates that specific interactions between the regulatory region of subunit γ and the C-terminal domain of subunit ε might be important for the modulation of chloroplast F_OF_1 activity. It should be noted, however, that chloroplast F_1 lacking subunit ε can still be activated by reduction (Richter et al. 1984; Duhe and Selman 1990), as well as the mutant $\varepsilon^{\Delta C}$ F_OF_1 (Nowak and McCarty 2004). Therefore, despite some interplay with the inhibition mediated by subunit ε C-terminal domain, the latter is not a prerequisite for inactivation of chloroplast F_OF_1 caused by oxidation of the γ subunit.

From the experiments on chloroplasts it was suggested that in vivo subunit γ is reduced by thioredoxin, which in turn is photoreduced in the chloroplasts by ferredoxin–thioredoxin reductase (Mills et al. 1980). Further experiments supported this suggestion and pointed out that thioredoxin-f rather than thioredoxin-m is responsible for F_OF_1 reduction in chloroplasts (Schwarz et al. 1997). An elegant biophysical study by Kramer and Crofts on leaves of intact plants provided evidence that light-dependent reduction by thioredoxin is indeed involved in the regulation of chloroplast F_OF_1 activity in vivo (Kramer and Crofts 1989). It was revealed that full reduction of F_OF_1 through the thioredoxin system occurs at a light intensity of $\sim$0.2% of the physiologically "normal" value that saturates primary photosynthetic proteins. Therefore, the thiol modulation is likely to be a "day–night" switch rather than being involved into daytime regulation of F_OF_1 activity (Kramer and Crofts 1989).

6
Mitochondrial Inhibitor Protein IF_1

Mitochondrial F_OF_1 has a more complicated subunit composition than bacterial and chloroplast enzymes. A special mitochondrial "inhibitor protein" (IF_1) that reversibly binds to F_OF_1 plays a role in regulation of ATP hydrolysis (for a review see Green and Grover 2000). The inhibitory effect of this small α-helical basic protein on ATPase activity of both isolated F_1 and of submitochondrial particles from beef heart mitochondria was reported in 1963 by Pullman and Monroy (Pullman and Monroy 1963). In the same study it was revealed that IF_1 does not inhibit ATP synthesis and that the inhibition of ATP hydrolysis is pH-dependent and occurs at pH below 8. Later, IF_1 was also found in yeast (Hashimoto et al. 1981) and rat (Cintron and Pedersen 1979) mitochondria. Bovine IF_1 was shown to inhibit F_OF_1 from yeast and vice verse (Cabezon et al. 2002; Ichikawa and Ogura 2003). In the case of yeast, it was reported that two other protein factors with molecular masses of 9 and 15 kDa interact in a complex manner to stabilize the F_1–IF_1 complex (Hashimoto et al. 1983).

The X-ray crystallographic studies clarified the structure of IF_1–F_1 complex from bovine mitochondria (Cabezon et al. 2003). It turned out that α-helical IF_1 N-terminus can insert itself into $\alpha_3\beta_3$ hexamer between the α and β subunits near their C-terminal regions and the βDELSEED region, which is involved in the subunit ε inhibitory effect in bacterial F_OF_1 (see Sect. 4).

The pH dependence of the IF_1-mediated inhibition (Pullman and Monroy 1963; Panchenko and Vinogradov 1985) was reported to correlate with the pH dependence of IF_1 oligomerization (Cabezon et al. 2000). At pH below neutral, IF_1 exists as a dimer that efficiently inhibits the ATPase activity of F_1, while at pH above neutral IF_1 forms a tetramer that has no inhibitory power. Such pH dependence was suggested to provide a feedback mechanism for preserving mitochondrial ATP in case of uncoupling or anoxia. When glycolysis becomes the only source of cellular ATP, it lowers the cytosolic pH, which is transmitted to the matrix and promotes the inhibition of ATP hydrolysis by IF_1 (Cabezon et al. 2000).

As mentioned above, similar to ADP-inhibition and inhibition mediated by subunit ε in bacterial and chloroplast F_OF_1, IF_1 inhibits ATP hydrolysis without detectable effect on ATP synthesis (Pullman and Monroy 1963; Asami et al. 1970; Iwatsuki et al. 2000). It has been demonstrated that IF_1 dissociates from F_OF_1 upon membrane energization (Schwerzmann and Pedersen 1981; Lippe et al. 1988), suggesting that rotation of subunit γ forces the release of bound IF_1. Experiments with mutant yeast strains lacking IF_1 revealed that in mitochondria it is responsible for prompt deactivation of ATP hydrolysis upon uncoupling (Mimura et al. 1993; Iwatsuki et al. 2000).

In vivo, the deletion of IF_1 in yeast does not affect the growth rate on non-fermentable carbon sources, but it is necessary to preserve mitochondrial and cellular ATP under starving conditions (Ichikawa et al. 2001).

7
Conclusions

F_OF_1 cannot be treated as a simple enzyme that merely accelerates a reversible reaction. Several mechanisms (ADP-inhibition, inhibition mediated by subunit ε in bacteria and chloroplasts, oxidation of subunit γ in chloroplasts, and binding of IF_1 in mitochondria) deactivate the enzyme upon dissipation of $\Delta\tilde{\mu}_{H^+}$ and prevent uncoupled ATP hydrolysis. Re-activation from the inhibited state might require $\Delta\tilde{\mu}_{H^+}$ higher than that necessary for ATP synthesis from thermodynamic considerations. Therefore, $\Delta\tilde{\mu}_{H^+}$ is necessary not only to provide energy for ATP synthesis, but also to maintain the F_OF_1 active state. High affinity to P_i in the presence of $\Delta\tilde{\mu}_{H^+}$ is a key feature of the active state maintenance, protecting the enzyme from ADP-inhibition. Such regulation supposedly prevents ATP waste upon membrane de-energization, but allows ATP-driven $\Delta\tilde{\mu}_{H^+}$ generation on well-coupled membranes.

Acknowledgements We thank Prof. T. Hisabori, Prof. A.D. Vinogradov, Prof. H. Akutsu, Prof. Y. Shirakihara, Dr. Y. Kato-Yamada, Dr. H. Konno, and Dr. T. Suzuki for stimulating discussions. We are also grateful to Dr. N. Sone, Dr. T. Suzuki, and Dr. N. Mitome for critical reading of the manuscript and for their helpful comments.

References

Abrahams JP, Leslie AG, Lutter R, Walker JE (1994) Structure at 2.4 Å resolution of F_1-ATPase from bovine heart mitochondria. Nature 370:621–628

Adachi K, Oiwa K, Nishizaka T, Furuike S, Noji H, Itoh H, Yoshida M, Kinosita K Jr (2007) Coupling of rotation and catalysis in F_1-ATPase revealed by single-molecule imaging and manipulation. Cell 130:309–321

Aggeler R, Capaldi RA (1996) Nucleotide-dependent movement of the epsilon subunit between alpha and beta subunits in the *Escherichia coli* F_1F_O-type ATPase. J Biol Chem 271:13888–13891

al Shawi MK, Ketchum CJ, Nakamoto RK (1997) The *Escherichia coli* F_OF_1 gammaM23K uncoupling mutant has a higher $K_{0.5}$ for P_i. Transition state analysis of this mutant and others reveals that synthesis and hydrolysis utilize the same kinetic pathway. Biochemistry 36:12961–12969

al Shawi MK, Senior AE (1992) Effects of dimethyl sulfoxide on catalysis in *Escherichia coli* F_1-ATPase. Biochemistry 31:886–891

Arana JL, Vallejos RH (1982) Involvement of sulfhydryl groups in the activation mechanism of the ATPase activity of chloroplast coupling factor 1. J Biol Chem 257:1125–1127

Ariga T, Muneyuki E, Yoshida M (2007) F_1-ATPase rotates by an asymmetric, sequential mechanism using all three catalytic subunits. Nat Struct Mol Biol 14:841–846

Asami K, Juniti K, Ernster L (1970) Possible regulatory function of a mitochondrial ATPase inhibitor in respiratory chain-linked energy transfer. Biochim Biophys Acta 205:307–311

Avron M, Jagendorf AT (1959) Evidence concerning the mechanism of adenosine triphosphate formation by spinach chloroplasts. J Biol Chem 234:967–972

Bakker-Grunwald T, Van Dam K (1974) On the mechanism of activation of the ATPase in chloroplasts. Biochim Biophys Acta 347:290–298

Bald D, Muneyuki E, Amano T, Kruip J, Hisabori T, Yoshida M (1999) The noncatalytic site-deficient alpha3beta3gamma subcomplex and F_OF_1-ATP synthase can continuously catalyse ATP hydrolysis when P_i is present. Eur J Biochem 262:563–568

Baltscheffsky M, Lundin A (1979) Flash-induced increase of ATPase activity in *Rhodospirillum rubrum* chromatophores. In: Mukohata Y, Packer L (eds) Cation flux across biomembranes. Academic, New York, pp 209–218

Boyer PD (1997) The ATP synthase – a splendid molecular machine. Annu Rev Biochem 66:717–749

Boyer PD (2002) Catalytic site occupancy during ATP synthase catalysis. FEBS Lett 512:29–32

Boyer PD, Cross RL, Momsen W (1973) A new concept for energy coupling in oxidative phosphorylation based on a molecular explanation of the oxygen exchange reactions. Proc Natl Acad Sci USA 70:2837–2839

Bulygin VV, Vinogradov AD (1991) Interaction of Mg^{2+} with F0.F1 mitochondrial ATPase as related to its slow active/inactive transition. Biochem J 276:149–156

Cabezon E, Butler PJ, Runswick MJ, Carbajo RJ, Walker JE (2002) Homologous and heterologous inhibitory effects of ATPase inhibitor proteins on F-ATPases. J Biol Chem 277:41334–41341

Cabezon E, Butler PJ, Runswick MJ, Walker JE (2000) Modulation of the oligomerization state of the bovine F_1-ATPase inhibitor protein, IF_1, by pH. J Biol Chem 275:25460–25464

Cabezon E, Montgomery MG, Leslie AG, Walker JE (2003) The structure of bovine F_1-ATPase in complex with its regulatory protein IF_1. Nat Struct Biol 10:744–750

Capaldi RA, Schulenberg B (2000) The epsilon subunit of bacterial and chloroplast F_1F_O ATPases. Structure, arrangement, and role of the epsilon subunit in energy coupling within the complex. Biochim Biophys Acta 1458:263–269

Carmeli C, Avron M (1972) Light-triggered and light-dependent ATPase activities in chloroplasts. Methods Enzymol 24:92–96

Carmeli C, Lifshitz Y (1972) Effects of Pi and ADP on ATPase activity in chloroplasts. Biochim Biophys Acta 267:86–95

Cintron NM, Pedersen PL (1979) A protein inhibitor of the mitochondrial adenosine triphosphatase complex of rat liver. Purification and characterization. J Biol Chem 254:3439–3443

Cipriano DJ, Dunn SD (2006) The role of the epsilon subunit in the *Escherichia coli* ATP synthase. The C-terminal domain is required for efficient energy coupling. J Biol Chem 281:501–507

Creczynski-Pasa TB, Graber P (1994) ADP binding and ATP synthesis by reconstituted H^+-ATPase from chloroplasts. FEBS Lett 350:195–198

Cross RL, Nalin CM (1982) Adenine nucleotide binding sites on beef heart F_1-ATPase. Evidence for three exchangeable sites that are distinct from three noncatalytic sites. J Biol Chem 257:2874–2881

Diez M, Zimmermann B, Borsch M, Konig M, Schweinberger E, Steigmiller S, Reuter R, Felekyan S, Kudryavtsev V, Seidel CA, Graber P (2004) Proton-powered subunit rotation in single membrane-bound F(0)F(1)-ATP synthase. Nat Struct Mol Biol 11:135–141

Dimroth P, Kaim G, Matthey U (1998) The motor of the ATP synthase. Biochim Biophys Acta 1365:87–92

Drobinskaya IY, Kozlov IA, Murataliev MB, Vulfson EN (1985) Tightly bound adenosine diphosphate, which inhibits the activity of mitochondrial F_1-ATPase, is located at the catalytic site of the enzyme. FEBS Lett 182:419–424

Duhe RJ, Selman BR (1990) The dithiothreitol-stimulated dissociation of the chloroplast coupling factor 1 epsilon-subunit is reversible. Biochim Biophys Acta 1017:70–78

Duncan TM, Bulygin VV, Zhou Y, Hutcheon ML, Cross RL (1995) Rotation of subunits during catalysis by *Escherichia coli* F_1-ATPase. Proc Natl Acad Sci USA 92:10964–10968

Dunham KR, Selman BR (1981a) Interactions of inorganic phosphate with spinach coupling factor 1. Effects on ATPase and ADP binding activities. J Biol Chem 256:10044–10049

Dunham KR, Selman BR (1981b) Regulation of spinach chloroplast coupling factor 1 ATPase activity. J Biol Chem 256:212–218

Elston T, Wang H, Oster G (1998) Energy transduction in ATP synthase. Nature 391:510–513

Etzold C, Deckers-Hebestreit G, Altendorf K (1997) Turnover number of *Escherichia coli* F_OF_1 ATP synthase for ATP synthesis in membrane vesicles. Eur J Biochem 243:336–343

Evron Y, Johnson EA, McCarty RE (2000) Regulation of proton flow and ATP synthesis in chloroplasts. J Bioenerg Biomembranes 32:501–506

Feldman RI, Boyer PD (1985) The role of tightly bound ADP on chloroplast ATPase. J Biol Chem 260:13088–13094

Feniouk BA, Cherepanov DA, Junge W, Mulkidjanian AY (2001) Coupling of proton flow to ATP synthesis in *Rhodobacter capsulatus*: F_OF_1-ATP synthase is absent from about half of chromatophores. Biochim Biophys Acta 1506:189–203

Feniouk BA, Junge W (2005) Regulation of the F_OF_1-ATP synthase: The conformation of subunit epsilon might be determined by directionality of subunit gamma rotation. FEBS Lett 579:5114–5118

Feniouk BA, Kozlova MA, Knorre DA, Cherepanov DA, Mulkidjanian AY, Junge W (2004) The proton driven rotor of ATP synthase: Ohmic conductance (10 fS), and absence of voltage gating. Biophys J 86:4094–4109

Feniouk BA, Mulkidjanian AY, Junge W (2005) Proton slip in the ATP synthase of *Rhodobacter capsulatus*: induction, proton conduction, and nucleotide dependence. Biochim Biophys Acta 1706:184–194

Feniouk BA, Suzuki T, Yoshida M (2007) Regulatory interplay between proton motive force, ADP, phosphate, and subunit epsilon in bacterial ATP synthase. J Biol Chem 282:764–772

Feniouk BA, Suzuki T, Yoshida M (2006) The role of subunit epsilon in the catalysis and regulation of F_OF_1-ATP synthase. Biochim Biophys Acta 1757:326–338

Fillingame RH (1975) Identification of the dicyclohexylcarbodiimide-reactive protein component of the adenosine 5'-triphosphate energy-transducing system of *Escherichia coli*. J Bacteriol 124:870–883

Fischer S, Graber P, Turina P (2000) The activity of the ATP synthase from *Escherichia coli* is regulated by the transmembrane proton motive force. J Biol Chem 275:30157–30162

Fitin AF, Vasilyeva EA, Vinogradov AD (1979) An inhibitory high affinity binding site for ADP in the oligomycin-sensitive ATPase of beef heart submitochondrial particles. Biochem Biophys Res Commun 86:434–439

Galkin MA, Vinogradov AD (1999) Energy-dependent transformation of the catalytic activities of the mitochondrial F_OF_1-ATP synthase. FEBS Lett 448:123–126

Garcia JJ, Capaldi RA (1998) Unisite catalysis without rotation of the gamma-epsilon domain in *Escherichia coli* F_1-ATPase. J Biol Chem 273:15940–15945

Gibbons C, Montgomery MG, Leslie AG, Walker JE (2000) The structure of the central stalk in bovine F(1)-ATPase at 2.4 Å resolution. Nat Struct Biol 7:1055–1061

Graber P, Schlodder E, Witt HT (1977) Conformational change of the chloroplast ATPase induced by a transmembrane electric field and its correlation to phosphorylation. Biochim Biophys Acta 461:426–440

Graber P, Witt HT (1976) Relations between the electrical potential, pH gradient, proton flux and phosphorylation in the photosynthetic membrane. Biochim Biophys Acta 423:141–163

Green DW, Grover GJ (2000) The IF_1 inhibitor protein of the mitochondrial F_OF_1-ATPase. Biochim Biophys Acta 1458:343–355

Grotjohann I, Graber P (2002) The H^+-ATPase from chloroplasts: effect of different reconstitution procedures on ATP synthesis activity and on phosphate dependence of ATP synthesis. Biochim Biophys Acta 1556:208–216

Hangarter RP, Grandoni P, Ort DR (1987) The effects of chloroplast coupling factor reduction on the energetics of activation and on the energetics and efficiency of ATP formation. J Biol Chem 262:13513–13519

Hara KY, Kato Y-Yamada, Kikuchi Y, Hisabori T, Yoshida M (2001) The role of the beta-DELSEED motif of F_1-ATPase: propagation of the inhibitory effect of the epsilon subunit. J Biol Chem 276:23969–23973

Hashimoto T, Negawa Y, Tagawa K (1981) Binding of intrinsic ATPase inhibitor to mitochondrial ATPase-stoichiometry of binding of nucleotides, inhibitor, and enzyme. J Biochem (Tokyo) 90:1151–1157

Hashimoto T, Yoshida Y, Tagawa K (1983) Binding properties of an intrinsic ATPase inhibitor and occurrence in yeast mitochondria of a protein factor which stabilizes and facilitates the binding of the inhibitor to F_1F_O-ATPase. J Biochem (Tokyo) 94:715–720

Hatefi Y, Yagi T, Phelps DC, Wong SY, Vik SB, Galante YM (1982) Substrate binding affinity changes in mitochondrial energy-linked reactions. Proc Natl Acad Sci USA 79:1756-1760

Hirono-Hara Y, Ishuzuka K, Kinosita K, Yoshida M, Noji H (2005) Activation of pausing F-1 motor by external force. Proc Natl Acad Sci USA 102:4288-4293

Hirono-Hara Y, Noji H, Nishiura M, Muneyuki E, Hara KY, Yasuda R, Kinosita K Jr, Yoshida M (2001) Pause and rotation of F_1-ATPase during catalysis. Proc Natl Acad Sci USA 98:13649-13654

Hisabori T, Konno H, Ichimura H, Strotmann H, Bald D (2002) Molecular devices of chloroplast F_1-ATP synthase for the regulation. Biochim Biophys Acta 1555:140-146

Hisabori T, Ueoka-Nakanishi H, Konno H, Koyama F (2003) Molecular evolution of the modulator of chloroplast ATP synthase: origin of the conformational change dependent regulation. FEBS Lett 545:71-75

Hong S, Pedersen PL (2003) ATP synthases: insights into their motor functions from sequence and structural analyses. J Bioenerg Biomembr 35:95-120

Humphrey W, Dalke A, Schulten K (1996) VMD - visual molecular dynamics. J Molec Graphics 14:33-38

Hyndman DJ, Milgrom YM, Bramhall EA, Cross RL (1994) Nucleotide-binding sites on *Escherichia coli* F_1-ATPase. Specificity of noncatalytic sites and inhibition at catalytic sites by MgADP. J Biol Chem 269:28871-28877

Ichikawa N, Karaki A, Kawabata M, Ushida S, Mizushima M, Hashimoto T (2001) The region from phenylalanine-17 to phenylalanine-28 of a yeast mitochondrial ATPase inhibitor is essential for its ATPase inhibitory activity. J Biochem (Tokyo) 130:687-693

Ichikawa N, Ogura C (2003) Overexpression, purification, and characterization of human and bovine mitochondrial ATPase inhibitors: comparison of the properties of mammalian and yeast ATPase inhibitors. J Bioenerg Biomembr 35:399-407

Iino R, Murakami T, Iizuka S, Kato Y-Y, Suzuki T, Yoshida M (2005) Real time monitoring of conformational dynamics of the epsilon subunit in F_1-ATPase. J Biol Chem 280:40130-40134

Itoh H, Takahashi A, Adachi K, Noji H, Yasuda R, Yoshida M, Kinosita K Jr (2004) Mechanically driven ATP synthesis by F_1-ATPase. Nature 427:465-468

Iwatsuki H, Lu YM, Yamaguchi K, Ichikawa N, Hashimoto T (2000) Binding of an intrinsic ATPase inhibitor to the F_1F_O-ATPase in phosphorylating conditions of yeast mitochondria. J Biochem (Tokyo) 128:553-559

Jagendorf AT, Avron M (1958) Cofactors and rates of photosynthetic phosphorylation by spinach chloroplasts. J Biol Chem 231:277-290

Jault JM, Allison WS (1993) Slow binding of ATP to noncatalytic nucleotide binding sites which accelerates catalysis is responsible for apparent negative cooperativity exhibited by the bovine mitochondrial F_1-ATPase. J Biol Chem 268:1558-1566

Jault JM, Allison WS (1994) Hysteretic inhibition of the bovine heart mitochondrial F_1-ATPase is due to saturation of noncatalytic sites with ADP which blocks activation of the enzyme by ATP. J Biol Chem 269:319-325

Jault JM, Matsui T, Jault FM, Kaibara C, Muneyuki E, Yoshida M, Kagawa Y, Allison WS (1995) The alpha 3 beta 3 gamma complex of the F_1-ATPase from thermophilic *Bacillus* PS3 containing the alpha D261N substitution fails to dissociate inhibitory MgADP from a catalytic site when ATP binds to noncatalytic sites. Biochemistry 34:16412-16418

Jault JM, Paik SR, Grodsky NB, Allison WS (1994) Lowered temperature or binding of pyrophosphate to sites for noncatalytic nucleotides modulates the ATPase activity of the beef heart mitochondrial F_1-ATPase by decreasing the affinity of a catalytic site for inhibitory MgADP. Biochemistry 33:14979–14985

Junesch U, Gräber P (1985) The rate of ATP synthesis as a function of delta pH in normal and dithiothreitol-modified chloroplasts. Biochim Biophys Acta 809:429–434

Junesch U, Graber P (1987) Influence of the redox state and the activation of the chloroplast ATP synthase on proton-transport-coupled ATP synthesis/hydrolysis. Biochim Biophys Acta 893:275–288

Junesch U, Graber P (1991) The rate of ATP-synthesis as a function of delta pH and delta psi catalyzed by the active, reduced H^+-ATPase from chloroplasts. FEBS Lett 294:275–278

Junge W (1987) Complete tracking of transient proton flow through active chloroplast ATP synthase. Proc Natl Acad Sci USA 84:7084–7088

Junge W, Lill H, Engelbrecht S (1997) ATP synthase: an electrochemical transducer with rotatory mechanics. Trends Biochem Sci 22:420–423

Kabaleeswaran V, Puri N, Walker JE, Leslie AG, Mueller DM (2006) Novel features of the rotary catalytic mechanism revealed in the structure of yeast F_1 ATPase. EMBO J 25:5433–5442

Kalashnikova TY, Milgrom YM, Murataliev MB (1988) The effect of inorganic pyrophosphate on the activity and Pi-binding properties of mitochondrial F_1-ATPase. Eur J Biochem 177:213–218

Kaplan JH, Jagendorf AT (1968) Further studies on chloroplast adenosine triphosphatase activation by acid-base transition. J Biol Chem 243:972–979

Kaplan JH, Uribe E, Jagendorf AT (1967) ATP hydrolysis caused by acid-base transition of spinach chloroplasts. Arch Biochem Biophys 120:365–370

Kasahara M, Penefsky HS (1978) High affinity binding of monovalent Pi by beef heart mitochondrial adenosine triphosphatase. J Biol Chem 253:4180–4187

Kato Y, Matsui T, Tanaka N, Muneyuki E, Hisabori T, Yoshida M (1997) Thermophilic F_1-ATPase is activated without dissociation of an endogenous inhibitor, epsilon subunit. J Biol Chem 272:24906–24912

Kato-Yamada Y (2005) Isolated epsilon subunit of *Bacillus subtilis* F(1)-ATPase binds ATP. FEBS Lett 579:6875–6878

Kato-Yamada Y, Bald D, Koike M, Motohashi K, Hisabori T, Yoshida M (1999) Epsilon subunit, an endogenous inhibitor of bacterial F_1-ATPase, also inhibits F_OF_1-ATPase. J Biol Chem 274:33991–33994

Kato-Yamada Y, Yoshida M (2003) Isolated epsilon subunit of thermophilic F_1-ATPase binds ATP. J Biol Chem 278:36013–36016

Kayalar C, Rosing J, Boyer PD (1976) 2,4-Dinitrophenol causes a marked increase in the apparent Km of Pi and of ADP for oxidative phosphorylation. Biochem Biophys Res Commun 72:1153–1159

Kayalar C, Rosing J, Boyer PD (1977) An alternating site sequence for oxidative phosphorylation suggested by measurement of substrate binding patterns and exchange reaction inhibitions. J Biol Chem 252:2486–2491

Keis S, Stocker A, Dimroth P, Cook GM (2006) Inhibition of ATP hydrolysis by thermoalkaliphilic F_1F_O-ATP synthase is controlled by the C terminus of the epsilon subunit. J Bacteriol 188:3796–3804

Kironde FA, Cross RL (1987) Adenine nucleotide binding sites on beef heart F_1-ATPase. Asymmetry and subunit location. J Biol Chem 262:3488–3495

Komatsu-Takaki M (1986) Interconversion of two distinct states of active CF_0-CF_1 (chloroplast ATPase complex) in chloroplasts. J Biol Chem 261:1116–1119

Komatsu-Takaki M (1989) Energy-dependent conformational changes in the epsilon subunit of the chloroplast ATP synthase (CF_0CF_1). J Biol Chem 264:17750–17753

Konno H, Murakami-Fuse T, Fujii F, Koyama F, Ueoka H-Nakanishi, Pack CG, Kinjo M, Hisabori T (2006) The regulator of the F_1 motor: inhibition of rotation of cyanobacterial F_1-ATPase by the epsilon subunit. EMBO J 25:4596-4604

Konno H, Suzuki T, Bald D, Yoshida M, Hisabori T (2004) Significance of the epsilon subunit in the thiol modulation of chloroplast ATP synthase. Biochem Biophys Res Commun 318:17-24

Kozlov IA, Vulfson EN (1985) Tightly bound nucleotides affect phosphate binding to mitochondrial F_1-ATPase. FEBS Lett 182:425-428

Kramer DM, Crofts AR (1989) Activation of the chloroplast ATP-ase measured by the electrochromic change in leaves of intact plants. Biochim Biophys Acta 976:28-41

Kramer DM, Sacksteder CA, Cruz JA (1999) How acidic is the lumen? Photosynth Res 60:151-163

Kuki M, Noumi T, Maeda M, Amemura A, Futai M (1988) Functional domains of epsilon subunit of *Escherichia coli* H^+-ATPase (F0F1). J Biol Chem 263:17437-17442

Laget PP, Smith JB (1979) Inhibitory properties of endogenous subunit epsilon in the *Escherichia coli* F_1 ATPase. Arch Biochem Biophys 197:83-89

Lefebvre-Legendre L, Balguerie A, Duvezin-Caubet S, Giraud MF, Slonimski PP, di Rago JP (2003) F_1-catalysed ATP hydrolysis is required for mitochondrial biogenesis in *Saccharomyces cerevisiae* growing under conditions where it cannot respire. Mol Microbiol 47:1329-1339

Lippe G, Sorgato MC, Harris DA (1988) The binding and release of the inhibitor protein are governed independently by ATP and membrane potential in ox-heart submitochondrial vesicles. Biochim Biophys Acta 933:12-21

Malyan A, Allison W (2002) Properties of noncatalytic sites of thioredoxin-activated chloroplast coupling factor 1. Biochim Biophys Acta 1554:153-158

Malyan AN (2005) Light-dependent incorporation of adenine nucleotide into noncatalytic sites of chloroplast ATP synthase. Biochemistry (Mosc) 70:1245-1250

Malyan AN (2006) ADP and ATP binding to noncatalytic sites of thiol-modulated chloroplast ATP synthase. Photosynth Res 88:9-18

Matsuyama S, Xu Q, Velours J, Reed JC (1998) The mitochondrial F_OF_1-ATPase proton pump is required for function of the proapoptotic protein Bax in yeast and mammalian cells. Mol Cell 1:327-336

McCarthy JE, Ferguson SJ (1983) The effects of partial uncoupling upon the kinetics of ATP synthesis by vesicles from *Paracoccus denitrificans* and by bovine heart submitochondrial particles. Implications for the mechanism of the proton-translocating ATP synthase. Eur J Biochem 132:425-431

Melandri AB, Fabbri E, Melandri BA (1975) Energy transduction in photosynthetic bacteria. VIII. Activation of the energy-transducing ATPase by inorganic phosphate. Biochim Biophys Acta 376:82-88

Mendel-Hartvig J, Capaldi RA (1991) Catalytic site nucleotide and inorganic phosphate dependence of the conformation of the epsilon subunit in *Escherichia coli* adenosinetriphosphatase. Biochemistry 30:1278-1284

Milgrom YM, Boyer PD (1990) The ADP that binds tightly to nucleotide-depleted mitochondrial F_1-ATPase and inhibits catalysis is bound at a catalytic site. Biochim Biophys Acta 1020:43-48

Milgrom YM, Cross RL (1993) Nucleotide binding sites on beef heart mitochondrial F1-ATPase. Cooperative interactions between sites and specificity of noncatalytic sites. J Biol Chem 268:23179-23185

Milgrom YM, Ehler LL, Boyer PD (1990) ATP binding at noncatalytic sites of soluble chloroplast F1-ATPase is required for expression of the enzyme activity. J Biol Chem 265:18725–18728

Milgrom YM, Ehler LL, Boyer PD (1991) The characteristics and effect on catalysis of nucleotide binding to noncatalytic sites of chloroplast F1-ATPase. J Biol Chem 266:11551–11558

Mills JD, Mitchell P (1982) Thiol modulation of CF_0-CF_1 stimulates acid/base-dependent phosphorylation of ADP by broken pea chloroplasts. FEBS 144:63–67

Mills JD, Mitchell P, Schurmann P (1980) Modulation of coupling factor ATPase activity in intact chloroplasts: The role of the thioredoxin system. FEBS 112:173–177

Mimura H, Hashimoto T, Yoshida Y, Ichikawa N, Tagawa K (1993) Binding of an intrinsic ATPase inhibitor to the interface between alpha- and beta-subunits of F_0F_1ATPase upon de-energization of mitochondria. J Biochem (Tokyo) 113:350–354

Minkov IB, Fitin AF, Vasilyeva EA, Vinogradov AD (1979) Mg^{2+}-induced ADP-dependent inhibition of the ATPase activity of beef heart mitochondrial coupling factor F1. Biochem Biophys Res Commun 89:1300–1306

Mitome N, Ono S, Suzuki T, Shimabukuro K, Muneyuki E, Yoshida M (2002) The presence of phosphate at a catalytic site suppresses the formation of the MgADP-inhibited form of F_1-ATPase. Eur J Biochem 269:53–60

Moyle J, Mitchell P (1975) Active/inactive state transitions of mitochondrial ATPase molecules influenced by Mg^{2+}, anions and aurovertin. FEBS Lett 56:55–61

Murataliev MB, Boyer PD (1992) The mechanism of stimulation of MgATPase activity of chloroplast F_1-ATPase by non-catalytic adenine-nucleotide binding. Acceleration of the ATP-dependent release of inhibitory ADP from a catalytic site. Eur J Biochem 209:681–687

Nalin CM, McCarty RE (1984) Role of a disulfide bond in the gamma subunit in activation of the ATPase of chloroplast coupling factor 1. J Biol Chem 259:7275–7280

Nelson N, Nelson H, Racker E (1972) Partial resolution of the enzymes catalyzing photophosphorylation. XII. Purification and properties of an inhibitor isolated from chloroplast coupling factor 1. J Biol Chem 247:7657–7662

Nishizaka T, Oiwa K, Noji H, Kimura S, Muneyuki E, Yoshida M, Kinosita K Jr (2004) Chemomechanical coupling in F_1-ATPase revealed by simultaneous observation of nucleotide kinetics and rotation. Nat Struct Mol Biol 11:142–148

Noji H, Yasuda R, Yoshida M, Kinosita K Jr (1997) Direct observation of the rotation of F_1-ATPase. Nature 386:299–302

Nowak KF, McCarty RE (2004) Regulatory role of the C-terminus of the epsilon subunit from the chloroplast ATP synthase. Biochemistry 43:3273–3279

Pacheco-Moises F, Garcia JJ, Rodriguez-Zavala JS, Moreno-Sanchez R (2000) Sulfite and membrane energization induce two different active states of the *Paracoccus denitrificans* F_0F_1-ATPase. Eur J Biochem 267:993–1000

Panchenko MV, Vinogradov AD (1985) Interaction between the mitochondrial ATP synthetase and ATPase inhibitor protein. Active/inactive slow pH-dependent transitions of the inhibitor protein. FEBS Lett 184:226–230

Penefsky HS (1977) Reversible binding of Pi by beef heart mitochondrial adenosine triphosphatase. J Biol Chem 252:2891–2899

Penefsky HS (2005) Pi binding by the F_1-ATPase of beef heart mitochondria and of the *Escherichia coli* plasma membrane. FEBS Lett 579:2250–2252

Perez JA, Ferguson SJ (1990a) Kinetics of oxidative phosphorylation in *Paracoccus denitrificans*. 1. Mechanism of ATP synthesis at the active site(s) of F_0F_1-ATPase. Biochemistry 29:10503–10518

Perez JA, Ferguson SJ (1990b) Kinetics of oxidative phosphorylation in *Paracoccus denitrificans*. 2. Evidence for a kinetic and thermodynamic modulation of F_OF_1-ATPase by the activity of the respiratory chain. Biochemistry 29:10518–10526

Petrack B, Craston A, Sheppy F, Farron F (1965) Studies on the hydrolysis of adenosine triphosphate by spinach chloroplasts. J Biol Chem 240:906–914

Pitard B, Richard P, Dunach M, Rigaud JL (1996) ATP synthesis by the F_OF_1 ATP synthase from thermophilic *Bacillus* PS3 reconstituted into liposomes with bacteriorhodopsin. 2. Relationships between proton motive force and ATP synthesis. Eur J Biochem 235:779–788

Pullman ME, Monroy GC (1963) A naturally occurring inhibitor of mitochondrial adenosine triphosphatase. J Biol Chem 238:3762–3769

Richard P, Pitard B, Rigaud JL (1995) ATP synthesis by the F_OF_1-ATPase from the thermophilic *Bacillus* PS3 co-reconstituted with bacteriorhodopsin into liposomes. Evidence for stimulation of ATP synthesis by ATP bound to a noncatalytic binding site. J Biol Chem 270:21571–21578

Richter ML (2004) Gamma-epsilon interactions regulate the chloroplast ATP synthase. Photosynth Res 79:319–329

Richter ML, McCarty RE (1987) Energy-dependent changes in the conformation of the epsilon subunit of the chloroplast ATP synthase. J Biol Chem 262:15037–15040

Richter ML, Patrie WJ, McCarty RE (1984) Preparation of the epsilon subunit and epsilon subunit-deficient chloroplast coupling factor 1 in reconstitutively active forms. J Biol Chem 259:7371–7373

Rodgers AJW, Wilce MCJ (2000) Structure of the gamma-epsilon complex of ATP synthase. Nat Struct Biol 7:1051–1054

Rondelez Y, Tresset G, Nakashima T, Kato Y-Yamada, Fujita H, Takeuchi S, Noji H (2005) Highly coupled ATP synthesis by F_1-ATPase single molecules. Nature 433:773–777

Rosen G, Gresser M, Vinkler C, Boyer PD (1979) Assessment of total catalytic sites and the nature of bound nucleotide participation in photophosphorylation. J Biol Chem 254:10654–10661

Rosing J, Kayalar C, Boyer PD (1977) Evidence for energy-dependent change in phosphate binding for mitochondrial oxidative phosphorylation based on measurements of medium and intermediate phosphate-water exchanges. J Biol Chem 252:2478–2485

Roveri OA, Muller JL, Wilms J, Slater EC (1980) The pre-steady state and steady-state kinetics of the ATPase activity of mitochondrial F_1. Biochim Biophys Acta 589:241–255

Sabbert D, Engelbrecht S, Junge W (1996) Intersubunit rotation in active F-ATPase. Nature 381:623–625

Schwartz M (1968) Light induced proton gradient links electron transport and phosphorylation. Nature 219:915–919

Schwarz O, Schurmann P, Strotmann H (1997) Kinetics and thioredoxin specificity of thiol modulation of the chloroplast H^+-ATPase. J Biol Chem 272:16924–16927

Schwerzmann K, Pedersen PL (1981) Proton–adenosinetriphosphatase complex of rat liver mitochondria: effect of energy state on its interaction with the adenosinetriphosphatase inhibitory peptide. Biochemistry 20:6305–6311

Senior AE, Nadanaciva S, Weber J (2002) The molecular mechanism of ATP synthesis by F_1F_O-ATP synthase. Biochim Biophys Acta 1553:188–211

Sherman PA, Wimmer MJ (1984) Activation of ATPase of spinach coupling factor 1. Release of tightly bound ADP from the soluble enzyme. Eur J Biochem 139:367–371

Shi XB, Wei JM, Shen YK (2001) Effects of sequential deletions of residues from the N- or C-terminus on the functions of epsilon subunit of the chloroplast ATP synthase. Biochemistry 40:10825–10831

Shimabukuro K, Yasuda R, Muneyuki E, Hara KY, Kinosita K Jr, Yoshida M (2003) Catalysis and rotation of F1 motor: cleavage of ATP at the catalytic site occurs in 1 ms before 40 degree substep rotation. Proc Natl Acad Sci USA 100:14731–14736

Shoshan V, Selman BR (1979) The relationship between light-induced adenine nucleotide exchange and ATPase activity in chloroplast thylakoid membranes. J Biol Chem 254:8801–8807

Slooten L, Vandenbranden S (1989) ATP-synthesis by proteoliposomes incorporating *Rhodospirillum rubrum* F_OF_1 as measured with firefly luciferase: dependence on delta and delta pH. Biochim Biophys Acta 976:150–160

Smith DJ, Stokes BO, Boyer PD (1976) Probes of initial phosphorylation events in ATP synthesis by chloroplasts. J Biol Chem 251:4165–4171

Smith JB, Sternweis PC (1977) Purification of membrane attachment and inhibitory subunits of the proton translocating adenosine triphosphatase from *Escherichia coli*. Biochemistry 16:306–311

Smith JB, Sternweis PC, Heppel LA (1975) Partial purification of active delta and epsilon subunits of the membrane ATPase from *Escherichia coli*. J Supramol Struct 3:248–255

Smith LT, Rosen G, Boyer PD (1983) Properties of ATP tightly bound to catalytic sites of chloroplast ATP synthase. J Biol Chem 258:10887–10894

Soteropoulos P, Suss KH, McCarty RE (1992) Modifications of the gamma subunit of chloroplast coupling factor 1 alter interactions with the inhibitory epsilon subunit. J Biol Chem 267:10348–10354

St Pierre J, Brand MD, Boutilier RG (2000) Mitochondria as ATP consumers: cellular treason in anoxia. Proc Natl Acad Sci USA 97:8670–8674

Strotmann H, Bickel S, Huchzermeyer B (1976) Energy-dependent release of adenine nucleotides tightly bound to chloroplast coupling factor CF1. FEBS Lett 61:194–198

Strotmann H, Thelen R, Muller W, Baum W (1990) A delta pH clamp method for analysis of steady-state kinetics of photophosphorylation. Eur J Biochem 193:879–886

Suzuki T, Murakami T, Iino R, Suzuki J, Ono S, Shirakihara Y, Yoshida M (2003) F_OF_1-ATPase/synthase is geared to the synthesis mode by conformational rearrangement of epsilon subunit in response to proton motive force and ADP/ATP balance. J Biol Chem 278:46840–46846

Suzuki T, Ueno H, Mitome N, Suzuki J, Yoshida M (2002) F_O of ATP synthase is a rotary proton channel: Obligatory coupling of proton translocation with rotation of *c*-subunit ring. J Biol Chem 277:13281–13285

Tomashek JJ, Glagoleva OB, Brusilow WS (2004) The *Escherichia coli* F_1F_O ATP synthase displays biphasic synthesis kinetics. J Biol Chem 279:4465–4470

Tsunoda SP, Rodgers AJ, Aggeler R, Wilce MC, Yoshida M, Capaldi RA (2001) Large conformational changes of the epsilon subunit in the bacterial F_1F_O ATP synthase provide a ratchet action to regulate this rotary motor enzyme. Proc Natl Acad Sci USA 98:6560–6564

Turina P, Melandri BA, Graber P (1991) ATP synthesis in chromatophores driven by artificially induced ion gradients. Eur J Biochem 196:225–229

Turina P, Rumberg B, Melandri BA, Graber P (1992) Activation of the H^+-ATP synthase in the photosynthetic bacterium *Rhodobacter capsulatus*. J Biol Chem 267:11057–11063

Ueno H, Suzuki T, Kinosita K Jr, Yoshida M (2005) ATP-driven stepwise rotation of F_OF_1-ATP synthase. Proc Natl Acad Sci USA 102:1333–1338

Uhlin U, Cox GB, Guss JM (1997) Crystal structure of the epsilon subunit of the proton-translocating ATP synthase from *Escherichia coli*. Structure 5:1219–1230

Vik SB (2000) What is the role of epsilon in the *Escherichia coli* ATP synthase? J Bioenerg Biomembr 32:485–491

Vik SB, Patterson AR, Antonio BJ (1998) Insertion scanning mutagenesis of subunit *a* of the F_1F_O ATP synthase near His^{245} and implications on gating of the proton channel. J Biol Chem 273:16229-16234

Weber J, Wilke S-Mounts, Grell E, Senior AE (1994) Tryptophan fluorescence provides a direct probe of nucleotide binding in the noncatalytic sites of *Escherichia coli* F_1-ATPase. J Biol Chem 269:11261-11268

Wilkens S, Capaldi RA (1994) Asymmetry and structural changes in ECF_1 examined by cryoelectronmicroscopy. Biol Chem Hoppe Seyler 375:43-51

Wilkens S, Dahlquist FW, McIntosh LP, Donaldson LW, Capaldi RA (1995) Structural features of the epsilon subunit of the *Escherichia coli* ATP synthase determined by NMR spectroscopy. Nat Struct Biol 2:961-967

Xiong H, Zhang D, Vik SB (1998) Subunit epsilon of the *Escherichia coli* ATP synthase: novel insights into structure and function by analysis of thirteen mutant forms. Biochemistry 37:16423-16429

Yagi H, Kajiwara N, Tanaka H, Tsukihara T, Kato-Yamada Y, Yoshida M, Akutsu H (2007) Structures of the thermophilic F_1 ATPase epsilon subunit suggesting ATP-regulated arm motion of its C-terminal domain in F_1. Proc Natl Acad Sci USA 104:11233-11238

Yasuda R, Noji H, Kinosita K Jr, Yoshida M (1998) F_1-ATPase is a highly efficient molecular motor that rotates with discrete 120 degree steps. Cell 93:1117-1124

Yoshida M, Allison WS (1983) Modulation by ADP and Mg^{2+} of the inactivation of the F_1-ATPase from the thermophilic bacterium, PS3, with dicyclohexylcarbodiimide. J Biol Chem 258:14407-14412

Yoshida M, Allison WS (1986) Characterization of the catalytic and noncatalytic ADP binding sites of the F1-ATPase from the thermophilic bacterium, PS3. J Biol Chem 261:5714-5721

Zharova TV, Vinogradov AD (2004) Energy-dependent transformation of F_OF_1-ATPase in *Paracoccus denitrificans* plasma membranes. J Biol Chem 279:12319-12324

Zhou JM, Xue ZX, Du ZY, Melese T, Boyer PD (1988) Relationship of tightly bound ADP and ATP to control and catalysis by chloroplast ATP synthase. Biochemistry 27:5129-5135

Zimmermann B, Diez M, Zarrabi N, Graber P, Borsch M (2005) Movements of the epsilon-subunit during catalysis and activation in single membrane-bound H^+-ATP synthase. EMBO J 24:2053-2063

Subject Index

Archaea 1, 73, 123, 185, 205, 224
–, diversity of 1
A_1A_0 ATP synthase
–, reversible ion pump 129
–, coupling ions 140
–, evolutionary relations 140
–, structure 141 ff
aceticlastic pathway 127
Acidianus amb. NDH-2
–, X-ray structure 194
Acidianus ambivalens 24
Anabena rhodopsin 80
antenna complex 34, 35, 40, 62
archaerhodopsin
–, crystal structure 76
autophosphorylation 105

bacteriorhodopsin (BR) 73,
–, conformational change of 100, 109
binuclear [NiFe] site 135
binuclear NiFe-center
–, non-protein metal ligands 225
biomethanation 124

carbon cycle 123 ff
cardiolipin 260
carotenoids 49, 63
cbb$_3$ oxidases 21
CF_1 thiol regulation, chloroplasts 295
che gens clusters 105
chemotaxis 110
chloride pump 74, 78, 87, 88
chlorophyll(s)
–, coordination of 39, 63
citrate fermentation, anaerobic 154
CO dehydrogenase 233
cobalamin 137,139
coenzyme B 125
coenzyme M 126, 132

complex I (respiratory)
–, evolution 136
conformational coupling 213
conformational transitions
–, in F_1 ATPase 293
cyanobcteria 33, 10 ff, 80
cyt bc$_1$ complex (complex III)
–, occurance, functions 253
–, purification 255
–, X-ray structures 256 ff
–, supernumerary subunits 257
–, inhibitors 259, 263, 267
–, redox centers 258
–, *cyt c* binding 268
cytochrome b_{559} 56
cytochrome b_6f 34, 254
cytochrome c 4, 7, 26
cytochrome c oxidase 2, 4

dacarboxalation phosphorylation
–, genaral occurance 152
D-channel 6, 19
DELSEED region of $F_1\beta$ subunit 294
dinucleotide binding motif(s) 187 ff
$\Delta\mu H^+$ regulation of ATP synthase 289
DoxB 24

electrochemical gradient(s) 2, 33, 35, 254
electron carriers
–, mobile 34
electron cryo-microscopy 74
electron transport chain
–, photosynthetic 34, 35 ff
–, in PS I 45
–, respiratory 185
electron tunneling 259
electron valve 242
endosymbiosis 33
excitation energy 42

F_1 ATPase motor   165
F_1 inhibitor protein, mitochondrial   297
F_1F_0 ATP synthase
–, structure and function   281
–, archaeal   139
–, catalytic mechanism   281–284
–, role of subunit γ 283
–, ADP inhibition of F_1 283 ff
–, P_i binding   286
–, function of subunit ε 291
$F_{420}H_2$-dehydrogenase   128, 227
–, structure of   134
FAD binding domains   188, 199
FeFe hydrogenases
–, eukaryotic   233
–, phylogenetic origin   235
FeS cluster   48
Fo motor   166
–, subunit c structures   167
–, *I. Tartaricus* subunit c   168
–, mechanism of motor   171
fumarate respiration   226 ff

gas channels, hydrophobic   235
glutarate fermentation, anaerobic   158

H^+ and Na^+ driven ATP synthases   162
H^+ versus Na^+ motors
–, torque generation   172, 176
H_2 sensors   228
H_2 uptake hydrogenases (Hup)   226
–, cytoplasmic, soluble   228
–, bidirectional   229
–, H_2 evolving   232, 240
–, respiratory   238
H_2:heterodisulfide-oxidoreductase   128, 136, 227
–, proton pumping   130
–, structure of   134
–, [4Fe-4S] cluster in   137
–, evolutionary relations   134, 137
Halobacterium salinarum   73
halorhodopsin (HR)   75
–, family   77
–, crystal structures   78
hem-Cu superfamilies   1, 26
–, classification of   2
–, phylogenetic analysis of   4 ff
–, binuclear center of   7
–, evolution of   26

heterodisulfide   128
human apoptosis factor (AIF)   193
hydrogenase
–, F_{420}-nonreducing   130, 135
–, proton pumping   130, 136
–, Ech–   131, 134, 136
–, hydrogen formation   132
hydrogenase(s)
–, general occurance   223
–, classification   224
–, X-ray structure   224
hydrogenosomes   235

ion channels,
–, light triggered   74, 83, 97 ff
ion gradient   74, 139
ion transfer (HR, BR)
–, mechanism   95

K-channel   6, 19

lateral gene transfer   28, 89, 140, 233, 235
light harvesting complex   42, 43
lipids
–, of photosystems   50, 64
lipoamide dehydrogenas   193

malonate fermentation, anaerobic   157
membrane potential   78, 82, 187, 241
methane hydrates   124
methanogenesis
–, general   125 ff
–, cofactors of   125
–, ion transduction in   125
–, substrates of   126, 240
methanogens
–, properties   123, 224
–, hyperthermophilic   141
–, genomes of   124
methanophenazine   129 f., 239, 241
methyl-CoM reductase   127, 131
methyl-transferase   137 ff
midpoint potentials   131, 258, 264
Mn cluster   51, 61, 62

Na^+-ion gradient   132, 137 ff
Na^+ motor,
–, ion path   173
–, driving force   176

Subject Index

Na$^+$ transport
–, linked to decarboxylation   159
Na$^+$/H$^+$ antiporters   140, 206, 207, 237
NADH dehydrogenases
–, families   185
NADH dehydrogenases Na$^+$-pumping (Nqr)   196
–, Nqr operons   197
–, subunits and cofactors   198 ff
–, inhibitors   202
–, Na$^+$/2e$^-$ stoichiometry   202
–, electron pathways   204
Natronomonas   16, 21
NDH-1 a sodium pump?   208 ff
NDH-1, proton pumping complex I (respiratory)
–, occurance   134, 205
–, FeS-clusters   206, 210
–, inhibitors   206
–, in Parkinson disease   205, 268
–, pump stoichiometry   205
–, evolutionary relations   136, 210
–, subunit composition   210
–, x-ray structure, partial   211 ff
–, pumping mechanism   213
NDH-2 (alternative NADH dehydrogenases)
–, characteristics   187
–, inhibitors of   187
–, physiological functions   188 ff
–, EF hand Ca- binding motif   188, 192
–, regulation and occurance   189
–, simiarity dentrogramms   191
–, homology modeling   194 ff
NiFe-hydrogenases   225
–, homology to complex I   229–231, 233, 236ff
nitric oxide (NO) reductases   2, 3 ff, 6, 8, 25
–, binuclear center of   7
Nobel prizes,
–, in bioenergetics   V–VIII

oxaloacetate decarboxylase
–, as sodium pump   160
oxidative phosphorylation   187, 239, 269
oxygen reductases
–, structure of   5, 19, 21
–, in archaea   8 ff, 19, 24

phosphorylation potential   144
photosynthesis   33, 111
–, evolution of   33, 38
–, oxygenic   33, 50, 58
photosystem(s)   33, 73
phototaxis   101 ff
phycobillisomes   34
phylloquinone   47
plant mitochondria   190
plastoquinone   34, 50 ff, 57, 59
PQ-pool   34
primary donor P680   59 ff
proteorhodopsin   79
–, genes   80
–, heterolg. expression   80
proton /ATP stoichiometry   144
proton channels   6
proton gradient   73, 132, 139, 239, 254
proton pumps
–, in algea   73, 82 ff, 88, 95
proton reduction   35
prton motive force   35
PS I
–, structure   35 ff
–, cofactors of   41
–, antenna system   48,   54
–, primary donors/acceptors   46 ff
PS II   50, 59
–, cofactors of   41, 51
–, antenna system of   62
–, primary donors/acceptors   54, 59
Pyrobaculum   20, 21

Q-cycle   34, 242, 253, 255 ff, 262, 264
Q$_i$ site (center N)   261
Q$_o$ site (center P)   254, 262
–, mutagenesis of   264
quinol oxidase(s)   30, 31
quinone pool (Q pool, PQ pool)   34, 185, 238

reaction center(s)   38 ff, 46, 53, 54
reactive oxygen species (ROS)   188, 205, 253, 267, 268
respirasomes   270
respiration(s)   2, 238
respiratory chain(s)   185, 239
retinal binding   88
rhodopsins
–, archaeal   75 ff

–, eubacterial 79
–, eukaryotic 82 ff, 85
–, structure of 85
–, evolution of type 1 89
–, photocycle 91
Rieske FeS protein 254, 257
–, domain movementss 256, 265
rotary catalysis
–, in F_1F_0ATPs synthase 281
rotor
–, of A_1A_0 ATP synthase 142 ff
–, subunit stoichiometry 143

salinixanthin 81
Schiff base region 96, 99
sensory rhodopsins (SR I, SR II) 73, 78, 79
–, evolution of 90
signal tranfer (SR II)
–, mechanism 95, 106 ff
sodium pumping
–, linked to decaarboxylation 161
SoxFG 8
SoxLN 24
SoxM 8
subunit coevolution 225

succinate and lactate fermentation 156
Sulfolobus acidocaldarius 8
supercomplexes 8, 24, 35
–, respiratory 24, 269

terminal oxidases 2 ff
–, evolution of 28
Thermoplasmales 25
Thermosynechococcus 35, 52
Thiobacillus ferrooxidans 25
thioredoxin reductase 193
transducer proteins 102 ff
two-component systems 104

ubiquinol oxidation
–, kinetics and mechanisms 266–268
ubiquinone "ramp" 212
ubiquinone-8 201

V_1V_0 ATPase 140

water oxidation 34, 51, 60

xanthorhodopsin 81

Printing: Krips bv, Meppel, The Netherlands
Binding: Stürtz, Würzburg, Germany

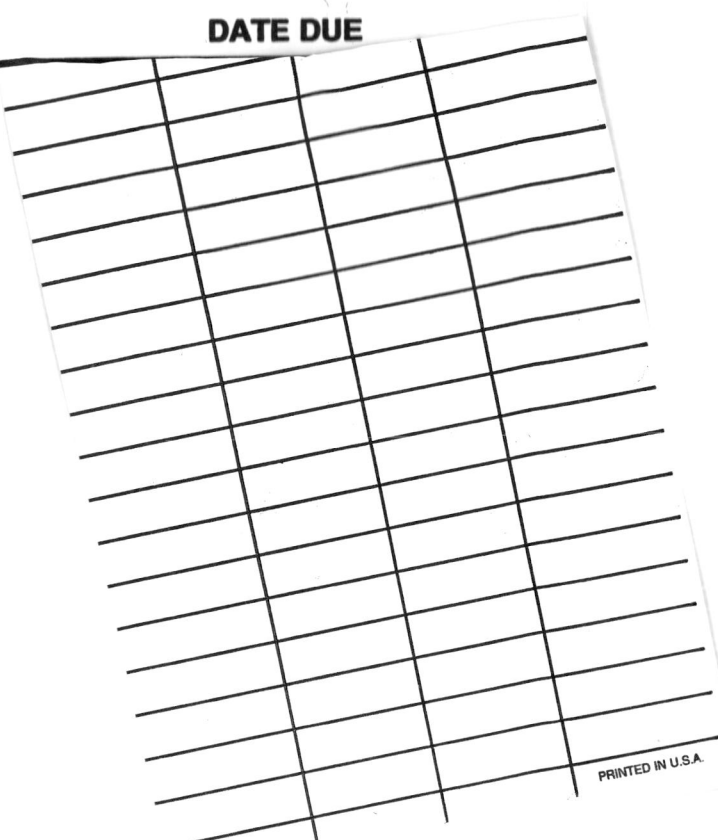